Genetics and Genomics of Livestock Health, Fertility and Product Quality

Genetics and Genomics of Livestock Health, Fertility and Product Quality

Guest Editor

Michael E. Davis

MDPI

Basel • Beijing • Wuhan • Barcelona • Belgrade • Novi Sad • Cluj • Manchester

Guest Editor
Michael E. Davis
Department of Animal Sciences
The Ohio State University
Columbus, OH
USA

Editorial Office
MDPI AG
Grosspeteranlage 5
4052 Basel, Switzerland

This is a reprint of the Special Issue, published open access by the journal *Animals* (ISSN 2076-2615), freely accessible at: https://www.mdpi.com/journal/animals/special_issues/Genetics_Genomics_Livestock_Health_Fertility_Product_Quality.

For citation purposes, cite each article independently as indicated on the article page online and as indicated below:

Lastname, A.A.; Lastname, B.B. Article Title. *Journal Name* **Year**, *Volume Number*, Page Range.

ISBN 978-3-7258-4445-6 (Hbk)
ISBN 978-3-7258-4446-3 (PDF)
https://doi.org/10.3390/books978-3-7258-4446-3

Contents

animals MDPI

Review

Genomic and Phenotypic Udder Evaluation for Dairy Cattle Selection: A Review

Miguel A. Gutiérrez-Reinoso [1,2], Pedro M. Aponte [3,4,5] and Manuel García-Herreros [6,*]

1 Carrera de Medicina Veterinaria, Facultad de Ciencias Agropecuarias y Recursos Naturales, Universidad Técnica de Cotopaxi (UTC), Latacunga 0501491, Ecuador; miguel.gutierrez@utc.edu.ec
2 Laboratorio de Biotecnología Animal, Departamento de Ciencia Animal, Facultad de Ciencias Veterinarias, Universidad de Concepción (UdeC), Chillán 3780000, Chile
3 Colegio de Ciencias Biológicas y Ambientales (COCIBA), Universidad San Francisco de Quito USFQ, Quito 170157, Ecuador; pmaponte@usfq.edu.ec
4 Colegio de Ciencias de la Salud, Escuela de Medicina Veterinaria, Universidad San Francisco de Quito USFQ, Quito 170157, Ecuador
5 Campus Cumbayá, Instituto de Investigaciones en Biomedicina "One-Health", Universidad San Francisco de Quito USFQ, Quito 170157, Ecuador
6 Instituto Nacional de Investigação Agrária e Veterinária (INIAV), 2005-048 Santarém, Portugal
* Correspondence: herrerosgm@gmail.com; Tel.: +351-243767 (ext. 330)

Simple Summary: Genomic and phenotypic selection criteria have been crucial in dairy cattle during the last decade. Udder health and milk production are important factors affecting productivity in dairy cattle. Furthermore, genomic and phenotypic selection are essential tools for increasing milk supply for human consumption, decreasing the use of antimicrobial products, improving animal health and welfare, and developing efficient dairy cattle production systems. The main aim of the present review is to explore the current advances, novel strategies, and future challenges in genomic and phenotypic udder evaluation traits for dairy cattle selection. Thus, the present review will explain the importance of genomic and phenotypic characterization of udder traits related to dairy cattle's health, production, and longevity. This review will encompass a comprehensive overview of the criteria for genomic and phenotypic udder evaluation and highlight potential future perspectives.

Abstract: The traditional point of view regarding dairy cattle selection has been challenged by recent genomic studies indicating that livestock productivity prediction can be redefined based on the evaluation of genomic and phenotypic data. Several studies that included different genomic-derived traits only indicated that interactions among them or even with conventional phenotypic evaluation criteria require further elucidation. Unfortunately, certain genomic and phenotypic-derived traits have been shown to be secondary factors influencing dairy production. Thus, these factors, as well as evaluation criteria, need to be defined. Owing to the variety of genomic and phenotypic udder-derived traits which may affect the modern dairy cow functionality and conformation, a definition of currently important traits in the broad sense is indicated. This is essential for cattle productivity and dairy sustainability. The main objective of the present review is to elucidate the possible relationships among genomic and phenotypic udder evaluation characteristics to define the most relevant traits related to selection for function and conformation in dairy cattle. This review aims to examine the potential impact of various udder-related evaluation criteria on dairy cattle productivity and explore how to mitigate the adverse effects of compromised udder conformation and functionality. Specifically, we will consider the implications for udder health, welfare, longevity, and production-derived traits. Subsequently, we will address several concerns covering the application of genomic and phenotypic evaluation criteria with emphasis on udder-related traits in dairy cattle selection as well as its evolution from origins to the present and future prospects.

Keywords: genomic analysis; phenotypic traits; evaluation criteria; milk production; health; welfare; longevity; dairy cattle

1. Introduction

For decades, dairy cattle breeding has been focused on increasing milk production [1]. Many functional traits have been described as having negative genomic correlations with milk production as well as reductions in genetic merit for health and fitness [2]. However, one of the constant goals and challenges in high-producing dairy cattle has been to balance fertility, udder health, and metabolic diseases without affecting milk production or compromising animal welfare [1–4].

Genomic and phenotypic evaluation methods provide accurate and efficient evaluations of udder traits in dairy cattle, allowing farmers to make more informed breeding decisions for animal selection. Phenotypic evaluation methods use physical measurements and observations to evaluate udder traits, providing a more holistic evaluation of the cow's overall health and productivity. On the other hand, genomic evaluation methods use DNA sequencing to identify specific genetic markers associated with desirable udder traits, providing a more direct and precise evaluation.

Combining genomic and phenotypic evaluation methods can provide a complete picture of a cow's udder health and milk production potential, allowing for more effective breeding and selection decisions. Furthermore, using genomic and phenotypic evaluation methods can ultimately lead to improved udder health, milk production, and overall profitability for dairy farmers and enhanced quality of dairy products for consumers.

In the dairy cow, different factors, such as udder conformation and size and environmental conditions, contribute to variations in cow performance and functionality, mainly in longevity and productivity [5]. In recent years, dairy cattle selection has been based on female-derived traits such as milk production performance and linear conformation traits [1]. Several of these linear conformation traits are negatively correlated with other functional traits, such as milk production, which has led to a reduction in health, body condition, and adaptability-derived traits [6]. Over the last 10–20 years, other traits related to udder health, such as somatic cell counts (SCCs), productive life (longevity), and other reproductive-related traits, such as daughter fertility, have been incorporated [3]. These traits have become increasingly important in the dairy industry, and recently, genomic selection has created new opportunities for increased accuracy of the selection of such traits in very young animals [3,7]. In other ruminant species, highly efficient selection schemes were established based on pyramidal population management with core breeders at the top. The selection schemes were based on pedigree and official milk records, artificial insemination results, controlled mating, and the estimation of breeding values with the aim of accelerating genetic progress [1]. On the other hand, the constant participation of young bulls in testing and breeding programs has been described in different dairy herds in the U.S. [8]. For bull selection, udder conformation and the expected difference in milk production were considered the most essential traits. However, traditionally, the most important traits for cow selection were milk production followed by udder conformation, feet and legs, and fat percentage [8]. Nowadays, the need to preserve particular traits of economic interest for milk production and to ensure positive genetic progress is one of the main objectives [9–11]. Therefore, current selection programs require specific breeding plans that take into account milk production and functional traits, including those related to udder conformation and functionality [12].

Moreover, the heritability assessment and genetic correlations between traits and selection indices allow identifying the most suitable individuals for selection in dairy breeds [11,13]. Poor udder and teat conformation has been reported to reduce profitability in dairy herds [14]. Furthermore, this has an impact on the incidence of mastitis at calving and leads to decreased productivity during the lifetime of the cow [9]. Therefore, it is necessary to evaluate the effects of udder conformation on performance and future progeny.

The objective of the present review is to provide a comprehensive description of the influence of genomic and phenotypic evaluation criteria in dairy cattle selection. It also aims to answer how these criteria may alleviate the adverse effects of defective udder conformation and function on different traits related to udder health, longevity, and milk production. Subsequently, we will address several concerns covering the application of genomic and phenotypic evaluation criteria in dairy cattle, its evolution from origins to the present day, and future prospects. This review is designed and organized as follows: The first part will briefly describe the udder evaluation criteria from the past to the present. In the second part, the use of the evaluation criteria based on udder-derived genomic and phenotypic traits as predictive tools for dairy cattle selection will be described, as well as several methodologies used for their application. In the third part, genomic and phenotypic evaluation criteria as evaluation tools for different traits associated with udder health, milk production (yield and quality), linear conformation, and longevity will be covered. Moreover, this part includes highly predictive genomic and phenotypic-derived methods that are key to the prognosis of other complex traits which directly or indirectly influence the udder conformation and its functionality. Incorporating both genomic and phenotypic evaluation criteria shows great promise in enhancing the accuracy of predicting udder-related traits and improving dairy cattle selection. However, this approach also presents several challenges, including integrating and optimizing these evaluation methods to develop the ideal selection process for enhancing udder conformation and functionality. Maximizing genetic gains across multiple traits associated with udder health, longevity, and milk production in dairy cattle has the potential to significantly boost livestock productivity and profitability in the future. By prioritizing these traits in breeding programs, farmers can improve the overall health and milk output of their herds, resulting in higher profits per animal and a more sustainable and efficient dairy industry.

2. Udder Evaluation Criteria: Past and Present

Udder evaluation has been a part of dairy cattle selection for centuries, with early breeders selecting cows based on visual inspection and manual palpation of the udder. In the early 1900s, standardized methods for udder evaluation were developed, including the linear score system and the udder index score. Dairy cattle breed associations have been developing classification systems since 1929, when the Holstein-Friesian Association of America introduced the first official system, intending to utilize these systems as phenotypic or genetic indicators of milk production ability, longevity, and udder health and milk production capacity over time [2–5,7–9].

In the mid-1900s, milk production testing became a common method for evaluating udder health and milk quality, with farmers tracking the amount and quality of milk produced by each cow. The establishment of modern livestock breeding programs dates back to the 1960s, when nucleus breeding initiatives emerged that utilized official pedigree data and milk production records to drive genetic progress [3]. This approach allowed registered dairy herds to make significant strides in genetic improvement, which could then be effectively disseminated to commercial herds [1]. As a result, these breeding programs have played a critical role in transforming the dairy industry and driving improvements in milk yield, udder health, and other key traits that impact profitability and sustainability.

While increasing milk production and improving quality (such as fat and protein content) continues to be the top priority for profitability, livestock breeders have also begun considering other important traits for selection. These may include mechanical milking ability, udder health, disease resistance (e.g., against mastitis and internal parasites), and even the nutritional value of milk (such as fatty acid composition) [1,15]. Given the importance of considering a range of traits beyond just milk production and quality, the dairy industry has increasingly focused on developing comprehensive selection criteria that consider factors such as udder health, disease resistance, and mechanical milking ability. One example of this is the Holstein Association of Canada's efforts to collect data on linear conformation traits in dairy cattle from the 1970s to the 1980s as a means of improving selection criteria based on type-derived traits [16]. The phenotypic linear traits considered were general appearance, milk form, body capacity, rump, feet and legs, mammary system, fore udder, and rear udder [16,17]. On the other hand, nine descriptive traits were also considered: stature, size, style, brisket, floor, loin strength, dorsal width, and pin placement [12,17]. Heritability estimates for type traits in dairy cattle ranged from 0.06 (style) to 0.33 (size). Phenotypic correlations between type traits and calving interval were essentially zero. The highest beneficial negative genetic correlations with calving interval were for breast depth (−0.42), rear udder (−0.37), and capacity (−0.34), and the highest antagonistic positive correlations were for milk yield (0.43) and milk quality traits (0.38) [17]. Regarding the selection of dairy cows for body size, they usually showed divergence associated with body weight, body dimensions, and calf birth weight but did not differ for milk yield. However, udder conformation in relation to body size differed between large and small cows [16].

In the late 1900s and early 2000s, genomic evaluation methods were developed, allowing for more accurate and efficient evaluations of udder traits. Today, a combination of visual inspection, manual palpation, milk production testing, and genomic and phenotypic evaluation methods are becoming widespread to evaluate udder traits in dairy cattle. Thus, current breeding programs based on the traditional quantitative approach have achieved appreciable genetic gains for milk production. They have implemented selection for other traits, including milk composition, udder morphology, somatic cell count, and mastitis resistance [1,18]. The implementation of novel selection strategies utilizing molecular-based data to enhance both conventional and high-value traits holds significant promise for advancing functional improvements in dairy breeds [1]. Recently, genome-wide association (GWAs)-based studies have demonstrated that udder conformation traits significantly affect animal health, milk production, and the profitability of Holstein cattle [19]. The release of complete genome sequences enabled the use of high-performance genetic markers improving udder trait predictions using genotyping platforms (DNA-derived arrays or SNP chips) [3]. In addition, DNA microarrays (MGEP) began to be widely used in functional genomics or transcriptomics [20]. In the coming years, genomics based on SNP chips or microarrays will possibly be replaced by next-generation sequencing (NGS) technologies [13,21,22]. Finally, the use of 'omic' technologies such as epigenomics, transcriptomics, proteomics, metabolomics, metagenomics, and meta-transcriptomics will open new horizons to accurately assess novel udder-derived traits for dairy cattle selection [23]. Overall, the integration of new technologies such as genomics, epigenomics, transcriptomics, proteomics, metabolomics, metagenomics, and meta-transcriptomics is crucial to accurately assess novel udder-derived traits for dairy cattle selection and, thus, improve the prediction of mammary gland characteristics related to udder health, longevity, welfare, and milk production.

3. Evaluation Criteria Based on Udder Phenotypic Traits

Phenotypic evaluation has long been a key component of dairy cattle breeding, allowing farmers and breeders to evaluate a cow's physical traits and characteristics to make informed breeding decisions. While more recent advancements in genomic evaluation have provided new tools for evaluating an animal's genetic potential, phenotypic evaluation remains a critical aspect of the breeding process, particularly when assessing traits that cannot be measured through DNA analysis. However, several intrinsic and extrinsic factors should be taken into account together with the phenotypic and genomic udder evaluation methods for dairy cattle selection (Figure 1). Understanding the criteria used to assess udder phenotypic traits is an essential aspect of dairy cattle breeding, as these traits are key indicators of a cow's milk production potential and overall udder health. The selection criteria for productive animals based on udder conformation include several traits associated with the general linear conformation of females, such as stature, strength, milk production, teat diameter, hind legs (lateral view), rump angle, rump width, anterior udder attachment, posterior udder height, posterior udder arch, udder depth, suspensory ligament, and teat placement [12]. The selection criteria based on udder conformation traits could differ depending on the livestock species, breed, or crossbreed evaluated. In one study, Montbéliarde (MO) × Holstein (HO) and Viking Red (VR) × HO crossbred cows were compared with first-calving pure HO cows to evaluate the conformation for production and reproductive traits [24]. It was observed that crossbred cows had less udder-to-hock clearance than HO cows. In addition, crossbred cows had more distance between the front and rear teats and longer teats than HO cows [24,25]. However, the frequency of first lactation cows that were discarded because of udder conformation issues was uniformly low across all groups (<1%) [24]. Although this review mainly focuses on dairy cattle, it is worth noting that other ruminant species, such as goats, also have their own linear evaluation systems for assessing individual traits based on their observed range. The American Dairy Goat Association's linear evaluation system analyzes individual traits in an example of other ruminant species such as caprine by assessing each trait based on its observed range in the animal being evaluated. This is carried out by starting from one biological end of the trait (such as head structure) and evaluating it until reaching the other biological end (such as leg structure). A score or grade ranging from 1 to 50 points is assigned to each trait, and a final type score between 50 and 99 points is calculated based on the importance given to each of the structural categories, which are general appearance (35%), dairy character (20%), body capacity (10%), and mammary system (35%) [12]. In several countries, particularly the United States, genetic evaluations are performed annually for the final type score and additional linear type traits, as in dairy goats [25].

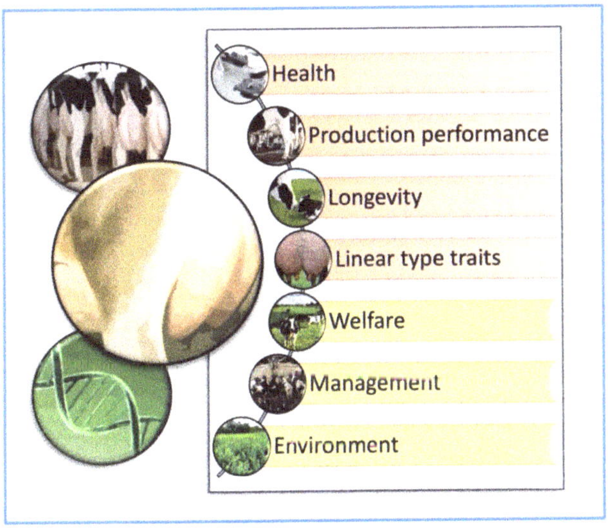

Figure 1. Brief scheme of the main elements that should be considered together with the phenotypic and genomic udder evaluation methods for dairy cattle selection. Orange: intrinsic components such as health, production, longevity, and linear (conformation) type traits. Yellow: extrinsic components such as welfare, management, and environment.

Among the selection criteria taken into account in cattle, the direct measuring and correlated responses to a single trait for milk production could be a convenient method in dairy production systems [26]. For example, several producers use a selection line consisting of artificial insemination (AI) bulls selected for their high estimated transmission capacity for milk production associated with udder conformation [27]. On the other hand, several studies currently target the genetic evaluation of novel traits based on methods to predict selection accuracy. These methods are based on reference populations of cows using indicator traits that could increase the accuracy of genomic selection for important traits such as udder health [3]. Moreover, in other ruminant species, linear or non-linear genetic relationships between functional productive life associated with anterior udder attachment, posterior udder height, posterior udder arch, udder depth, suspensory ligament, and teat location have been described in U.S. dairy goats [12].

Furthermore, heritabilities and correlations between traits were estimated through linear mixed models with additive pedigree genetic relationships and ASREML software (version 2023) [12]. Therefore, udder conformation traits can be used as a selection criterion to increase longevity in dairy goats and can be extrapolated to other species, such as cattle. This consideration of non-linear relationships can help design more efficient breeding programs using conformation traits [12,28]. However, udder conformation in cows with large funnel teats, pendulous udders after calving, and blind quarters presented a higher risk of subclinical mastitis or intramammary infections [29].

Generally, breeders and technicians have considered that there are several traits of importance to consider, such as milk value and udder conformation [30]. Significant interactions indicated that dairy aptitude is the most crucial trait [30]. Subjective visual assessment of dairy cattle by appraisers is carried out to determine various distinguishing characteristics, including linear type traits related to udder conformation [31]. Thus, genetic links between routine observations carried out by appraisers during the evaluation process can be used to validate the classifications from individual appraisers [31,32]. In general, udder central ligament, teat length, udder (in general), and teat placement receive different scores that relate to milk production ability in several cattle breeds [31,33,34].

4. Evaluation Criteria Based on Udder Genomic Traits

As genomic evaluation methods have become more prevalent in dairy cattle breeding, understanding the criteria used to evaluate udder genomic traits has become increasingly crucial for breeders seeking to make informed decisions about animal selection. The progress of genetic and genomic evaluation in dairy cattle is based on discovering new traits of interest for predicting selection accuracy. Thus, the genetic and genomic assessment aims to improve how using indicator traits could increase the accuracy of genomic selection [3]. To develop effective selection programs for new traits, it is necessary to create large databases based on highly reliable estimated genetic values (EGVs) [2,35,36]. For evaluating the most costly traits, it is possible to perform extensive phenotyping in combination with genotyping of female dairy cattle to record and characterize their functionality [2]. These novel traits include udder health, hoof health, feed efficiency, methane emissions, and others [3]. There are various deterministic methods to estimate the reliability of genomic evaluations. These depend on factors such as the number of cows used as a reference, the trait's heritability, and other population characteristics. This is conducted by analyzing the impact of individual genetic markers called single nucleotide polymorphisms (SNPs) [3].

When analyzing genetic data, a positive correlation was found between Somatic Cell Score (SCS) and milk quality and composition traits, highlighting the interconnected nature of these characteristics [11,20,23]. The genetic traits for milk composition were negatively correlated with the udder conformation (-0.40) and with rear muscling and udder volume (-0.28) [11]. These results were obtained by using the current selection index, which mainly focuses on dairy strength and could negatively influence all other traits [11]. Moreover, genetic correlations between udder traits are favorable, indicating that selection for a given trait will positively affect the evolution of overall udder conformation. In particular, the degree of udder suspension was highly correlated with udder depth (0.82) [26]. Genetic correlations with milk production were low and negative, except for udder depth (-0.48) [26]. Accordingly, more appropriate selection indices based on genetically derived traits should also be considered, together with functional traits [37]. Recently, a more detailed report on genomic methods and general conformation traits applied for dairy cattle selection was published [13]. However, this report did not thoroughly consider the specific udder conformation and functional traits and their relationship with udder-related genomic traits. Udder conformation is not only crucial for displaying visual characteristics but also for high milk production and low incidence of mastitis. A recent study measured different udder type traits in Sahiwal (*Bos indicus*) and Karan Fries (*Bos taurus* × *Bos indicus*) and their association with SNPs in vitamin D receptor and protein tyrosine phosphatase, receptor type (*R*) genes. The *GG* genotype of SNP *rs454303072* was associated with a wider rear udder, larger udder circumference, and longer distance between the front and rear teats and left and right teats in Karan Fries cattle. On the other hand, in Sahiwal cattle, the *AA* genotype of this SNP was found to be associated with a higher and wider rear udder, larger udder circumference, and wider udder. In addition, the *AA* genotype of the SNP *rs382671389* was associated with longer front teats in Karan Fries cattle. The *TT* and *CC* genotypes of SNP *rs435289107* were associated with udder-type traits in Karan Fries and Sahiwal cattle, respectively. These results suggest that *BTA5* harbors genomic regions associated with udder traits in *Bos indicus* and *Bos indicus x Bos taurus* cattle [38].

All data based on genome sequences from previous generations provides comprehensive information on the polymorphic loci that characterize a given breed and differentiate it from others. Association analysis with imputed sequences, mainly when applied to multiple traits simultaneously, is a powerful approach to detecting candidate causal variants underlying complex phenotypes. Twelve quantitative trait loci (QTL) affecting different morphological characteristics of the mammary gland were detected in the German Fleckvieh cattle population. Most of the QTLs were located in non-coding regions of the genome but in close proximity to candidate genes that could be involved in mammary gland morphology (*SP5, GC, NPFFR2, CRIM1, RXFP2, TBX5, RBM19,* and *ADAM12*) [39]. In Holstein cattle, genetic analyses of udder conformation traits have been developed [19]. A genome-wide association has been described with five udder conformation traits, including anterior udder attachment, central suspensory ligament, posterior udder attachment height, posterior udder attachment width, and udder depth [19]. Heritability and standard errors for these five udder traits ranged from 0.04 ± 0.00 to 0.49 ± 0.03. In this study, phenotypic data were measured on 1000 Holstein cows, and the GeneSeek Genomic Profiler (GGP) Bovine 100 K SNP chip was used to analyze genotypic data [19]. Numerous candidate genes were identified within 200 kb of significant SNPs. Among all significant SNPs associated with udder conformation traits, most of them were located within the following genes: Microsomal Glutathione S-Transferase 1 (*MGST1*), Microsomal Glutathione S-Transferase 2 (*MGST2*), Microtubule-associated scaffold protein 1 (*MTUS1*), *LOC101903734*, *LOC112447118*, Structural maintenance of chromosomes protein 5 and 6 (*SMC5-SMC6*), Parkin RBR E3 ubiquitin-protein ligase (*PRKN*), Syntaxin-Binding Protein 6 (*STXBP6*), glutamate ionotropic receptor delta-type subunit 2 (*GRID2*), E2F transcription factor 8 (*E2F8*), Cadherin 11 (*CDH11*), Forkhead Box P1 (*FOXP1*), Localisation factor complex 1 (*SLF1*), transmembrane protein 117 (*TMEM117*), SET binding factor 2 (*SBF2*), CGPVitamin D binding protein (*GC*), galectin 2 (*LGALS2*), adhesion G protein-coupled receptor B3 (*ADGRB3*), and Glutamate-Cysteine Ligase Catalytic Subunit (*GCLC*) as well as *KEGG* pathway genes. While one of the SNPs on *Chr6* was close (50 kb) to the ubiquitin-conjugating enzyme E2 K gene (*UBE2K*), three SNPs on *Chr21*, *Chr29*, and *Chr18* were located close (100 kb) to *STXBP6*, *E2F8*, and *CDH11*, respectively [19]. These results could provide valuable biological information for the genetic structure of udder conformation traits in Holstein dairy cattle [19].

Several studies have identified quantitative trait loci (QTL) to identify and understand genetic variants associated with economically relevant phenotypes. However, to identify QTL and genomic regions associated with phenotypes in Holstein cattle, the GWAS method has become the most widely used, proving to be an effective tool for identifying genetic variants associated with economically important traits [40].

5. Evaluation Criteria Based on Genomic and Phenotypic Traits: Udder Health, Production, and Longevity

5.1. Genomic, Genotypic, and Phenotypic Udder Traits: Impact on Mammary Gland Health

The development of national health registration programs began several decades ago. Accordingly, the first reports of genetic results based on health traits in the United States started in 2004 [41]. A few years later, in 2007, Canada reported large amounts of data on mastitis incidence in cows [42,43]. Available data based on the mastitis incidence can be analyzed by multi-trait models using SCS data and other derived traits indicative of SCS in early lactation, partial SCS, test day SCS, anterior udder attachment, udder depth, and BCS to perform traditional and genomic evaluations for mastitis resistance [35,43,44]. Udder health traits associated with bacteriological testing of quarters maximize the information on infection; however, these tests are not practical on a population scale [45]. In addition, somatic cell counts (SCCs) are highly variable, difficult to interpret, and not sensitive indicators of subclinical infections [45].

Nevertheless, several milk enzymes are possible indicators of tissue damage at the mammary gland level [45]. The heritabilities are around 0.2 for SCCs and 0.1 for other traits, reflecting genetic variations in the immunological defenses and their effect on the teat, phagocytosis, and immune response [45]. There is a positive relationship between somatic cell count (SCC) and milk yield at the genetic level but a strong negative relationship between them at the phenotypic level [45]. Genetic correlation refers to the degree to which the same genes affect two different traits, in this case, SCC and milk yield. A positive genetic correlation between SCC and milk yield means that the genes contributing to higher milk yield also contribute to higher SCC.

In several countries, the genomic values for mastitis resistance have been adopted as part of national selection indices [29,36,42,46,47]. A recent study based on milk production traits aimed to identify genes associated with mammary conformation and health phenotypes in Holstein, Montbéliarde, and Normande breeds [48]. Cows were genotyped with medium (50 k) or low (7 to 10 k) density chips and imputed to 50 k, resulting in 1013 genes associated with udder morphology, mastitis, and production phenotypes (e.g., *ESR1*, *FGF2*, *FGFR2*, *FGFR2*, *GLI2*, *IQGAP3*, *PGR*, *PRLR*, *RREB1*, *BTRC*, and *TGFBR2*, among others) [48]. Another study identified potential candidate genes for novel udder health traits targeting specific mastitis pathogens and Holstein Friesian cows' differential somatic cell fractions. Thirty-five significant SNPs were detected on 14 different chromosomes for cell fractions and pathogens. In this case, depending on environmental stressors, the HEMK1 gene plays a role in developing the immune system.

Furthermore, it was observed that the *CHL1* gene is up-regulated with stress levels and influences the mechanisms of the immune system [49]. Therefore, the activity interactions between the genes may fluctuate depending on environmental stressors. On the other hand, other authors report that *CFAP69*, *STEAP2*, and *ITGB3BP* genes located on chromosome 4 are related to the mechanisms of the mammary gland immune system [50,51]. The objective measures of mammary gland conformation and linear type scores are used to predict mastitis in experimental linear classification programs [1]. However, the relationships between conformation and mastitis are often inconsistent due to moderate or low correlations between mastitis indicators [46]. Selection to reduce the occurrence of cows with deep udders, especially low rear udders, open teats, teats that are set back, and also short and wide teats, can improve efforts to reduce the incidence of mastitis, along with better health control, therapeutic managing, and proper milking procedures [46]. A study in the U.S. reported genetic analyses performed for health traits, including reproductive problems, metabolic diseases, lameness, and the incidence of mastitis [36]. This study concluded that genetic selection for health traits using recorded phenotypic data and including genomic data would substantially improve selection accuracy [36]. In addition to the mentioned traits, others, such as milk electrical conductivity, are available to help predict mastitis incidence [52]. Conformation traits in Viking Red (VR), Montbéliarde (MO), and Holstein (HO) crossbred cows in commercial dairy herds were associated with significantly lower health indices during the first (−23%), second (−29%), and third (−21%) lactation comparing the VR and MO group with the HO group [14]. Crossbred cows had significantly less udder to hock spacing but significantly wider rear teat width and longer teat length compared to the HO group [14]. It is important to note that in the least-developed countries, the national phenotypic, genetic, and genomic evaluations for udder and other health traits are not currently being planned. It is, therefore, essential to establish national udder health registration programs to facilitate future selection protocols [35,36,53].

There are factors associated with the distribution of clinical bovine mastitis between the hind and forequarters, in addition to risk factors related to certain aspects of lactation, udder conformation, and management practices [54–56]. Thus, programs to select sires whose progeny have the lowest SCCs should be carefully planned, taking into account the interpretation of SCCs as an immunological defense mechanism [11,27,57,58]. High SCCs decrease the likelihood of subsequent infections; however, there is a heritable variation in SCCs in dairy cattle. Therefore, using SCCs as a defense marker should be studied in greater depth [45]. Clinical mastitis in the hindquarters was found to be more common among primiparous cows, with a prevalence rate of 61.9% in affected cases [54]. Moreover, udder conformation does not seem to play an essential role in mastitis incidence and distribution [37,46,59]. In another study, the genetic traits of milking speed (subjectively scored) and SCS estimation using Restricted Maximum Likelihood (REML) were evaluated [58]. The results showed that heritabilities were 0.15 for milking speed and 0.14 and 0.16 for mean SCS in the first and second lactation, respectively [58]. Genetic correlations between milking speed and SCS were 0.41 and 0.25 for the first and second lactation, respectively, indicating that faster milking was associated with increased SCS [58]. The genetic correlations were higher between udder depth and SCS (−0.26) and between rear udder width and milking speed (−0.24). Therefore, an udder health index applied to sire selection was developed for an aggregated genotype, including subclinical mastitis in 1 and ≥2 lactations and clinical mastitis in 1 and ≥2 lactations concerning milking speed [58]. Finally, the genetic and phenotypic correlations between type traits, heritability indices, and arithmetic mean SCCs during lactation were determined using a minimum norm quadratic unbiased estimator (MINQUE) and two multiple trait REML methods [57]. The heritabilities obtained for all traits were low. Regarding SCCs, heritabilities ranged from 0.09 to 0.11, and for type traits, from 0.08 to 0.14. The genetic correlations between SCCs and type traits ranged from −0.22 to 0.30; however, the phenotypic correlations were very low. In general, the correlations indicate an association between SCCs and udder conformation traits; that is to say, a desirable type score would be associated with a low SCC [57]. In summary, the traits included in the health index were milking speed, udder conformation, and SCS in the first and subsequent lactations [58]. The accuracy of the health index was found to be 0.776. This was a 15% increase compared to an index that was based on SCS count alone [58].

Prior studies anticipated that variations in udder infection rates could be linked to udder conformation and milkability [60]. The number of infected quarters was 1.53 for the low infection rate group and 2.53 for the high infection rate group [60]. Poor udder and teat conformation were associated with high levels of intramammary infection, as indicated by increased SCCs [9,55,56,61]. In addition, both udder physical traits and SCCs were related to the productive life in sheep and cattle [61]. Udder conformation and teat end lesions have been associated with risk factors such as elevated SCCs and intramammary infections in dairy cows [31,46,55]. Cows with large udder hindquarters had significantly fewer *Staphylococcus aureus*- and *Streptococcus uberis*-derived infections [54,55]. There was a similar, although less significant, negative association with *Escherichia coli*-derived infection [55,62]. *Streptococcus agalactiae*- and *Streptococcus dysgalactiae*-derived infections were more frequent in cows with large pendulous or small udder conformations [55]. Moreover, anatomical teat characteristics (length, cylinder diameter, and tip) in dairy cows were associated with intramammary infections caused by teat skin bacterial populations [56,61,63,64]. Because of this, linear conformation data have been evaluated for udder type in dairy cattle, including teat position (front or rear), which is associated with the occurrence of clinical mastitis [56]. In addition, *GAS6* negatively regulates the *Staphylococcus aureus*-induced inflammatory response in the mouse mammary gland [65]. Finally, the *NCAM1* gene, also known as *CD56*, is a member of the immunoglobulin superfamily and is the phenotypic marker of natural killer cells [66].

Other udder-related pathologies, such as udder cleft dermatitis (UCD), could be considered a trait affecting udder health and conformation [47]. Dermatitis is an inflammatory skin condition associated with mastitis and digital dermatitis that affects the anterior parts of the udder in dairy cows [47]. The factors with the most significant effects on such lesions were days in milk, milk production, and early calving [47]. The etiology of UCD is probably multifactorial, involving udder conformation traits, other cow-related aspects, and herd-related risk factors [47]. The high prevalence of severe UCD lesions in dairy cows from Nordic countries such as Sweden emphasizes the need for preventive genomic and phenotypic evaluations together with efficient treatments to avoid the occurrence of UCD in future dairy cattle [47].

Recently, eight different machine learning methods were compared to predict the udder health status of cows based on somatic cell counts. The study found that all methods had prediction accuracies above 75%. Neural Network, Random Forest, and linear methods best predicted udder health classes based on recorded milk traits. The findings suggest that machine learning algorithms can be a promising tool to improve farmers' decision-making in identifying cows with high somatic cell counts, which can improve surveillance methods and potentially reduce the economic and health impact of bovine mastitis in dairy farms [67,68]. The use of machine learning and data mining techniques in genomic and phenotypic udder evaluation for dairy cattle selection has become an essential tool in improving udder health. Several research groups have been investigating the implementation of these methods [67,69–71]. Furthermore, integrating both genomic and phenotypic information has been shown to significantly improve the accuracy of predictions regarding udder health in dairy cattle.

5.2. Genomic, Genotypic, and Phenotypic Udder Traits: Effects on Milk Production and Quality

Traditionally, dairy cattle have been selected for their ability to produce milk in quantity and quality [20]. The traditional approach to dairy cattle selection has changed, and secondary traits are now included in the selection indices, with less emphasis on milk production and more interest in milk quality and other non-productive traits [20,53,63]. Greater emphasis on non-productive traits is reflected in the industry's desire to breed less productive but more functional dairy cattle [20]. According to some studies, there is a low but statistically significant relationship between conformation traits and the lifetime production efficiency of cows (Guilford scale), being slightly higher for milk production than longevity [72]. Type and conformation traits seem to be more suitable for predicting the lifetime production efficiency of cows compared to other more detailed traits [17,25,72]. For example, lifetime performance was most strongly related to overall linear score and udder linear score (r = 0.22), followed by type and conformation scores, feet and hooves (r = 0.13), and, finally, other detailed traits such as udder width and dairy character (r = 0.14) [72]. Therefore, type and conformation traits seem to be more suitable than genomic traits for predicting the lifetime production efficiency of cows [11,72].

The phenotypic and genomic traits related to milk yield, protein, and fat are included in the performance tests in young bulls [8,11]. Several linear-type traits have been added to the factor analysis. Heritability (h2) estimates for milk yield and milk quality-derived traits ranged from 0.125 to 0.219 [11,45]. Moreover, genetic estimates for milk yield and milk quality traits were negatively correlated with traits explaining udder conformation (−0.40) and rear muscularity [11]. The latter genetic traits also showed a negative correlation with udder volume (−0.28). Additionally, head typicality and rear leg traits were not correlated with milk yield and milk quality but were negatively correlated with meat-related traits such as rear muscularity (−0.32) [11]. The consequence of these results is that the use of the current selection index, which is mainly focused on milk production traits, may lead to a deterioration of all other traits. Therefore, more appropriate selection indices should consider an association between genetic and functional traits [11,23,37,57]. A previous study revealed that 51 significant SNPs overlapped with 51 protein-coding genes together with five QTL for SCC. On chromosome *BTA18*, three QTL for SCC overlapped with SNP *rs41622425*, which is located within the intron of a gene encoding a protein called Pleckstrin Homology-Like Domain Family B Member 3 (*PHLDB3*). In addition, on the same chromosome, *rs110437995* overlapped with QTL for SCC. They also detected three overlapping genes, GAS6, PKD2, and NCAM1, previously associated with inflammation in humans and mice [73].

The selection and correlation of different traits for milk production based on the selection of AI bulls preferred for high transmission capacity is now possible [3,27,57]. The estimates regarding sire selection criteria may vary; however, the selection for milk production traits effectively increases milk production [8]. Nonetheless, in a research project evaluating the direct and correlated effects of single-trait selection on milk yield all selective breeding groups increased productivity, but also undesirable responses correlated with selection for milk production were detected [27]. Sometimes, bull sires from different genetic lines were selected by progeny testing based on the first lactation productivity and milk quality by associating traits for fat-corrected milk production, percentage of daughters discarded at the first lactation, and daughters' udder conformation. However, one study showed no differences regarding milk yield among lines [57]. Moreover, negative correlations existed for milk production traits and quality traits without affecting udder conformation traits for the selection lines considered. Additionally, cows selected for the higher production traits tend to have higher health costs, specifically for mammary treatment [27,74]. Finally, other factors, such as chronic stress, can affect cow welfare, and acute stress during milking can decrease milk production [6,37]. Therefore, it is important to quantify whether long-term stress persistence would impact milk production depending on the genetic line studied [6]. Another study compared the reproductive performance, milk yield and composition, udder health, and conformation traits of Holstein, F1, and R1 crossbred Holstein × Simmental cows [75]. Milk yield and fat-protein yield did not differ between groups [75]. Furthermore, udder conformation traits lacked high scores in crossbred cows. Finally, Holstein cows had significantly shallower udders and higher udder spacing than the other groups [75].

It has been observed by RNA sequencing that *DDIT3*, *RPL23A*, *SESN2*, and *NR4A1* genes are significantly and differentially expressed between mammary glands of lactating Holstein cows with extremely high or low protein and fat percentages. Therefore, these four genes could affect milk production and composition traits [76]. In addition, the genes identified in another study, such as *PGM1* and *ARL4A*, were related to milk production traits, lactose synthesis, glucose metabolism [77], and milk production or composition [78]. Finally, the *ROR1* gene located on chromosome 3 has been considered a causal determinant gene for somatic cell count [79]. These studies shed light on the importance of genetic factors in determining milk production and composition traits and highlight the need for further research in this area to improve the efficiency and profitability of the dairy industry.

Linear evaluation plays a vital role in estimating the milk production of dairy cows; however, these evaluations, sometimes taken through subjective methods, can show variations [5,31]. On the other hand, objective methods for estimating milk productivity provide more accurate and reliable data but require more sophisticated technologies [5]. In one study, sire-derived type traits were analyzed in cows with low, medium, or high production performance during the first lactation [80]. The factors examined in the model included herd-year-season, age at calving, the month of calving, recording status interaction, change in herd size, and season. In addition, fat and protein production and type-based sire linear regressions were analyzed within each group of cows with similar production traits [80]. The impact of leg traits on the number of lactations was almost the same as for udder traits. In addition, interactions between traits were significant. Conformation-type traits were of little importance for herd life in low-producing cows. However, few differences were observed in the relationships between herd life and type in medium- versus high-producing cows, indicating that there is no need to increase emphasis on type in response to current trends for higher production [80]. These studies reinforce the idea that a comprehensive and detailed analysis is necessary to estimate milk production in dairy cows accurately. Furthermore, a primary goal of modern cattle breeding is to attain an equilibrium between various phenotypic traits and overall functions linked to mammary gland conformation, which can effectively manifest desirable, productive traits in diverse environments [1,3,11,59,80].

5.3. Genomic, Genotypic, and Phenotypic Udder Traits: Influence on True and Functional Longevity

Longevity-related traits have garnered growing interest among various milk-producing species, including sheep, cattle, goats, and other animals, as efforts to enhance the longevity of these animals continue to gain importance. Therefore, understanding the impact of genomic, genotypic, and phenotypic udder traits is crucial in determining dairy cows' true and functional longevity. Functional longevity is defined as the number of days between first calving and culling, that is to say, the length of cows´ productive life [81]. Functional longevity is an economically important trait to increase the profitability of dairy management [82]. In different bovine dairy herds, the reasons for culling cows can be voluntary (mainly because of low productivity) or involuntary (mainly because of health and low fertility) [81,83]. The longevity-derived traits include (i) true longevity (all reasons for culling, including productivity) and (ii) functional longevity (all reasons for culling except productivity) [82]. Conformation traits and functional longevity have been related by survival analysis (Cox proportional hazards models) in first-lactation Holstein cows [81]. The dairy character had the strongest correlation between a composite trait and functional longevity, followed by the udder final score [81,84]. Among the descriptive traits, the highest correlations were observed for traits related to longevity vs. udder attachment and longevity vs. udder depth [81,83]. Cows with deep udders had significantly lower functional survival than cows with shallow udders [75]. The highest positive effect on longevity was exerted by udder score and legs and hooves ($r = 0.11$), udder placement ($r = 0.14$), and anterior udder attachment ($r = 0.10$) [72]. In addition, central ligament weakness was associated with significantly reduced cow longevity [81].

A more accurate methodology for improving animal longevity is necessary to decrease the involuntary culling rates rather than extending traits that influence the herd life [82]. Therefore, the proportional hazard model is helpful for assessing genetic fitness for the traits influencing herd life. However, the differences between estimates made with the proportional hazards models and those made with linear animal models for one or more traits are unclear. Productive traits, udder traits, and leg and hoof traits are genetically correlated with longevity; consequently, these traits are used to assess longevity indirectly [12,16]. The reliability of genetic fitness-related estimates for longevity is increased by combining direct and indirect estimates [82]. Therefore, these genetic correlations should be reviewed periodically in different dairy cattle production systems as they vary according to the year of birth [82]. In light of this, QTLs affecting economically relevant traits were studied for eight US Holstein cattle genetic lines [85]. A marker on chromosome 14 associated with differences in fat yield, fat percentage, and milk yield was observed in two genetic lines. Other markers located on chromosomes 16 and 20 were related to differences in udder depth and anterior udder attachment, respectively. A marker on chromosome 27 was associated with a difference in milkability index. These additional markers complete the quantitative trait locus mapping to identify QTLs affecting economically important traits in a selected commercial Holstein population [85]. Another study in dairy cattle demonstrated the association of SNPs located within selected genes with traits associated with longevity [86]. In summary, the polymorphism with the most significant effect on functional longevity is leptin (*LEP-R25C*) in dairy cattle [86]. In addition, four potential candidate genes, *ETV1*, *ONECUT1*, *MACROD2*, and *SIRT1*, have been identified in Holstein cows that directly (through disease resistance mechanisms) or indirectly (through milk productivity) influence longevity [87]. Finally, the three bovine selectin-related genes identified as *SELP*, *SELL*, and *SELE* have been described as relevant genes for udder traits, with *SELP* being the most polymorphic of these genes. It is suggested that these polymorphisms in *SELP* and *SELE* are associated with fertility, milk production traits, and longevity in Holstein Friesian dairy cows [88].

The relationship between different conformation traits and functional longevity is essential in dairy cows and has been evaluated for years using survival analyses [80,81]. As mentioned, the highest correlations among descriptive traits were observed for longevity-udder attachment and longevity-udder depth [26,35,43,46,59]. However, functional longevity decreased with decreasing body condition in dairy cows [81]. Other relevant factors affecting longevity could be related to chronic stress in dairy cattle. For example, environmental factors can affect genomic and phenotypic traits in the dairy cow [6]. The transition of a dairy herd to an automatic milking system from a conventional parlor system can also be stressful for the cow, as many changes occur during this process [6,37]. The robotic milking methods applied to dairy cows will be essential tools that will probably be introduced, as a general rule, in all dairy farms [89]. However, there is a substantial possibility that the robotic methods may not be able to be adapted adequately to the spatial location of the teats [89]. It has been found that robotic attachment is more successful in cows whose udder conformation is more suitable. Moreover, the cows showed some signs of discomfort after the omission of milking when robotic milking systems were used [6,89]. In 60% of cases, there is milk loss after robotic milking, which indicates ineffectiveness and additional risk for mastitis, affecting longevity [89]. Therefore, special attention should be paid to the conformational udder-related traits in cows destined for robotic milking that will be useful for dairy cows with high performance, good udder conformation, health, and longevity [3,9,81]. The use of mechanical milking has led to the selection of animals based on udder morphology (linear scoring) [12,25]. In order to improve longevity, teat placement, degree of udder suspension, udder depth, and the degree of separation of the two halves are evaluated using nine-point linear scales [26,90]. It has also been observed that the linear combination of individual linear traits is crucial for longevity and is highly correlated with teat placement, degree of udder suspension, and udder depth, with repeatability across lactation showing a value of 0.76 [26].

Dairy cows have a short productive life, so longevity-derived traits are essential from an economic perspective [53]. Combining a long lifespan with high functionality could help improve dairy cow productivity [2,91]. Furthermore, genetic values for milk production, fertility, udder health, and leg health are indicators that should be positively associated with the probability of achieving long-lifetime productivity [53,91]. In conclusion, to increase the lifetime of milk production in dairy herds, producers should select heifers with high scores for conformation traits involving udder, feet, and hooves. In addition, they need to consider high genetic values for milk production, fertility, and udder health [53,92].

5.4. Relationship between Genomic, Genotypic, and Phenotypic Udder Traits: Health, Production, and Longevity

The intricate interplay between genomic, genotypic, and phenotypic udder traits is fundamental to understanding the complex relationship between health, production, and longevity in dairy cattle (Figure 2). The dairy cattle sector is significantly impacted by the economic costs associated with the high incidence and prevalence of clinical and subclinical mastitis, including expenses related to treatment, production losses, and reduced animal welfare [59,93]. The large databases generated have allowed for assessing the incidence of this health problem and investigating the genetic background of clinical mastitis and its relationships with other udder-derived traits of interest for the dairy industry [59]. There is persistent controversy about low milk SCCs and susceptibility to mastitis [94]. However, high SCCs in milk may indicate inflammation or infection of the mammary gland [94]. Several studies described that cow selection for mastitis resistance is often based on genetically correlated indicator traits, such as SCCs, udder depth, and attachment [59,94]. Moreover, genetic correlations showed a significant relationship between clinical mastitis, SCCs, and udder depth during early lactation [59]. Therefore, SCCs are also a reference trait for selecting dairy ruminants less prone to mastitis incidence [59]. The selection programs favoring animals with fewer SCCs are still widely debated [59]. However, the divergent selection of animals based on lower SCCs could solve future udder-related health problems [94,95]. It is important to note that the susceptibility to mastitis linked to SCCs is a phenotypic trait that may be related to immunomodulation but not to selection [75]. Genetic correlations between clinical mastitis and other economically important traits indicated that selection would also improve resistance against other diseases, improving fertility and longevity [51,74]. However, milk production remains positively correlated with clinical mastitis, highlighting the importance of including health traits in phenotypic selection programs to achieve genetic progress for all relevant traits [51,74]. Thus, implementing new genomic evaluation models in phenotypic-related studies is expected to enhance the accuracy and efficiency of selecting dairy cattle with improved mastitis resistance [22,34,39]. The genetic variability of mastitis resistance is well established in dairy cattle, with many studies focused on the estimation of heritabilities and genetic correlation between mastitis-related phenotypic traits such as SCCs and clinical cases [36,77–82].

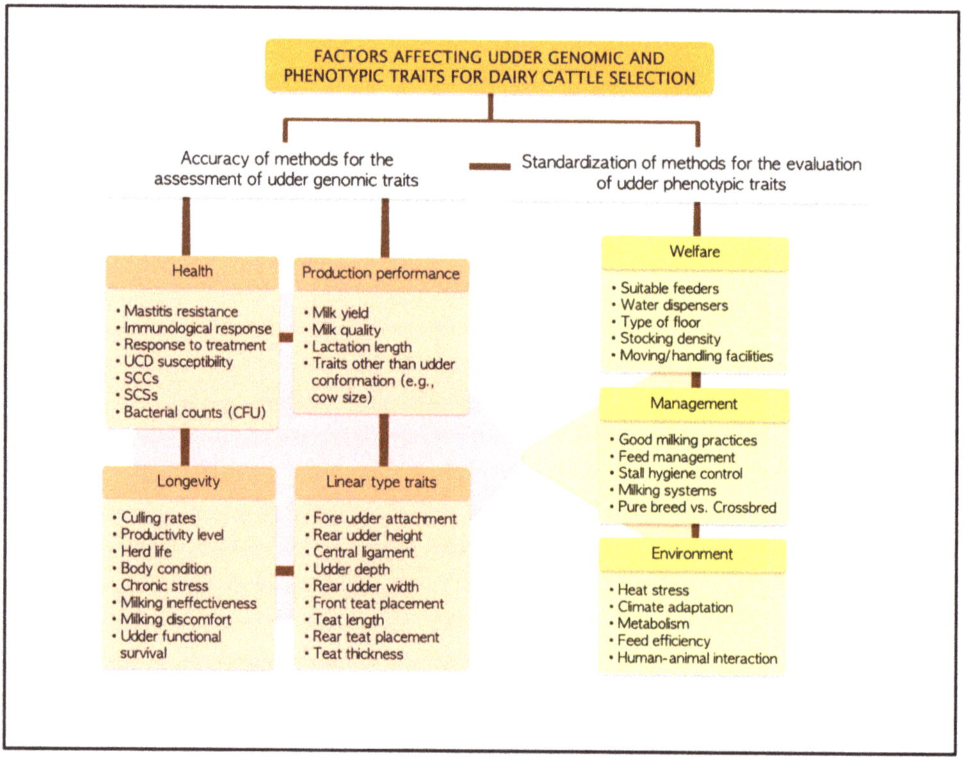

Figure 2. Overview of the factors affecting udder genomic, genotypic, and phenotypic traits for dairy cattle selection. Orange squares on the left: factors affecting udder health, production, and longevity depending on linear (conformation) type traits. Yellow squares on the right: factors affecting udder function depending on welfare, management, and environment-related traits. Both direct and indirect factors are interconnected as they depend on each other and vice versa. Improving dairy herd selection in the medium- and long-term requires the development of cutting-edge prediction methods for udder conformation and function, whether genomic or phenotypic, which underscores their critical importance. Under each factor topic (Health, Production, Longevity, Linear type, Welfare, Management, and Environment), examples of relevant factors are listed. UCD: Udder Cleft Dermatitis; SCCs: Somatic Cell Counts; SCSs: Somatic Cell Scores; CFU: Colony-Forming Units.

Other studies involved molecular genome mapping results which provided information on quantitative trait loci (QTL) related to mastitis resistance and provided a better understanding of the genetic relevance of the traits [37]. Many countries have implemented selection programs based on a linear decrease in SCCs for increasing mastitis resistance [37]. Improving the selection accuracy for mastitis resistance includes advances in modeling, an optimal combination of mastitis-related traits, and associated udder predictors [37]. In addition, the definition of the overall breeding objective that includes udder-related conformational and functional traits and the inclusion of molecular-based information is now available from QTL [37,96]. These cutting-edge studies will lead to a better understanding of the genetic background of mastitis resistance and allow for more accurate selection, improving udder health, animal welfare, and profitability in the future modern dairy industry [59,97]. Other potential future studies on mastitis resistance will be based on bacteriology as new indicator traits and tools for dairy cattle selection [59]. The relationship between cows and microorganisms has traditionally been related to intramammary infections [64]. However, recent metagenomic studies indicate that mammary secretions from clinically healthy quarters may harbor genomic markers for detecting several bacterial groups, the majority of which were not associated with mastitis but were defined as "commensal mammary microbiota" [64,98,99]. Whether disruption of udder microbiota diversity (dysbiosis) plays a role in determining susceptibility to mastitis is still unknown, and little is known about the contributions of various biotic and abiotic factors on the composition of udder microbiota diversity [64,100,101]. Therefore, there is a need to study the teat end, teat canal, and mammary secretions-related microbiota as interconnected niches of a highly dynamic microbial ecosystem [64,102]. In addition, it is necessary to analyze host-associated factors, including physiological and anatomical traits, as well as genetic traits that may affect the udder microbiota [64,103].

Within the domain of conformation traits, a study was conducted to assess the impact of udder morphological characteristics on milk production in *Bos indicus* cows [104,105]. First, the udder diameter and height, teat length and diameter, and milk production were measured, and finally, the study determined the values of udder morphological characteristics in local zebu cows. The results showed that the udder size was highly and positively correlated with milk production. These findings will be instrumental in genetic improvement programs for zebu cows [106]. In studies related to *Bos taurus*, it has been observed that udder morphology, as well as teat tip length and diameter, play a significant role in the incidence of intramammary infection among Jersey crossbred cows in warm and humid climates [63]. Moreover, pendulous udder, flat and inverted teat tip, and very long and thick teat characteristics were associated with higher susceptibility to intramammary infection in Jersey crossbred cows. Therefore, these traits should be considered relevant when selecting dairy cows [63].

On the other hand, the influence of udder and teat conformation and size on the teat-seeking behavior of newborn calves is a determining factor for calves suckling for the first time [23,30,107]. A shorter distance between the udder and ground (low udders) increased the time spent searching for teats and the time of first suckling, which also significantly affected calf survival [107]. Finally, longevity is an increasingly important trait in dairy and beef cattle cows [61,62]. Increased longevity reduces costs and raises incomes for the producers [63]. The phenotypic relationship between conformation-type traits and longevity in cattle estimates the effect of voluntary culling [61,63]. It has been described that udder conformation had no impact on longevity in beef cattle [63]. However, muscle-derived traits were the most important factors for longevity in beef cows [63]. Therefore, relevant udder-derived traits in dairy cattle differ from those applicable to beef cattle.

Dairy cattle offer an attractive model for investigating the genes responsible for the substantial diversity observed both within and between mammalian species concerning their milk volume, protein, and fat composition, highlighting the potential significance of genomic traits. Many phenotypes for these traits and the complete genome sequence of the key founders of modern dairy cattle populations are available. Association tests were conducted on Holstein and Jersey cattle with exceptional phenotypes to identify variants within the target regions, while gene expression data were analyzed to pinpoint potential candidate genes such as *BTRC, MGST1, SLC37A1, STAT5A, STAT5B, PAEP, VDR, CSF2RB, MUC1, NCF4*, and *GHDC* associated with milk production [108]. In *Bos taurus*, 141 and five novel genes related to milk production and SCS have been identified, respectively. These novel genes were also found to be functionally related to heat tolerance (e.g., *SLC45A2, IRAG1*, and *LOC101902172*), longevity (e.g., *SYT10* and *LOC101903327*), and fertility [109].

In a previous study, polymorphisms in the gene regulating the osteopontin OPN (*SPP1*) were found to be associated with milk yield traits and somatic cell score (SCS), a trait used to estimate the genetic value of udder health in dairy cattle [110]. Several genomic regions have been found to contain genes, including *ZNF683, DHX9, CUX1, TNNT1*, and *SPRY1*, that may biologically contribute to mastitis development and whose functions range from regulation of cell proliferation to immune system signaling [111]. Genetic investigation of the composite risk trait involved a new locus and candidate genes that have potentially pleiotropic effects related to mastitis [111]. On the other hand, several studies have identified genetic factors related to mastitis resistance or susceptibility in dairy cows. Classical traits related to mastitis include somatic cell counts, electrical conductivity, and clinical cases of the disease. However, new traits such as acute phase proteins, immunological assays, and milk flow patterns are considered by discovering new biological pathways, e.g., the role of the mammary epithelium and the nervous system and their relationship to disease development [112]. Furthermore, the usefulness of these traits for genetic improvement has been determined [113]. An additional study detected genomic regions that hold potential candidate genes linked to mastitis resistance and susceptibility, immune system and inflammation, milk properties, udder morphology, and various cancer types situated on a segment of chromosome 8 (SNP rs41609496). Among the genes within this segment are AKNA, MIR455, ORM1, and GNG10, which play crucial roles in mammary gland immunity and are relevant to milk production [112].

6. Perspectives and Future Challenges for Genomic, Genotypic, and Phenotypic Udder Evaluation

Several secondary traits have been included in the selection indices, decreasing the emphasis on the milk yield-derived trait [114]. The dairy cattle industry has recognized the importance of including non-production traits in the selection process of functional dairy cattle. These traits include type, growth, size, body composition, nutritional efficiency, disease resistance, udder health, reproduction, and management (Figure 3). Most of these traits can be found within selection indices worldwide, although the relative importance of each trait is variable [115]. For example, the non-lactation traits mentioned above are quantitative, where the whole animal's phenotype represents the sum of minor traits [20].

The development and implementation of genetic evaluations for health traits have led to an accelerated pace for incorporating new functional traits for dairy cattle selection. Since 2018, the Council on Dairy Cattle Breeding (CDCB, Bowie, MD, USA) has made routine official genomic evaluations available to producers for six health-related traits in Holstein cattle [23]. These traits include resistance to milk fever, abomasal displacement, ketosis, clinical mastitis, metritis, and retained placenta. The genetic merit indices include these health-related traits [23]. Improving cow health is mainly carried out through changes in management practices and genetic selection based on SCSs, productive life, and longevity [23,116]. Genomic testing allows an accelerated improvement of traits with low heritability, such as health traits; however, phenotypic traits remain essential for successful genomic evaluations associated with data where appropriate quality control standards have been applied [23,117]. Mastitis control evaluations are submitted to Interbull together with SCSs for international validation and udder health evaluation [23]. In addition, the evaluation includes data from different breeds regarding health traits, using multiple trait models and additional functional evaluations of new traits such as calf health and feed efficiency [23,118].

In the present era, a wealth of information can facilitate informed decisions for the selection of udder-derived traits with greater accuracy than ever before; however, the development of future decisions will require new and ongoing cooperation among livestock industry stakeholders [59]. Therefore, it is crucial to provide accurate and unbiased information related to udder-derived and other production-related traits [23,119]. Functional traits based on direct information about cow health have also become more critical due to increasing concerns about animal welfare and consumer demands for healthy and natural products [2]. Genetic selection can be applied to quantitative traits, but it is challenging to link successful genetic selection to physiological traits [20]. The bovine genome sequence will have an essential role that should not be underestimated in the future of dairy cattle [20,120]. Completing the bovine genome sequence was the first step towards modernizing our approach to dairy cattle genetics [20,121].

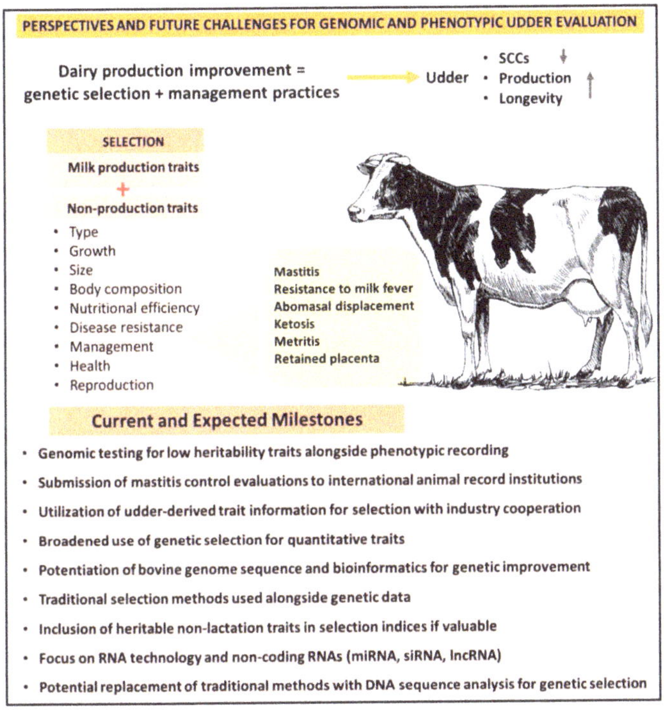

Figure 3. Scheme of the perspectives and future challenges for genomic, genotypic, and phenotypic udder evaluation. Current and expected milestones should be considered for the improvement of dairy cattle selection. miRNA: micro-RNA; siRNA: small interfering RNA; long non-coding RNA

Nevertheless, finding functional genes is complex, and scanning billions of DNA base pairs is expensive and time-consuming [20,122]. Moreover, the function of most genes is unknown or incompletely understood [20]. However, combining all information in a usable format, known as bioinformatics, opens new doors for the future selection of high-production dairy cattle [20,122]. The critical information lies in the subtle changes in gene expression and the cumulative effect that these changes may have [20]. Traditional genetic selection methods in dairy cattle will complement genetic data in the near future [20,123]. The long-term prognosis for genome science is good, but progress will take time. Genetic selection in the genome era will be different because DNA sequence analysis may replace traditional methods for dairy genetic selection [7,13,20,124–127].

The study of microRNAs (miRNAs) remains a challenge. miRNAs are non-coding RNAs that play an essential role in post-transcriptional regulation [128]. *miRNA-148th* is highly expressed and plays a crucial role in epigenetic regulation by decreasing the expression of DNA methyltransferase 1 [103]. Another important miRNA in milk is *miRNA-125b*, which regulates p53 [128]. Therefore, milk-derived miRNAs could be considered potential biomarkers for different mammary gland traits because changes in immunoregulatory miRNAs have been observed in quarters with chronic subclinical mastitis, with *bta-miR-223-3p* being the most upregulated miRNA [129]. Furthermore, recent advances report an essential role of exosomal miRNAs in cell signaling pathways regulating mammary gland immune response in ruminants [130]. In the immediate future, understanding miRNAs will enable their application as genomic markers in dairy cattle as a screening tool for udder health. For example, the miR-200 family of miRNAs plays an essential role in inhibiting the growth and progression of future mammary tumors. Thus, the miR-200 family could also offer a preventive strategy. In addition, increased miR-200 expression in late gestation and early lactation could be a determinant for the development of mammary gland alveoli or the regulation of milk production [131]. The regulatory role of miRNAs in response to most mastitis-causing pathogens is not yet well understood. However, discovering the function of miRNAs and their post-transcriptional regulation would aid in understanding the response of the bovine mammary gland to pathogens and improve lactation efficiency, milk quality, and longevity [132,133].

Besides genomic selection through genetic markers to predict an animal's breeding value for udder health and milk productivity, other emerging technologies include automated milking systems [69], which allow for the real-time data collection on milk yield and udder health. Additionally, new imaging technologies, such as ultrasound [134] and thermography [135], can be used to assess udder health and identify potential issues early on. Machine learning methods, such as neural networks and decision trees, have shown promising results in predicting udder health status based on somatic cell counts. These methods can help identify cows at risk of mastitis and enable earlier interventions. Artificial intelligence, including deep learning algorithms, will likely be explored in the context of udder health prediction [67]. As data collection and analysis technologies continue to improve, it is likely that machine learning and artificial intelligence will play an increasingly important role in predicting and managing udder health in dairy cattle. In the near future, it is likely that these technologies will continue to advance and become more widely used in the dairy industry, potentially leading to more accurate and efficient selection for udder-related traits.

7. Conclusions

Implementing genomic, genotypic, and phenotypic udder evaluation criteria for dairy cattle selection provides solid predictions for improving udder health, milk productivity, and longevity characteristics, thus generating more sustainable genetic lines. The estimation criteria of the different udder-derived traits regarding functionality and conformation should be considered as a whole. Furthermore, the association of both genomic and phenotypic traits significantly improves the accuracy of the prediction values. Thus, udder-derived estimation criteria should be based on different genomic and phenotypic databases to make predictions more reliable. Selection criteria based on genomic and phenotypic udder-derived traits provide a better understanding of the cattle's health, milk yield and quality, welfare, and herd life. This insight becomes highly relevant as estimation criteria derived from genomic and phenotypic data repositories regarding udder-derived traits point the way for selecting future herds and accelerating the process of genetic improvement in dairy cattle.

Author Contributions: All the authors listed in this work have contributed to the preparation and execution of this review. M.A.G.-R., P.M.A.: Conceptualization, methodology, critical reading, writing—original draft, and funding acquisition; and M.G.-H.: conceptualization, methodology, writing—original draft, critical reading, funding acquisition, and writing—review and editing. All authors have read and agreed to the published version of the manuscript.

Funding: The authors received financial support from Agencia Nacional de Investigación y Desarrollo (ANID); Programa de Becas/Doctorado Nacional/Government of Chile: 2020–21201280; National Institute for Agricultural and Veterinary Research (INIAV, Portugal); and Ministerio de Economia y Finanzas/Universidad Técnica de Cotopaxi (UTC)/Proyectos Generativos/Government of Ecuador: DIRGI-CP2022-005.

Institutional Review Board Statement: Not applicable.

Informed Consent Statement: Not applicable.

Data Availability Statement: Data sharing is not applicable to this article.

Acknowledgments: We would like to thank the people and laboratories involved in the research related to the present review topic.

Conflicts of Interest: The authors participating in this review declare not to have conflict of interest.

References

1. Carta, A.; Casu, S.; Salaris, S. Invited Review: Current State of Genetic Improvement in Dairy Sheep. *J. Dairy Sci.* **2009**, *92*, 5814–5833. [CrossRef] [PubMed]
2. Egger-Danner, C.; Cole, J.B.; Pryce, J.E.; Gengler, N.; Heringstad, B.; Bradley, A.; Stock, K.F. Invited Review: Overview of New Traits and Phenotyping Strategies in Dairy Cattle with a Focus on Functional Traits. *Animal* **2014**, *9*, 191–207. [CrossRef] [PubMed]
3. Chesnais, J.P.; Cooper, T.A.; Wiggans, G.R.; Sargolzaei, M.; Pryce, J.E.; Miglior, F. Using Genomics to Enhance Selection of Novel Traits in North American Dairy Cattle. *J. Dairy Sci.* **2016**, *99*, 2413–2427. [CrossRef] [PubMed]
4. Gutiérrez-Reinoso, M.A.; Aponte, P.M.; Cabezas, J.; Rodriguez-Alvarez, L.; Garcia-Herreros, M. Genomic Evaluation of Primiparous High-Producing Dairy Cows: Inbreeding Effects on Genotypic and Phenotypic Production-Reproductive Traits. *Animals* **2020**, *10*, 1704. [CrossRef] [PubMed]
5. Flower, F.C.; Weary, D.M. Gait Assessment in Dairy Cattle. *Animal* **2009**, *3*, 87–95. [CrossRef] [PubMed]
6. Jacobs, J.A.; Siegford, J.M. Lactating Dairy Cows Adapt Quickly to Being Milked by an Automatic Milking System. *J. Dairy Sci.* **2012**, *95*, 1575–1584. [CrossRef] [PubMed]
7. Cecchinato, A.; Macciotta, N.P.P.; Mele, M.; Tagliapietra, F.; Schiavon, S.; Bittante, G.; Pegolo, S. Genetic and Genomic Analyses of Latent Variables Related to the Milk Fatty Acid Profile, Milk Composition, and Udder Health in Dairy Cattle. *J. Dairy Sci.* **2019**, *102*, 5254–5265. [CrossRef]
8. Shanks, R.D.; Rooney, K.A.; Hutjens, M.F. Breeding Practices on Illinois Holstein Farms. *J. Dairy Sci.* **1983**, *66*, 1209–1217. [CrossRef]
9. Beard, J.K.; Musgrave, J.A.; Funston, R.N.; Travis Mulliniks, J. The Effect of Cow Udder Score on Cow/Calf Performance in the Nebraska Sandhills. *Transl. Anim. Sci.* **2019**, *3*, 14–19. [CrossRef]
10. Gutiérrez-Reinoso, M.A.; Aponte, P.M.; García-Herreros, M. A Review of Inbreeding Depression in Dairy Cattle: Current Status, Emerging Control Strategies, and Future Prospects. *J. Dairy Res.* **2022**, *89*, 3–12. [CrossRef]
11. Mancin, E.; Sartori, C.; Guzzo, N.; Tuliozi, B.; Mantovani, R. Selection Response Due to Different Combination of Antagonistic Milk, Beef, and Morphological Traits in the Alpine Grey Cattle Breed. *Animals* **2021**, *11*, 1340. [CrossRef] [PubMed]
12. Castañeda-Bustos, V.J.; Montaldo, H.H.; Valencia-Posadas, M.; Shepard, L.; Pérez-Elizalde, S.; Hernández-Mendo, O.; Torres-Hernández, G. Linear and Nonlinear Genetic Relationships between Type Traits and Productive Life in US Dairy Goats. *J. Dairy Sci.* **2017**, *100*, 1232–1245. [CrossRef] [PubMed]
13. Gutierrez-Reinoso, M.A.; Aponte, P.M.; Garcia-Herreros, M. Genomic Analysis, Progress and Future Perspectives in Dairy Cattle Selection: A Review. *Animals* **2021**, *11*, 599. [CrossRef]
14. Hazel, A.R.; Heins, B.J.; Hansen, L.B. Health Treatment Cost, Stillbirth, Survival, and Conformation of Viking Red-, Montbéliarde-, and Holstein-Sired Crossbred Cows Compared with Pure Holstein Cows during Their First 3 Lactations. *J. Dairy Sci.* **2020**, *103*, 10917–10939. [CrossRef] [PubMed]
15. Heimes, A.; Brodhagen, J.; Weikard, R.; Hammon, H.M.; Meyerholz, M.M.; Petzl, W.; Zerbe, H.; Engelmann, S.; Schmicke, M.; Hoedemaker, M.; et al. Characterization of Functional Traits with Focus on Udder Health in Heifers with Divergent Paternally Inherited Haplotypes on BTA18. *BMC Vet. Res.* **2019**, *15*, 241. [CrossRef]
16. Hansen, L.B.; Cole, J.B.; Marx, G.D.; Seykora, A.J. Productive Life and Reasons for Disposal of Holstein Cows Selected for Large versus Small Body Size. *J. Dairy Sci.* **1999**, *82*, 795–801. [CrossRef] [PubMed]
17. Dadati, E.; Kennedy, B.W.; Burnside, E.B. Relationships between Conformation and Calving Interval in Holstein Cows. *J. Dairy Sci.* **1986**, *69*, 3112–3119. [CrossRef]
18. Strandberg, E.; Shook, G.E. Genetic and Economic Responses to Breeding Programs That Consider Mastitis. *J. Dairy Sci.* **1989**, *72*, 2136–2142. [CrossRef]
19. Nazar, M.; Abdalla, I.M.; Chen, Z.; Ullah, N.; Liang, Y.; Chu, S.; Xu, T.; Mao, Y.; Yang, Z.; Lu, X. Genome-Wide Association Study for Udder Conformation Traits in Chinese Holstein Cattle. *Animals* **2022**, *12*, 2542. [CrossRef]
20. Lucy, M.C. Non-Lactational Traits of Importance in Dairy Cows and Applications for Emerging Biotechnologies. *N. Z. Vet. J.* **2005**, *53*, 406–415. [CrossRef]
21. Pérez-Enciso, M. Genomic Relationships Computed from Either Next-Generation Sequence or Array SNP Data. *J. Anim. Breed. Genet.* **2014**, *131*, 85–96. [CrossRef] [PubMed]
22. Silpa, M.V.; König, S.; Sejian, V.; Malik, P.K.; Nair, M.R.R.; Fonseca, V.F.C.; Maia, A.S.C.; Bhatta, R. Climate-Resilient Dairy Cattle Production: Applications of Genomic Tools and Statistical Models. *Front. Vet. Sci.* **2021**, *8*, 625189. [CrossRef] [PubMed]
23. Parker Gaddis, K.L.; VanRaden, P.M.; Cole, J.B.; Norman, H.D.; Nicolazzi, E.; Dürr, J.W. Symposium Review: Development, Implementation, and Perspectives of Health Evaluations in the United States. *J. Dairy Sci.* **2020**, *103*, 5354–5365. [CrossRef] [PubMed]

24. Hazel, A.R.; Heins, B.J.; Hansen, L.B. Fertility, Survival, and Conformation of Montbéliarde × Holstein and Viking Red × Holstein Crossbred Cows Compared with Pure Holstein Cows during First Lactation in 8 Commercial Dairy Herds. *J. Dairy Sci.* **2017**, *100*, 9447–9458. [CrossRef]
25. Wiggans, G.R.; Hubbard, S.M. Genetic Evaluation of Yield and Type Traits of Dairy Goats in the United States. *J. Dairy Sci.* **2001**, *84*, 69–73. [CrossRef]
26. Casu, S.; Pernazza, I.; Carta, A. Feasibility of a Linear Scoring Method of Udder Morphology for the Selection Scheme of Sardinian Sheep. *J. Dairy Sci.* **2006**, *89*, 2200–2209. [CrossRef]
27. Kelm, S.C.; Freeman, A.E.; Brundage, A.L.; Pearson, R.E.; Martin, T.G.; McGilliard, L.D.; Hansen, L.B.; Young, C.W.; Voelker, H.H.; Shook, G.E.; et al. Direct and Correlated Responses to Selection for Milk Yield: Results and Conclusions of Regional Project NC-2, "Improvement of Dairy Cattle through Breeding, with Emphasis on Selection". *J. Dairy Sci.* **2000**, *83*, 2721–2732. [CrossRef]
28. Varona, L.; Moreno, C.; Altarriba, J. Genetic Correlation of Longevity with Growth, Post-Mortem, Docility and Some Morphological Traits in the Pirenaica Beef Cattle Breed. *Animal* **2012**, *6*, 873–879. [CrossRef]
29. Persson Waller, K.; Persson, Y.; Nyman, A.K.; Stengärde, L. Udder Health in Beef Cows and Its Association with Calf Growth. *Acta Vet. Scand.* **2014**, *56*, 9. [CrossRef]
30. Just, A.; Wellmann, R.; Bennewitz, J. Estimation of Relative Economic Weights and the Marginal Willingness to Pay for Breeding Traits of Brown Swiss Cattle Using Discrete Choice Experiments. *J. Dairy Sci.* **2018**, *101*, 5207–5213. [CrossRef]
31. Veerkamp, R.F.; Gerritsen, C.L.M.; Koenen, E.P.C.; Hamoen, A.; De Jong, G. Evaluation of Classifiers That Score Linear Type Traits and Body Condition Score Using Common Sires. *J. Dairy Sci.* **2002**, *85*, 976–983. [CrossRef] [PubMed]
32. Buaban, S.; Prempree, S.; Sumreddee, P.; Duangjinda, M.; Masuda, Y. Genomic Prediction of Milk-Production Traits and Somatic Cell Score Using Single-Step Genomic Best Linear Unbiased Predictor with Random Regression Test-Day Model in Thai Dairy Cattle. *J. Dairy Sci.* **2021**, *104*, 12713–12723. [CrossRef] [PubMed]
33. Dahiya, S.; Kumar, S.; Kumar, M. Current Status of Research on Linear Type Traits in Indian Cattle and Future Strategies. *Trop. Anim. Health Prod.* **2020**, *52*, 2221–2232. [CrossRef] [PubMed]
34. Manafiazar, G.; Goonewardene, L.; Miglior, F.; Crews, D.H.; Basarab, J.A.; Okine, E.; Wang, Z. Genetic and Phenotypic Correlations among Feed Efficiency, Production and Selected Conformation Traits in Dairy Cows. *Animal* **2015**, *10*, 381–389. [CrossRef] [PubMed]
35. Miglior, F.; Koeck, A.; Jamrozik, J.; Schenkel, F.; Kelton, D.; Kistemaker, G.; Van Doormaal, B. Index for Mastitis Resistance and Use of BHBA for Evaluation of Health Traits in Canadian Holsteins. *Interbull Bull.* **2014**, *48*, 73–78.
36. Parker Gaddis, K.L.; Cole, J.B.; Clay, J.S.; Maltecca, C. Genomic Selection for Producer-Recorded Health Event Data in US Dairy Cattle. *J. Dairy Sci.* **2014**, *97*, 3190–3199. [CrossRef]
37. Rupp, R.; Boichard, D. Genetics of Resistance to Mastitis in Dairy Cattle. *Vet. Res.* **2003**, *34*, 671–688. [CrossRef]
38. Sinha, R.; Sinha, B.; Kumari, R.; Vineeth, M.R.; Sharma, N.; Verma, A.; Gupta, I.D. Association of Udder Type Traits with Single Nucleotide Polymorphisms in Sahiwal (Bos indicus) and Karan Fries (*Bos taurus* × *Bos indicus*) Cattle. *Anim. Biotechnol.* **2022**, 1–12. [CrossRef]
39. Pausch, H.; Emmerling, R.; Schwarzenbacher, H.; Fries, R. A Multi-Trait Meta-Analysis with Imputed Sequence Variants Reveals Twelve QTL for Mammary Gland Morphology in Fleckvieh Cattle. *Genet. Sel. Evol.* **2016**, *48*, 14. [CrossRef]
40. Jiang, J.; Ma, L.; Prakapenka, D.; VanRaden, P.M.; Cole, J.B.; Da, Y. A Large-Scale Genome-Wide Association Study in U.S. Holstein Cattle. *Front. Genet.* **2019**, *10*, 412. [CrossRef]
41. Zwald, N.R.; Weigel, K.A.; Chang, Y.M.; Welper, R.D.; Clay, J.S. Genetic Selection for Health Traits Using Producer-Recorded Data. II. Genetic Correlations, Disease Probabilities, and Relationships with Existing Traits. *J. Dairy Sci.* **2004**, *87*, 4295–4302. [CrossRef] [PubMed]
42. Heringstad, B.; Klemetsdal, G.; Steine, T. Selection Responses for Disease Resistance in Two Selection Experiments with Norwegian Red Cows. *J. Dairy Sci.* **2007**, *90*, 2419–2426. [CrossRef] [PubMed]
43. Jamrozik, J.; Koeck, A.; Miglior, F.; Kistemaker, G.; Schenkel, F.; Kelton, D.; Doormaal, B. Van Genetic and Genomic Evaluation of Mastitis Resistance in Canada. *Interbull Bull.* **2013**, *47*.
44. Koeck, A.; Miglior, F.; Kelton, D.F.; Schenkel, F.S. Alternative Somatic Cell Count Traits to Improve Mastitis Resistance in Canadian Holsteins. *J. Dairy Sci.* **2012**, *95*, 432–439. [CrossRef] [PubMed]
45. Miller, R.H. Traits for Sire Selection Related to Udder Health and Management. *J. Dairy Sci.* **1984**, *67*, 459–471. [CrossRef]
46. Thomas, C.L.; Vinson, W.E.; Pearson, R.E.; Dickinson, F.N.; Johnson, L.P. Relationships between Linear Type Scores, Objective Type Measures, and Indicators of Mastitis. *J. Dairy Sci.* **1984**, *67*, 1281–1292. [CrossRef]
47. Ekman, L.; Nyman, A.K.; Landin, H.; Magnusson, U.; Waller, K.P. Mild and Severe Udder Cleft Dermatitis-Prevalence and Risk Factors in Swedish Dairy Herds. *J. Dairy Sci.* **2018**, *101*, 556–571. [CrossRef]
48. Marete, A.; Lund, M.S.; Boichard, D.; Ramayo-Caldas, Y. A System-Based Analysis of the Genetic Determinism of Udder Conformation and Health Phenotypes across Three French Dairy Cattle Breeds. *PLoS ONE* **2018**, *13*, e0199931. [CrossRef]
49. Wagner, P.; Yin, T.; Brügemann, K.; Engel, P.; Weimann, C.; Schlez, K.; König, S. Genome-Wide Associations for Microscopic Differential Somatic Cell Count and Specific Mastitis Pathogens in Holstein Cows in Compost-Bedded Pack and Cubicle Farming Systems. *Animals* **2021**, *11*, 1839. [CrossRef]
50. Yang, C.; Zhao, X.; Cui, N.; Liang, Y. Cadherins Associate with Distinct Stem Cell-Related Transcription Factors to Coordinate the Maintenance of Stemness in Triple-Negative Breast Cancer. *Stem Cells Int.* **2017**, *2017*, 5091541. [CrossRef]

51. Li, Y.; Han, B.; Liu, L.; Zhao, F.; Liang, W.; Jiang, J.; Yang, Y.; Ma, Z.; Sun, D. Genetic Association of DDIT3, RPL23A, SESN2 and NR4A1 Genes with Milk Yield and Composition in Dairy Cattle. *Anim. Genet.* **2019**, *50*, 123–135. [CrossRef] [PubMed]
52. Fernando, R.S.; Spahr, S.L.; Jaster, E.H. Comparison of Electrical Conductivity of Milk with Other Indirect Methods for Detection of Subclinical Mastitis. *J. Dairy Sci.* **1985**, *68*, 449–456. [CrossRef] [PubMed]
53. Van Eetvelde, M.; Verdru, K.; de Jong, G.; van Pelt, M.L.; Meesters, M.; Opsomer, G. Researching 100 t Cows: An Innovative Approach to Identify Intrinsic Cows Factors Associated with a High Lifetime Milk Production. *Prev. Vet. Med.* **2021**, *193*, 105392. [CrossRef] [PubMed]
54. Lancelot, R.; Paye, B.; Lescourret, F. Factors Affecting the Distribution of Clinical Mastitis among Udder Quarters in French Dairy Cows. *Vet. Res.* **1997**, *28*, 45–53. [PubMed]
55. Bhutto, A.L.; Murray, R.D.; Woldehiwet, Z. Udder Shape and Teat-End Lesions as Potential Risk Factors for High Somatic Cell Counts and Intra-Mammary Infections in Dairy Cows. *Vet. J.* **2010**, *183*, 63–67. [CrossRef]
56. Guarín, J.F.; Baumberger, C.; Ruegg, P.L. Anatomical Characteristics of Teats and Premilking Bacterial Counts of Teat Skin Swabs of Primiparous Cows Exposed to Different Types of Bedding. *J. Dairy Sci.* **2017**, *100*, 1436–1444. [CrossRef]
57. Miller, R.H.; Pearson, R.E.; Rothschild, M.F.; Fulton, L.A. Comparison of Single and Multiple-Trait Selected Sires. Response in Mastitis Traits. *J. Dairy Sci.* **1981**, *64*, 832–837. [CrossRef]
58. Boettcher, P.J.; Dekkers, J.C.M.; Kolstad, B.W. Development of an Udder Health Index for Sire Selection Based on Somatic Cell Score, Udder Conformation, and Milking Speed. *J. Dairy Sci.* **1998**, *81*, 1157–1168. [CrossRef]
59. Martin, P.; Barkema, H.W.; Brito, L.F.; Narayana, S.G.; Miglior, F. Symposium Review: Novel Strategies to Genetically Improve Mastitis Resistance in Dairy Cattle. *J. Dairy Sci.* **2018**, *101*, 2724–2736. [CrossRef]
60. Van der Geer, D.; Grommers, F.J.; van Houten, M. Comparison of Dairy Cows with Low or High Rate of Udder Infection. *Tijdschr. Diergeneeskd.* **1979**, *104*. [CrossRef]
61. Huntley, S.J.; Cooper, S.; Bradley, A.J.; Green, L.E. A Cohort Study of the Associations between Udder Conformation, Milk Somatic Cell Count, and Lamb Weight in Suckler Ewes. *J. Dairy Sci.* **2012**, *95*, 5001–5010. [CrossRef] [PubMed]
62. Alain, K.; Karrow, N.A.; Thibault, C.; St-Pierre, J.; Lessard, M.; Bissonnette, N. Osteopontin: An Early Innate Immune Marker of Escherichia Coli Mastitis Harbors Genetic Polymorphisms with Possible Links with Resistance to Mastitis. *BMC Genom.* **2009**, *10*, 444. [CrossRef] [PubMed]
63. Bharti, P.; Bhakat, C.; Pankaj, P.K.; Bhat, S.A.; Arul Prakash, M.; Thul, M.R.; Puhle Japheth, K. Relationship of Udder and Teat Conformation with Intra-Mammary Infection in Crossbred Cows under Hot-Humid Climate. *Vet. World* **2015**, *8*, 898–901. [CrossRef] [PubMed]
64. Derakhshani, H.; Fehr, K.B.; Sepehri, S.; Francoz, D.; De Buck, J.; Barkema, H.W.; Plaizier, J.C.; Khafipour, E. Invited Review: Microbiota of the Bovine Udder: Contributing Factors and Potential Implications for Udder Health and Mastitis Susceptibility. *J. Dairy Sci.* **2018**, *101*, 10605–10625. [CrossRef] [PubMed]
65. Zahoor, A.; Yang, Y.; Yang, C.; Akhtar, M.; Guo, Y.; Shaukat, A.; Guo, M.Y.; Deng, G. Gas6 Negatively Regulates the Staphylococcus Aureus-Induced Inflammatory Response via TLR Signaling in the Mouse Mammary Gland. *J. Cell. Physiol.* **2020**, *235*, 7081–7093. [CrossRef]
66. Van Acker, H.H.; Capsomidis, A.; Smits, E.L.; Van Tendeloo, V.F. CD56 in the Immune System: More Than a Marker for Cytotoxicity? *Front. Immunol.* **2017**, *8*, 892. [CrossRef]
67. Bobbo, T.; Biffani, S.; Taccioli, C.; Penasa, M.; Cassandro, M. Comparison of Machine Learning Methods to Predict Udder Health Status Based on Somatic Cell Counts in Dairy Cows. *Sci. Rep.* **2021**, *11*, 13642. [CrossRef]
68. Bobbo, T.; Matera, R.; Pedota, G.; Manunza, A.; Cotticelli, A.; Neglia, G.; Biffani, S. Exploiting Machine Learning Methods with Monthly Routine Milk Recording Data and Climatic Information to Predict Subclinical Mastitis in Italian Mediterranean Buffaloes. *J. Dairy Sci.* **2023**, *106*, 1942–1952. [CrossRef]
69. Rodenburg, J. Robotic Milking: Technology, Farm Design, and Effects on Work Flow. *J. Dairy Sci.* **2017**, *100*, 7729–7738. [CrossRef]
70. Ebrahimi, M.; Mohammadi-Dehcheshmeh, M.; Ebrahimie, E.; Petrovski, K.R. Comprehensive Analysis of Machine Learning Models for Prediction of Sub-Clinical Mastitis: Deep Learning and Gradient-Boosted Trees Outperform Other Models. *Comput. Biol. Med.* **2019**, *114*, 103456. [CrossRef]
71. Zhou, X.; Xu, C.; Wang, H.; Xu, W.; Zhao, Z.; Chen, M.; Jia, B.; Huang, B. The Early Prediction of Common Disorders in Dairy Cows Monitored by Automatic Systems with Machine Learning Algorithms. *Animals* **2022**, *12*, 1251. [CrossRef] [PubMed]
72. Sawa, A.; Bogucki, M.; Krężel-Czopek, S.; Neja, W. Relationship between Conformation Traits and Lifetime Production Efficiency of Cows. *ISRN Vet. Sci.* **2013**, *2013*, 124690. [CrossRef] [PubMed]
73. Jeretina, J.; Skok, D.J. Genes Associated with Somatic Cell Count Index in Brown Swiss Cattle. *J. Anim. Sci.* **2020**, *98*, skaa330. [CrossRef] [PubMed]
74. Mekonnen, S.A.; Koop, G.; Getaneh, A.M.; Lam, T.J.G.M.; Hogeveen, H. Failure Costs Associated with Mastitis in Smallholder Dairy Farms Keeping Holstein Friesian × Zebu Crossbreed Cows. *Animal* **2019**, *13*, 2650–2659. [CrossRef] [PubMed]
75. Knob, D.A.; Scholz, A.M.; Alessio, D.R.M.; Mendes, B.P.B.; Perazzoli, L.; Kappes, R.; Thaler Neto, A. Reproductive and Productive Performance, Udder Health, and Conformation Traits of Purebred Holstein, F1, and R1 Crossbred Holstein × Simmental Cows. *Trop. Anim. Health Prod.* **2020**, *52*, 1639–1647. [CrossRef] [PubMed]
76. Li, T.; Gao, J.; Zhao, X.; Ma, Y. Digital Gene Expression Analyses of Mammary Glands from Meat Ewes Naturally Infected with Clinical Mastitis. *R. Soc. Open Sci.* **2019**, *6*, 181604. [CrossRef]

77. Lemay, D.G.; Hovey, R.C.; Hartono, S.R.; Hinde, K.; Smilowitz, J.T.; Ventimiglia, F.; Schmidt, K.A.; Lee, J.W.S.; Islas-Trejo, A.; Silva, P.I.; et al. Sequencing the Transcriptome of Milk Production: Milk Trumps Mammary Tissue. *BMC Genom.* **2013**, *14*, 872 [CrossRef]
78. Raschia, M.A.; Nani, J.P.; Maizon, D.O.; Beribe, M.J.; Amadio, A.F.; Poli, M.A. Single Nucleotide Polymorphisms in Candidate Genes Associated with Milk Yield in Argentinean Holstein and Holstein x Jersey Cows. *J. Anim. Sci. Technol.* **2018**, *60*, 31 [CrossRef]
79. Lashneva, I.; Sermyagin, A.A.; Ignatieva, L.P.; Gladyr, E.; Ermilov, A.; Zinovieva, N.A. PSXII-7 Milk Somatic Cells Monitoring in Russian Holstein Cattle Population as a Base for Determining Genetic and Genomic Variability. *J. Anim. Sci.* **2021**, *99*, 252. [CrossRef]
80. Boettcher, P.J.; Jairath, L.K.; Koots, K.R.; Dekkers, J.C.M. Effects of Interactions between Type and Milk Production on Survival Traits of Canadian Holsteins. *J. Dairy Sci.* **1997**, *80*, 2984–2995. [CrossRef]
81. Zavadilová, L.; Němcová, E.; Štípková, M. Effect of Type Traits on Functional Longevity of Czech Holstein Cows Estimated from a Cox Proportional Hazards Model. *J. Dairy Sci.* **2011**, *94*, 4090–4099. [CrossRef]
82. Sasaki, O. Estimation of Genetic Parameters for Longevity Traits in Dairy Cattle: A Review with Focus on the Characteristics of Analytical Models. *Anim. Sci. J.* **2013**, *84*, 449–460. [CrossRef] [PubMed]
83. Forabosco, F.; Groen, A.F.; Bozzi, R.; Van Arendonk, J.A.M.; Filippini, F.; Boettcher, P.; Bijma, P. Phenotypic Relationships between Longevity, Type Traits, and Production in Chianina Beef Cattle. *J. Anim. Sci.* **2004**, *82*, 1572–1580. [CrossRef] [PubMed]
84. Hocking, P.M.; McAllister, A.J.; Wolynetz, M.S.; Batra, T.R.; Lee, A.J.; Lin, C.Y.; Roy, G.L.; Vesely, J.A.; Wauthy, J.M.; Winter, K.A. Factors Affecting Length of Herdlife in Purebred and Crossbred Dairy Cattle. *J. Dairy Sci.* **1988**, *71*, 1011–1024. [CrossRef]
85. Ashwell, M.S.; Van Tassell, C.P.; Sonstegard, T.S. A Genome Scan to Identify Quantitative Trait Loci Affecting Economically Important Traits in a US Holstein Population. *J. Dairy Sci.* **2001**, *84*, 2535–2542. [CrossRef] [PubMed]
86. Szyda, J.; Morek-Kopeć, M.; Komisarek, J.; Zarnecki, A. Evaluating Markers in Selected Genes for Association with Functional Longevity of Dairy Cattle. *BMC Genet.* **2011**, *12*, 30. [CrossRef] [PubMed]
87. Shabalina, T.; Yin, T.; König, S. Survival Analyses in Holstein Cows Considering Direct Disease Diagnoses and Specific SNP Marker Effects. *J. Dairy Sci.* **2020**, *103*, 8257–8273. [CrossRef]
88. Chen, X.; Zhang, S.; Cheng, Z.; Cooke, J.S.; Werling, D.; Wathes, D.C.; Pollott, G.E. Polymorphisms in the Selectin Gene Cluster Are Associated with Fertility and Survival Time in a Population of Holstein Friesian Cows. *PLoS ONE* **2017**, *12*, e0175555. [CrossRef]
89. Stefanowska, J.; Plavsic, M.; Ipema, A.H.; Hendriks, M.M.W.B. The Effect of Omitted Milking on the Behaviour of Cows in the Context of Cluster Attachment Failure during Automatic Milking. *Appl. Anim. Behav. Sci.* **2000**, *67*, 277–291. [CrossRef]
90. Kern, E.L.; Cobuci, J.A.; Costa, C.N.; Pimentel, C.M.M.M. Factor Analysis of Linear Type Traits and Their Relation with Longevity in Brazilian Holstein Cattle. *Asian-Australas. J. Anim. Sci.* **2014**, *27*, 784–790. [CrossRef]
91. Short, T.H.; Lawlor, T.J. Genetic Parameters of Conformation Traits, Milk Yield, and Herd Life in Holsteins. *J. Dairy Sci.* **1992**, *75*, 1987–1998. [CrossRef] [PubMed]
92. Williams, M.; Sleator, R.D.; Murphy, C.P.; McCarthy, J.; Berry, D.P. Re-Assessing the Importance of Linear Type Traits in Predicting Genetic Merit for Survival in an Aging Holstein-Friesian Dairy Cow Population. *J. Dairy Sci.* **2022**, *105*, 7550–7563. [CrossRef] [PubMed]
93. Malchiodi, F.; Jamrozik, J.; Christen, A.M.; Fleming, A.; Kistemaker, G.J.; Richardson, C.; Daniel, V.; Kelton, D.F.; Schenkel, F.S.; Miglior, F. Symposium Review: Multiple-Trait Single-Step Genomic Evaluation for Hoof Health. *J. Dairy Sci.* **2020**, *103*, 5346–5353. [CrossRef]
94. Rainard, P.; Foucras, G.; Boichard, D.; Rupp, R. Invited Review: Low Milk Somatic Cell Count and Susceptibility to Mastitis. *J. Dairy Sci.* **2018**, *101*, 6703–6714. [CrossRef] [PubMed]
95. Cole, J.B.; VanRaden, P.M. Symposium Review: Possibilities in an Age of Genomics: The Future of Selection Indices. *J. Dairy Sci.* **2018**, *101*, 3686–3701. [CrossRef]
96. Cole, J.B.; Wiggans, G.R.; Ma, L.; Sonstegard, T.S.; Lawlor, T.J.; Crooker, B.A.; Van Tassell, C.P.; Yang, J.; Wang, S.; Matukumalli, L.K.; et al. Genome-Wide Association Analysis of Thirty One Production, Health, Reproduction and Body Conformation Traits in Contemporary U.S. Holstein Cows. *BMC Genom.* **2011**, *12*, 408. [CrossRef]
97. Sahana, G.; Lund, M.S.; Andersson-Eklund, L.; Hastings, N.; Fernandez, A.; Iso-Touru, T.; Thomsen, B.; Viitala, S.; Sørensen, P.; Williams, J.L.; et al. Fine-Mapping QTL for Mastitis Resistance on BTA9 in Three Nordic Red Cattle Breeds. *Anim. Genet.* **2008**, *39*, 354–362. [CrossRef]
98. Cremonesi, P.; Severgnini, M.; Romanò, A.; Sala, L.; Luini, M.; Castiglioni, B. Bovine Milk Microbiota: Comparison among Three Different DNA Extraction Protocols to Identify a Better Approach for Bacterial Analysis. *Microbiol. Spectr.* **2021**, *9*, e00374-21. [CrossRef]
99. Gryaznova, M.V.; Syromyatnikov, M.Y.; Dvoretskaya, Y.D.; Solodskikh, S.A.; Klimov, N.T.; Mikhalev, V.I.; Zimnikov, V.I.; Mikhaylov, E.V.; Popov, V.N. Microbiota of Cow's Milk with Udder Pathologies. *Microorganisms* **2021**, *9*, 1974. [CrossRef]
100. Rainard, P. Mammary Microbiota of Dairy Ruminants: Fact or Fiction? *Vet. Res.* **2017**, *48*, 25. [CrossRef]
101. Hoque, M.N.; Istiaq, A.; Clement, R.A.; Sultana, M.; Crandall, K.A.; Siddiki, A.Z.; Hossain, M.A. Metagenomic Deep Sequencing Reveals Association of Microbiome Signature with Functional Biases in Bovine Mastitis. *Sci. Rep.* **2019**, *9*, 13536. [CrossRef] [PubMed]

102. Zadoks, R.N.; Middleton, J.R.; McDougall, S.; Katholm, J.; Schukken, Y.H. Molecular Epidemiology of Mastitis Pathogens of Dairy Cattle and Comparative Relevance to Humans. *J. Mammary Gland Biol. Neoplasia* **2011**, *16*, 357–372. [CrossRef] [PubMed]
103. Doyle, C.J.; Gleeson, D.; O'Toole, P.W.; Cotter, P.D. Impacts of Seasonal Housing and Teat Preparation on Raw Milk Microbiota: A High-Throughput Sequencing Study. *Appl. Environ. Microbiol.* **2016**, *83*, AEM-02694. [CrossRef] [PubMed]
104. Riley, D.G.; Sanders, J.O.; Knutson, R.E.; Lunt, D.K. Comparison of F1 Bos Indicus x Hereford Cows in Central Texas: II. Udder, Mouth, Longevity, and Lifetime Productivity. *J. Anim. Sci.* **2001**, *79*, 1439–1449. [CrossRef]
105. Tolleson, M.W.; Gill, C.A.; Herring, A.D.; Riggs, P.K.; Sawyer, J.E.; Sanders, J.O.; Riley, D.G. Association of Udder Traits with Single Nucleotide Polymorphisms in Crossbred Bos Indicus- Bos Taurus Cows. *J. Anim. Sci.* **2017**, *95*, 2399–2407. [CrossRef]
106. Mingoas, K.J.P.; Awah-Ndukum, J.; Dakyang, H.; Zoli, P.A. Effects of Body Conformation and Udder Morphology on Milk Yield of Zebu Cows in North Region of Cameroon. *Vet. World* **2017**, *10*, 901. [CrossRef]
107. Ventorp, M.; Michanek, P. The Importance of Udder and Teat Conformation for Teat Seeking by the Newborn Calf. *J. Dairy Sci.* **1992**, *75*, 262–268. [CrossRef]
108. Raven, L.A.; Cocks, B.G.; Kemper, K.E.; Chamberlain, A.J.; Vander Jagt, C.J.; Goddard, M.E.; Hayes, B.J. Targeted Imputation of Sequence Variants and Gene Expression Profiling Identifies Twelve Candidate Genes Associated with Lactation Volume, Composition and Calving Interval in Dairy Cattle. *Mamm. Genome* **2016**, *27*, 81–97. [CrossRef]
109. Buaban, S.; Lengnudum, K.; Boonkum, W.; Phakdeedindan, P. Genome-Wide Association Study on Milk Production and Somatic Cell Score for Thai Dairy Cattle Using Weighted Single-Step Approach with Random Regression Test-Day Model. *J. Dairy Sci.* **2022**, *105*, 468–494. [CrossRef]
110. Dudemaine, P.L.; Thibault, C.; Alain, K.; Bissonnette, N. Genetic Variations in the SPP1 Promoter Affect Gene Expression and the Level of Osteopontin Secretion into Bovine Milk. *Anim. Genet.* **2014**, *45*, 629–640. [CrossRef]
111. Miles, A.M.; Posbergh, C.J.; Huson, H.J. Direct Phenotyping and Principal Component Analysis of Type Traits Implicate Novel QTL in Bovine Mastitis through Genome-Wide Association. *Animals* **2021**, *11*, 1147. [CrossRef] [PubMed]
112. Abbasi-Moshaii, B.; Moradi, M.H.; Yin, T.; Rahimi-Mianji, G.; Nejati-Javaremi, A.; König, S. Genome-Wide Scan for Selective Sweeps Identifies Novel Loci Associated with Resistance to Mastitis in German Holstein Cattle. *J. Anim. Breed. Genet.* **2023**, *140*, 92–105. [CrossRef] [PubMed]
113. Detilleux, J.C. Genetic Factors Affecting Susceptibility to Udder Pathogens. *Vet. Microbiol.* **2009**, *134*, 157–164. [CrossRef] [PubMed]
114. Brito, L.F.; Bedere, N.; Douhard, F.; Oliveira, H.R.; Arnal, M.; Peñagaricano, F.; Schinckel, A.P.; Baes, C.F.; Miglior, F. Review: Genetic Selection of High-Yielding Dairy Cattle toward Sustainable Farming Systems in a Rapidly Changing World. *Animal* **2021**, *15*, 100292. [CrossRef] [PubMed]
115. Hagan, B.A.; Moro-Mendez, J.; Cue, R.I. Realized Genetic Selection Differentials in Canadian Ayrshire, Jersey, and Brown Swiss Dairy Cattle Populations. *J. Dairy Sci.* **2021**, *104*, 1951–1966. [CrossRef]
116. González, M.E.; González, V.M.; Montaño, M.F.; Medina, G.E.; Mahadevan, P.; Villa, C.; Villa, R. Genome-Wide Association Analysis of Body Conformation Traits in Mexican Holstein Cattle Using a Mix of Sampled and Imputed SNP Genotypes. *Genet. Mol. Res.* **2017**, *16*. [CrossRef]
117. Lund, M.S.; Guldbrandtsen, B.; Buitenhuis, A.J.; Thomsen, B.; Bendixen, C. Detection of Quantitative Trait Loci in Danish Holstein Cattle Affecting Clinical Mastitis, Somatic Cell Score, Udder Conformation Traits, and Assessment of Associated Effects on Milk Yield. *J. Dairy Sci.* **2008**, *91*, 4028–4036. [CrossRef]
118. Wu, X.; Lund, M.S.; Sahana, G.; Guldbrandtsen, B.; Sun, D.; Zhang, Q.; Su, G. Association Analysis for Udder Health Based on SNP-Panel and Sequence Data in Danish Holsteins. *Genet. Sel. Evol.* **2015**, *47*, 50. [CrossRef]
119. Fessenden, B.; Weigel, D.J.; Osterstock, J.; Galligan, D.T.; Di Croce, F. Validation of Genomic Predictions for a Lifetime Merit Selection Index for the US Dairy Industry. *J. Dairy Sci.* **2020**, *103*, 10414–10428. [CrossRef]
120. Cai, Z.; Guldbrandtsen, B.; Lund, M.S.; Sahana, G. Weighting Sequence Variants Based on Their Annotation Increases the Power of Genome-Wide Association Studies in Dairy Cattle. *Genet. Sel. Evol.* **2019**, *51*, 20. [CrossRef]
121. Mesbah-Uddin, M.; Guldbrandtsen, B.; Capitan, A.; Lund, M.S.; Boichard, D.; Sahana, G. Genome-Wide Association Study with Imputed Whole-Genome Sequence Variants Including Large Deletions for Female Fertility in 3 Nordic Dairy Cattle Breeds. *J. Dairy Sci.* **2022**, *105*, 1298–1313. [CrossRef] [PubMed]
122. VanRaden, P.M.; Tooker, M.E.; O'Connell, J.R.; Cole, J.B.; Bickhart, D.M. Selecting Sequence Variants to Improve Genomic Predictions for Dairy Cattle. *Genet. Sel. Evol.* **2017**, *49*, 32. [CrossRef]
123. VanRaden, P.M. Efficient Methods to Compute Genomic Predictions. *J. Dairy Sci.* **2008**, *91*, 4414–4423. [CrossRef] [PubMed]
124. Schrooten, C.; Bovenhuis, H.; Coppieters, W.; Van Arendonk, J.A.M. Whole Genome Scan to Detect Quantitative Trait Loci for Conformation and Functional Traits in Dairy Cattle. *J. Dairy Sci.* **2000**, *83*, 795–806. [CrossRef]
125. Glantz, M.; Lindmark Månsson, H.; Paulsson, M.; Stålhammar, H. Genomic Selection in Relation to Bovine Milk Composition and Processability. *J. Dairy Res.* **2012**, *79*, 53–59. [CrossRef]
126. Jiménez-Montero, J.A.; González-Recio, O.; Alenda, R. Genotyping Strategies for Genomic Selection in Small Dairy Cattle Populations. *Animal* **2012**, *6*, 1216–1224. [CrossRef]
127. Van der Heide, E.M.M.; Veerkamp, R.F.; van Pelt, M.L.; Kamphuis, C.; Ducro, B.J. Predicting Survival in Dairy Cattle by Combining Genomic Breeding Values and Phenotypic Information. *J. Dairy Sci.* **2020**, *103*, 556–571. [CrossRef]

128. Melnik, B.C.; Schmitz, G. MicroRNAs: Milk's Epigenetic Regulators. *Best Pract. Res. Clin. Endocrinol. Metab.* **2017**, *31*, 427–442. [CrossRef]
129. Saenz-de-Juano, M.D.; Silvestrelli, G.; Bauersachs, S.; Ulbrich, S.E. Determining Extracellular Vesicles Properties and MiRNA Cargo Variability in Bovine Milk from Healthy Cows and Cows Undergoing Subclinical Mastitis. *BMC Genom.* **2022**, *23*, 189. [CrossRef]
130. Cintio, M.; Polacchini, G.; Scarsella, E.; Montanari, T.; Stefanon, B.; Colitti, M. MicroRNA Milk Exosomes: From Cellular Regulator to Genomic Marker. *Animals* **2020**, *10*, 1126. [CrossRef] [PubMed]
131. Roth, M.J.; Moorehead, R.A. The MiR-200 Family in Normal Mammary Gland Development. *BMC Dev. Biol.* **2021**, *21*, 12. [CrossRef] [PubMed]
132. Tabashiri, R.; Sharifi, S.; Pakdel, A.; Bakhtiarizadeh, M.R.; Pakdel, M.H.; Tahmasebi, A.; Hercus, C. Genome-Wide Post-Transcriptional Regulation of Bovine Mammary Gland Response to Streptococcus Uberis. *J. Appl. Genet.* **2022**, *63*, 771–782. [CrossRef] [PubMed]
133. Jin, X.L.; Yang, J.X.; Li, Z.; Liu, H.Y.; Liu, J.X. Progress on the MiRNA Related with Mammary Gland Development and Lactation. *Yi Chuan* **2013**, *35*, 695–702. [CrossRef] [PubMed]
134. Schwarz, T.; Scheeres, N.; Małopolska, M.M.; Murawski, M.; Agustin, T.D.; Ahmadi, B.; Strzałkowska, N.; Rajtar, P.; Micek, P.; Bartlewski, P.M. Associations between Mammary Gland Echotexture and Milk Composition in Cows. *Animals* **2020**, *10*, 2005. [CrossRef]
135. Zaninelli, M.; Redaelli, V.; Luzi, F.; Bronzo, V.; Mitchell, M.; Dell'Orto, V.; Bontempo, V.; Cattaneo, D.; Savoini, G. First Evaluation of Infrared Thermography as a Tool for the Monitoring of Udder Health Status in Farms of Dairy Cows. *Sensors* **2018**, *18*, 862. [CrossRef]

animals

MDPI

Article

Improving Genomic Prediction Accuracy in the Chinese Holstein Population by Combining with the Nordic Holstein Reference Population

Zipeng Zhang [1,†], Shaolei Shi [1,†], Qin Zhang [2], Gert P. Aamand [3], Mogens S. Lund [4], Guosheng Su [4,*] and Xiangdong Ding [1,*]

1 National Engineering Laboratory for Animal Breeding, College of Animal Science and Technology, China Agricultural University, Beijing 100193, China
2 Shandong Provincial Key Laboratory of Animal Biotechnology and Disease Control and Prevention, Shandong Agricultural University, Taian 271018, China
3 Nordic Cattle Genetic Evaluation, 8200 Aarhus, Denmark
4 Center for Quantitative Genetics and Genomics, Aarhus University, 8830 Tjele, Denmark
* Correspondence: guosheng.su@qgg.au.dk (G.S.); xding@cau.edu.cn (X.D.)
† These authors contributed equally to this work.

Simple Summary: The size of the reference population is critical to the accuracy of genomic prediction. In addition, joining the reference populations from different breeding organizations is a convenient and effective method by which to enlarge reference populations. By adding the Nordic Holstein reference population to the Chinese Holstein reference population, we found that the accuracy of genomic prediction in the Chinese Holstein population was improved substantially for the traits with high or moderate genetic correlation between the two populations; however, the low-genetic-correlation traits did not improve. These findings are important for the purposes of multi-country joint genomic evaluation.

Abstract: The size of the reference population is critical in order to improve the accuracy of genomic prediction. Indeed, improving genomic prediction accuracy by combining multinational reference populations has proven to be effective. In this study, we investigated the improvement of genomic prediction accuracy in seven complex traits (i.e., milk yield; fat yield; protein yield; somatic cell count; body conformation; feet and legs; and mammary system conformation) by combining the Chinese and Nordic Holstein reference populations. The estimated genetic correlations between the Chinese and Nordic Holstein populations are high with respect to protein yield, fat yield, and milk yield—whereby these correlations range from 0.621 to 0.720—and are moderate with respect to somatic cell count (0.449), but low for the three conformation traits (which range from 0.144 to 0.236). When utilizing the joint reference data and a two-trait GBLUP model, the genomic prediction accuracy in the Chinese Holsteins improves considerably with respect to the traits with moderate-to-high genetic correlations, whereas the improvement in Nordic Holsteins is small. When compared with the single population analysis, using the joint reference population for genomic prediction in younger animals, results in a 2.3 to 8.1 percent improvement in accuracy. Meanwhile, 10 replications of five-fold cross-validation were also implemented in order to evaluate the performance of joint genomic prediction, thereby resulting in a 1.6 to 5.2 percent increase in accuracy. With respect to joint genomic prediction, the bias was found to be quite low. However, for traits with low genetic correlations, the joint reference data do not improve the prediction accuracy substantially for either population.

Keywords: genomic prediction; joint reference population; genotype by environment interaction; multi-trait GBLUP

1. Introduction

Genomic selection (GS) [1] is a new landmark method of genetic evaluation that follows the BLUP method [2]. It has already become a tool for routine genetic evaluation in regard to dairy cattle breeding. The principle of GS is that at least one genetic marker in the whole genome is in linkage disequilibrium with the quantitative trait loci of the trait of interest [1]. These genetic markers are used to estimate the effect of each QTL and then calculate the genomic estimated breeding value for the target trait. In addition, genetic markers can capture the relationship of individuals who cannot be traced to a common ancestor in the pedigree. When compared with the traditional BLUP method, genomic selection substantially improves the prediction accuracy in selection candidates and shortens the generation interval [3,4]. In the United States, seven years after the introduction of genomic selection in dairy cattle, the annual rates of genetic gain increased from 50% to 100% for yield traits, and from three-fold to four-fold for lowly heritable traits—including female fertility, herd life, and somatic cell count [5].

The reference population size is one of the most important factors for the purposes of improving genomic prediction accuracy [6,7]. A cost-effective approach is to use genotype imputation as a strategy in order to reduce cost or to increase/maintain the prediction accuracy for selection candidates and to render genomic selection economically feasible. Another cost-effective approach to increase the reference population size is to combine with other reference populations. Furthermore, it has been reported that combining multiple reference populations can increase genomic prediction accuracy [8–10]. Indeed, with the help of frozen semen technology, the genetic material of Holstein dairy cattle has been delivered globally. As such, this fact has provided a genetic basis for the joint genomic evaluation of dairy cattle.

When combining multiple populations in order to enlarge the reference population, the consistency of linkage disequilibrium (LD) among the populations requires consideration. If high LD consistency exists among the respective populations, then combining the different populations is expected to improve the genomic prediction accuracy considerably. With respect to this, the EuroGeomics Consortium [9] and the North American Consortium [11] combined multiple, genetically related populations into a large reference population, thereby resulting in a significant improvement with respect to genomic prediction accuracy. In addition, the combination of the Holstein populations from different countries for the purposes of cross-country joint assessment resulted in good outcomes for the countries who possessed relatively small reference populations. For instance, Zhou et al. combined both Chinese and Nordic Holstein populations, which possessed highly consistent LD (0.97), as a joint reference population. As such, they found a substantial improvement in the genomic prediction accuracy of milk production traits in Chinese populations [12]. In addition, Li et al. found that prediction reliabilities for the Brazilian Holstein population could be greatly increased by including the data of Nordic and French Holstein populations [8]. Meanwhile, certain studies have, instead, shown that multi-population or multi-breed genomic selection, via the combination of distantly related populations, could not improve prediction accuracy and even, in certain circumstances, reduced it [13–15].

Genotype-by-environment interaction (G $\times$ E) is also a key factor that influences the accuracy of multi-population genomic selection. This is due to the fact that many economically important traits are influenced by a combination of genetics, environment, and G $\times$ E. Moreover, different climates, housing conditions, nutritional levels, disease pressure, and feeding densities may also cause G $\times$ E. Furthermore, when phenotypes of different reference populations are measured in different environments, ignoring the influence of G $\times$ E decreases the gain in genomic prediction accuracy when combining reference populations. As such, the biases may increase [16] due to the fact that the LD differences can lead to an increased bias in the estimation of SNP effects.

It must be noted that genomic selection has been carried out in Chinese Holstein breeding since 2012. While the reference population mainly comprised cows [17], the number of progeny-tested bulls in the reference population has been increasing in recent years. It is well-known that in a cattle population, a breeding bull possesses a large number of daughters; thus, the phenotypic information of a bull is of greater importance than that of a cow. The Chinese Holsteins were originally imported from Europe and North America, whereby frozen semen and embryos were imported from abroad in recent decades. Zhou et al. reported very high LD concordance with respect to the adjacent markers between the Chinese and Nordic Holstein populations, as based on 54 K marker data and with a correlation of 0.97 [12]. Therefore, after including the bull reference population of the Nordic Holstein, which consisted of progeny-tested bulls, the genomic prediction accuracy for the Chinese population is expected to improve considerably.

The objective of this study, therefore, was to investigate the improvement of genomic prediction in the Chinese Holstein population, with a joint reference population that consists of Chinese and Nordic Holsteins. We also tested the genomic prediction accuracy using cross-population prediction, i.e., the Chinese Holstein reference population was used for the prediction of Nordic Holsteins, and the Nordic Holstein reference population was used for prediction of Chinese Holsteins.

2. Materials and Methods

2.1. Phenotypes

The dataset utilized in this study was derived from 9206 Chinese Holsteins with genotypes born between 1984 and 2018 and 7084 Holsteins with genotypes born between 1977 and 2012 in certain Nordic countries (including Denmark, Finland, and Sweden). The Chinese Holstein population comprises 5333 bulls and 3873 cows, while the Nordic Holstein population consists of all bulls. Seven traits—milk yield (MY), fat yield (FY), protein yield (PY), somatic cell score (SCS) (with respect to the Chinese Holsteins) or somatic cell count (SCC) (regarding the Nordic Holsteins), body conformation (CONF), feet and legs (FL), and mammary system conformation (MS)—were analyzed. In the Chinese population, 6000 individuals with records were available for MY, FY, and PY, as well as 7000 individuals that possessed records for the other traits. In regard to the Nordic population, the number of individuals with phenotypes for MY, FY, and PY was the same as in the Chinese population. However, the 6495 Holsteins did possess phenotypes for CONF, FL, and MS, and 6347 Holsteins possessed phenotypes for SCC. In the two populations, the numbers of individuals for the three milk-production traits were equal. In addition, the numbers of individuals for the three types of traits were also the same. The de-regressed proof (DRP) was a standardized DRP with respect to the Nordic population, but the original DRP was revealed as a deviation from the base population in the Chinese population. The summary statistics of the DRP for each trait are listed in Table 1. Among them, in regard to the Nordic Holstein population, the averaged reliabilities of the DRP for both milk-production traits and the SCC were found to be higher than or equal to 0.914; further, the reliabilities of the DRP on type traits ranged from 0.710 to 0.844. In addition, the averaged reliability of the DRP in the Chinese Holstein population was found to be lower. Moreover, the milk-production traits and SCS possessed a reliability of 0.502 to 0.660 on average, and the reliabilities of DRP on all type traits ranged from 0.401 to 0.409.

Table 1. Descriptive statistics of DRP [a] regarding the Chinese and Nordic Holstein populations.

		Milk Yield (kg)	Fat Yield (kg)	Protein Yield (kg)	Body Conformation	Feet and Legs	Mammary System Conformation	Somatic Cell Score (10^3/mL)
Number of Records	Nordic	6000	6000	6000	6495	6495	6495	6347
	Chinese	6000	6000	6000	7000	7000	7000	7000
Cut-off date [b]	Nordic		December 2006			January 2017		April 2007
	Chinese		July 2013			September 2013		June 2013
DRP reliability	Nordic	0.945	0.945	0.945	0.844	0.710	0.803	0.914
	Chinese	0.656	0.660	0.658	0.401	0.409	0.403	0.502
Max	Nordic	132.8	130.3	146.8	159.2	201.4	160.3	150.4
	Chinese	4907.8	192.5	130.0	51.5	48.5	48.4	389.5
Min	Nordic	45.0	36.7	7.7	56.8	−60.0	−7.3	51.1
	Chinese	−5390.9	−185.9	−182.0	−43.8	−37.2	−59.4	215.3
Mean	Nordic	93.6	93.4	90.7	100.0	94.8	91.1	94.7
	Chinese	261.3	3.6	7.7	−1.8	−1.3	−1.5	300.1
S.D	Nordic	12.7	12.3	13.8	12.4	15.4	14.9	11.7
	Chinese	1237.1	47.0	39.5	10.1	9.7	10.9	18.7

[a] The DRP is the standardized DRP in the Nordic population, but the original DRP is understood as the deviation from the base population with respect to the Chinese population. [b] Cut-off date in order to divide the whole data into reference and validation sets.

2.2. *Genotypes and Quality Control*

All the individuals in both populations were genotyped with Illumina BovineSNP50 BeadChip (Illumina Inc., San Diego, CA, USA). For each population, the SNPs with a minor allele frequency (MAF) of less than 0.01 were deleted. Further, SNPs with a call rate of less than 0.90 were removed. After genotype quality control was conducted, the individuals with a call rate of less than 0.90 were excluded.. For the remaining individuals, the SNPs with missing genotypes were imputed via using Beagle 3.31 software [18]. The SNPs with an imputation accuracy (R square) less than 0.95 were also removed. As such, after genotype quality control was performed, a total of 43,613 autosomal SNPs, but no animals with a call rate of less than 0.90.

2.3. *Phenotypes and Quality Control*

We analyzed the following three scenarios:

(1) A single-trait GBLUP model was used in order to predict direct genomic breeding values (DGV) while using thereference data of the same population;

(2) A single-trait GBLUP model was used to predict the DGV while using the reference data of a population in order to predict another population (i.e., using the Nordic reference population in order to predict the Chinese validation population);

(3) A two-trait GBLUP model was used for the purposes of genomic prediction while using the joint Chinese–Nordic population, in which the same trait in the different populations was treated as different, but as genetically correlated traits.

The variance components and DGV were estimated using DMU [19]. Furthermore, the algorithm used to estimate variance components was determined by the average information that restricted the maximum likelihood (REML) [20,21].

2.3.1. Single-Trait GBLUP

The single-trait GBLUP model was defined as

$$y = \mathbf{1}\mu + \mathbf{Z}g + e$$

where y is the vector of n pseudo-phenotypes (i.e., the de-regressed proof (DRP)); **1** is a vector of 1 s; μ is the overall mean; g is the vector of additive genetic values with respect to the individuals with a genotype; Z is the design matrix linking the phenotype to the genetic values; and e is the vector of the random residuals. The assumptions of the random effects are: $g \sim N\left(0, G\sigma_g^2\right)$ and $e \sim N\left(0, D\sigma_e^2\right)$. Furthermore, σ_g^2 and σ_e^2 are the addictive genetic variance and residual variance, respectively. In addition, G represents the genomic relationship matrix, which was calculated with method 1, as was described by VanRaden [22]. Lastly, D is the diagonal matrix of the weights for the residuals [23], where the diagonal elements were calculated as $d_{ii} = \frac{1-r_i^2}{r_i^2}$, and r_i^2 is the reliability of the DRP with respect to the individual i.

2.3.2. Two-Trait GBLUP

Two-trait GBLUP model is:

$$\begin{bmatrix} y_1 \\ y_2 \end{bmatrix} = \begin{bmatrix} \mathbf{1}_{n_1} & 0 \\ 0 & \mathbf{1}_{n_2} \end{bmatrix} \begin{bmatrix} \mu_1 \\ \mu_2 \end{bmatrix} + \begin{bmatrix} Z_1 & 0 \\ 0 & Z_2 \end{bmatrix} \begin{bmatrix} g_1 \\ g_2 \end{bmatrix} + \begin{bmatrix} e_1 \\ e_2 \end{bmatrix}$$

where y_1 and y_2 are the vectors of DRP for trait 1 and trait 2, thus, corresponding to the pseudo observations in the Chinese Holstein and Nordic Holstein populations, respectively. Furthermore, μ_1 and μ_2 are the overall means of the two traits, and g_1 and g_2 are the vectors of the additive genetic values of the two traits. Moreover, it was assumed that $\begin{bmatrix} g_1 \\ g_2 \end{bmatrix} \sim N(0, G_0 \otimes G)$, whereby G_0 is the genetic variance and the covariance matrix for the two traits $G_0 = \begin{bmatrix} \sigma_{g_1}^2 & \sigma_{g_1 g_2} \\ \sigma_{g_1 g_2} & \sigma_{g_2}^2 \end{bmatrix}$. Then, G is the genomic relationship matrix of all genotyped animals, and $\sigma_{g_1}^2$ and $\sigma_{g_2}^2$ are the additive genetic variance with respect to the two populations. Further, $\sigma_{g_1 g_2}$ is the genetic variance-covariance matrix between the two populations. Next, e_1 and e_2 are the vectors of the random residuals of the two populations, which were assumed to be independent of each other with respect to $e_1 \sim N\left(0, D_1\sigma_e^2\right)$ and $e_2 \sim N\left(0, D_2\sigma_e^2\right)$. The construction of G, D_1, and D_2 were the same as G and D, which were described in the above single-trait GBLUP model.

2.3.3. Genomic Prediction Accuracy

The genomic prediction accuracy and prediction unbiasedness were obtained through two ways. The first was in terms of predicting the younger individuals that were using the older animals. All the individuals from the Chinese and Nordic Holstein populations were divided into reference and validation population sets, with a ratio around 4:1 in population size and according to the cut-off date as shown in Table 1. The 1/5 younger individuals were also used as the validation population, thereby called the Chinese validation population and the Nordic validation population. The remaining 4/5 individuals of each population were used as the reference population, thus, called the Chinese reference population and the Nordic reference population, respectively. Alternatively, a five-fold cross-validation was also implemented. The prediction accuracy and prediction unbiasedness were obtained through five-fold cross-validation. For each trait, the individuals in both populations were randomly split into five groups, in which the numbers of individuals from the Chinese population and the Nordic population were nearly the same among the five groups. In each round of cross-validation, one group was taken as the validation population, the other four groups were used as the reference population. Similarly, either with respect to the Chinese population or the Nordic population, the five-fold cross-validation of a single population was carried out via using their own five groups. Furthermore, the five-fold cross-validation was replicated a total of 10 times.

The prediction accuracy of the GBLUP was calculated as

$$acc = \frac{cor(DRP,DGV)}{\bar{r}},$$

where $cor(DRP, DGV)$ is the correlation between the DRP and the DGV for the animals in the validation population. In addition, $\bar{r}$ is the averaged accuracy of the DRP in the validation population.

Next, the prediction unbiasedness of GBLUP was calculated as:

$$b = \frac{cov(DRP, DGV)}{var(DGV)} \quad (1)$$

where $cov(DRP,DGV)$ is the covariance between the DRP and the DGV for the animals in the validation population. Further, the $var\ (DGV)$ is the variance of the DGV in the validation population.

3. Results

3.1. Genetic Correlation of Traits in Two Holstein Populations

Table 2 shows the genetic correlations between the Chinese and Nordic Holstein populations, which were estimated with a two-trait GBLUP model. Regarding the three milk-production traits (i.e., MY, FY, and PY), the genetic correlations are high and range between 0.621 to 0.720. Moreover, the SCS possess a genetic correlation of 0.449. Regarding the type traits (i.e., CONF, FL, and MS), the genetic correlations from the Chinese and Nordic Holstein populations are low (i.e., less than or equal to 0.236). Moreover, the genetic correlation between the same trait in the two populations indicates that G×E exists [24].

Table 2. Genetic correlation of traits between Chinese and Nordic Holstein populations.

Traits	Variance		Covariance	Correlation
	Chinese	Nordic		
Milk yield	470,936.770	102.627	5005.297	0.720 (0.030)
Fat yield	634.546	175.126	94.554	0.715 (0.031)
Protein yield	422.192	124.551	95.1678	0.621 (0.035)
Body conformation	29.540	12.190	102.790	0.221 (0.043)
Feet and legs	40.368	9.601	109.742	0.144 (0.044)
Mammary system conformation	31.657	14.707	123.003	0.236 (0.044)
Somatic cell score	58.536	34.502	101.069	0.449 (0.046)

Regarding the individuals with milk-production trait records, the average genomic relationships withinthe Chinese and Nordic Holstein populations are 0.030 (±0.043) and 0.033 (±0.038), respectively. The average genomic relationship between the two populations is 0.022 (±0.022), which is lower than the average genomic relationships within the populations. With respect to the individuals with type trait records, the average genomic relationships withinthe Chinese and Nordic Holstein populations are 0.029 (±0.039) and 0.033 (±0.037), respectively. In addition, the average genomic relationship between the two populations is the same as that found in the milk-production traits. Regarding the individuals with SCC/SCS records, the average genomic relationship of the Chinese and Nordic Holstein populations is 0.029 (±0.041) and 0.033 (±0.037), respectively. However, the average genomic relationship between the two populations is 0.022 (±0.021).

3.2. Comparison between Single Population and Joint Population Prediction Accuracy

Table 3 shows the prediction accuracy and unbiasedness of the younger validation individuals with respect to both the Chinese and Nordic Holstein populations, which was achieved using the own- and joint-reference populations. In regard to the milk-production traits (i.e., MY, FY, and PY) and the SCS in the Chinese Holstein population, the genomic prediction accuracies using the single Chinese reference population (i.e., as per the single-trait model) are 0.402, 0.429, 0.394, and 0.253, respectively. Indeed, the genomic prediction accuracy via using the joint reference population (i.e., the two-trait model) reaches 0.464, 0.510, 0.433, and 0.276, respectively. Moreover, the joint reference population gains 2.3 to 8.1 percentage points over the single reference population with respect to the accuracy of genomic selection for the Chinese Holstein population. However, in regard to the Nordic Holstein population, the prediction accuracy is improved very little when compared with the own reference population. Further, the MY trait gains the greatest improvement in terms of prediction accuracy, albeit only from 0.683 to 0.703. With respect to the type traits, the prediction accuracy, with a negligible change of −0.4 to 0.2 percentage points when compared to their single own reference groups, is not improved in either of the Chinese orNordic Holstein populations.

Table 3. Prediction accuracy and unbiasedness regarding the younger individuals with respect to the seven traits in the Chinese and Nordic Holstein populations, as predicted by the single/joint reference populations.

Reference	Validation		Milk Yield	Fat Yield	Protein Yield	Body Conformation	Feet and Legs	Mammary System Conformation	Somatic Cell Score
Single	Nordic	Accuracy	0.683	0.681	0.654	0.752	0.589	0.703	0.682
		Unbiasedness	0.823	0.807	0.722	0.993	0.898	0.974	0.987
	Chinese	Accuracy	0.402	0.429	0.394	0.439	0.493	0.507	0.253
		Unbiasedness	0.937	0.948	0.860	0.839	0.800	0.989	0.726
Joint	Nordic	Accuracy	0.703	0.696	0.665	0.753	0.591	0.704	0.687
		Unbiasedness	0.842	0.815	0.731	1.002	0.89	0.973	0.983
	Chinese	Accuracy	0.464	0.510	0.433	0.441	0.489	0.507	0.276
		Unbiasedness	0.976	0.983	0.898	0.82	0.773	0.946	0.927

For the Chinese Holstein population, the unbiasedness of genomic prediction ofthe milk-production traits and the SCS is significantly improved when using the joint reference population. In particular, the unbiasedness of genomic prediction regarding SCS dramatically improves from 0.726 to 0.927. However, for the type traits, the joint reference population leads to a bigger bias. For example, the the genomic prediction unbiasedness of MS is 0.989 when using the single Chinese reference population, but decreases to 0.946 when using the joint reference population. However, the Nordic Holstein population does not benefit from the joint reference population with respect to the prediction unbiasedness, as is the case in regards to accuracy.

Table 4 further details the prediction accuracy and unbiasedness, from the single and joint reference populations in the 10 replicates of the five-fold cross-validation, for both the Chinese and Nordic Holstein populations. When compared with the single population, the impact of the joint reference data regarding the genomic prediction from the cross-validation is consistent with that of the younger validation individuals. Nonetheless, the improvement in the prediction accuracy of the milk-production traits and the SCS for the Chinese population is found to be lower, ranging from 1.6 to 5.2 percentage points, than the prediction accuracy of the younger validation individuals. Indeed, the changes in unbiasedness for all traits are small, and the genomic prediction unbiasedness obtained from the cross-validation is generally found to be better than that from the younger validation individuals, which is closer to 1.0.

Table 4. Prediction accuracy and unbiasedness regarding the seven traits in the Chinese and Nordic Holstein populations, as predicted by the single/joint reference populations in the 10 replicates of the five-fold cross-validation.

Reference	Validation		Milk Yield	Fat Yield	Protein Yield	Body Conformation	Feet and Legs	Mammary System Conformation	Somatic Cell Score
Single	Nordic	Accuracy	0.860 (±0.007)	0.854 (±0.009)	0.897 (±0.007)	0.783 (±0.013)	0.707 (±0.024)	0.853 (±0.016)	0.763 (±0.015)
		Unbiasedness	1.006 (±0.021)	1.004 (±0.018)	1.009 (±0.016)	0.989 (0.033)	1.002 (±0.046)	1.016 (±0.027)	0.982 (±0.030)
	Chinese	Accuracy	0.502 (±0.024)	0.475 (±0.026)	0.507 (±0.024)	0.590 (±0.033)	0.680 (±0.035)	0.590 (±0.037)	0.413 (±0.031)
		Unbiasedness	1.086 (±0.080)	1.072 (±0.079)	0.897 (±0.007)	0.943 (±0.069)	0.904 (±0.054)	0.989 (±0.085)	1.212 (±0.111)
Joint	Nordic	Accuracy	0.864 (±0.007)	0.860 (±0.023)	0.902 (±0.019)	0.783 (±0.013)	0.708 (±0.024)	0.853 (±0.016)	0.766 (±0.015)
		Unbiasedness	1.006 (±0.022)	1.006 (±0.020)	1.010 (±0.019)	0.988 (±0.033)	1.002 (±0.046)	1.016 (±0.028)	0.982 (±0.030)
	Chinese	Accuracy	0.547 (±0.026)	0.527 (±0.055)	0.536 (±0.023)	0.593 (±0.032)	0.680 (±0.034)	0.594 (±0.037)	0.429 (±0.033)
		Unbiasedness	1.070 (±0.075)	1.058 (±0.023)	1.067 (±0.068)	0.943 (±0.066)	0.904 (±0.053)	0.987 (±0.084)	1.200 (±0.114)

3.3. The Genomic Prediction Accuracy of Cross Population

In addition, we analyzed the genomic prediction with respect to the cross population, as demonstrated in Tables 5 and 6. Regarding the younger validation individuals, the prediction accuracy from the cross population (e.g., when predicting with respect to the Chinese Holstein population using the Nordic reference population) is found to be worse than the prediction that was obtained using their own, respective, reference population. More specifically, there is a large decrease in the prediction accuracy of the type traits. For example, the prediction accuracies of FL, for the validation population of the Chinese and Nordic Holsteins, are found to be 0.493 and 0.589 (Table 3) when compared with their own reference populations, respectively. However, the prediction accuracies of the FL, when estimated from the cross population, are only 0.088 and 0.212 (Table 5) for the Chinese and Nordic Holstein populations, respectively. This, therefore, implies the importance of the own reference population. Similar results are also found in regard to the other traits, but the decrease in accuracy when using the foreign reference population varies. For example, the decrease in genomic prediction accuracy for milk-production traits is found to be relatively small. However, the prediction accuracy of the MY trait in the Chinese validation population predicted by the Chinese reference population is 0.402, while the accuracy of prediction when using the Nordic reference population is 0.354. Similarly, as shown in Table 6, the performance of the cross population in regard to the prediction accuracy obtained in the 10 replicates of five-fold cross-validation is found to be worse, even when compared with the own or joint reference population.

Table 5. Accuracy of the younger individuals in the cross population prediction.

Reference	Validation	Milk Yield	Fat Yield	Protein Yield	Body Conformation	Feet and Legs	Mammary System Conformation	Somatic Cell Score
Chinese	Nordic	0.302	0.280	0.258	0.101	0.088	0.028	0.132
Nordic	Chinese	0.354	0.323	0.303	0.198	0.212	0.142	0.194

Table 6. Accuracy of the cross-population prediction in the 10 replicates of the five-fold cross-validation.

Reference	Validation	Milk Yield	Fat Yield	Protein Yield	Body Conformation	Feet and Legs	Mammary System Conformation	Somatic Cell Score
Chinese	Nordic	0.378 (±0.030)	0.390 (±0.029)	0.364 (±0.025)	0.332 (±0.037)	0.068 (±0.035)	0.349 (±0.030)	0.142 (±0.033)
Nordic	Chinese	0.444 (±0.021)	0.468 (±0.028)	0.500 (±0.023)	0.323 (±0.035)	0.122 (±0.040)	0.303 (±0.033)	0.216 (±0.036)

4. Discussion

In this study, we investigated whether enlarging the Holstein reference population, by combining the Chinese and Nordic Holstein reference populations, could improve the accuracy of genomic selection in the two Holstein populations. The results show that the genomic prediction of the milk-production traits (i.e., MY, FY, and PY) and SCC in the Chinese population is improved substantially using the joint reference population, albeit this is not the case for the type traits. However, the Nordic Holstein population does not gain much improvement in terms of the genomic prediction of all the traits. Although the reference population size of the Chinese and Nordic Holstein is nearly equal (Table 1), the Nordic reference population provides much more information. This is due to the fact that the Nordic reference population consists of progeny-tested bulls that possess a higher DRP reliability with respect to the concerned traits. Therefore, the joint reference population is found to be more helpful regarding the Chinese Holstein population. Regarding the aforementioned, Lund et al. also demonstrate that the benefit that is obtained from the joint reference population is influenced by the DRP reliability of the foreign reference population [9]. Moreover, in the joint genomic prediction of the Holstein and Jersey populations (which was conducted in Lund et al.'s study), it was also indicated that the Jersey population, which possessed a smaller reference population, gained more of an improvement in terms of genomic prediction, while the Holstein population acquired only a slight improvement due to its larger reference population [15].

On the other hand, the accuracies of the genomic selection for all traits in the Nordic Holstein population are higher than in those in the Chinese Holstein population, when their own reference population (i.e., the progeny-tested bulls) is used only to further demonstrate the composition of the reference population, which is essential for the purposes of genomic selection. Nevertheless, it is feasible to add cows with genotypic information in order to enlarge the reference population for the purposes of improving the genomic prediction accuracy when the number of bulls is insufficient [25]. Indeed, Ding et al. [17] investigated if it is plausible to use cows as the Holstein reference population. When 3084 cows were used as the reference population, the correlation between the genomic EBV and the conventional EBV, regarding the validation population of the five milk-production traits, went as high as from 0.594 to 0.760. Moreover, Mc Hugh et al. also showed that the genomic information that is obtained from cows could improve the genomic prediction accuracy [26]. In addition, cows can be genotyped using low-density chips and then imputed with high-density chips in order to achieve a low-cost increase in the cow reference population size [27].

Indeed, the size of the reference population is a critical factor regarding genome prediction [6,7]. Combining populations is a simple and practical cost-efficient strategy in order to improve genomic prediction accuracy. Moreover, Li et al. found that adding Nordic and French bull data in order to enlarge the Brazilian reference population could improve the genomic prediction reliability of the three milk-production traits by 0.030 to 0.055 [8]. Similar results were obtained by adding 870 foreign Brown Swiss cows to the US Brown Swiss reference group. Further, the genomic prediction accuracy was improved by 3.2%, on average, with a single-trait model, and by 4.6% with a two-trait model [28]. In our study, only a two-trait model was used to estimate the genomic breeding values. This was decided upon due to the fact that the single-trait model could not be carried out due to the different scales with respect to the DRP in the Chinese and Nordic populations. Moreover, the two-trait model could also account for G × E regarding the two populations [29]. Meanwhile, our results also indicate that the joint reference population is helpful with respect to improving the genomic prediction accuracy of the milk-production traits and the SCS/SCC, while no such improvements are gained regarding the type traits. The reason for this could be that the genetic correlations of the three type traits (i.e., CONF, FL, and MS) are much lower than those of the milk-production traits and the SCS/SCC (Table 2). Indeed, the high genetic correlations mean that the foreign population could provide more information and increase the efficiency of using the joint reference population. This point was confirmed in the study of dairy cattle in the EuroGenomics project that was reported by Lund et al. A 12–19% increase, in the project, with respect to the genomic prediction accuracy was gained on udder depth, as this trait possessed the highest genetic correlation (0.98) among the collaborating countries. In contrast, the genetic correlation for the SCS was relatively low (0.88), thereby resulting in an increase in accuracy of 8–15% [9]. In our study, the genetic correlations of the type traits between the Chinese and Nordic Holstein populations are quite low (0.144–0.236), which may be due to the different trait definitions or measurement methods for these type traits in the two populations. Consequently, combining the two reference populations does not provide more information than using a single population. Thus, no improvements in genomic prediction accuracy are obtained.

Although the Chinese Holstein cows mainly originated from Europe and North America, the differences in climate, feeding environment, selection criterion, etc., between China and the Nordic countries resulted in different performances with respect to the two Holstein populations. This could be explained by the existence of G × E in these two populations. Robertson proposed that a genetic correlation less than 0.80 between the same trait in two different environments indicates the existence of G × E a [24]. According to this proposal, all of the traits in this study showed G × E in the Chinese and Nordic populations, e.g., milk-production traits possess a high genetic correlation (0.621–0.720), SCC/SCS are moderately genetically correlated (0.449), and the type traits are lowly genetically correlated (0.144–0.236). In the dairy cattle populations in the Eurogenomics project, the G × E could be ignored in most traits, e.g., udder depth and SCC, as their genetic correlations were greater than 0.8. Furthermore, these traits could, thus, be treated as one trait, i.e., a single-trait model using the joint reference population could be a good approach [9]. However, in the scenario with G × E, it is not reasonable to implement a single-trait model in terms of genomic selection, particularly with respect to the traits with low genetic correlations. In such situations, the single-trait model may generate large bias using a combined reference population, while the two-trait model could be of better use due to its accounting for G × E [30]. When treating the same trait of two populations as different traits, it enables one to capture G × E as a covariance between the populations, which, in turn, allows one to account for G × E in the model [31]. In addition, the two-trait model is more flexiblewhen the scales used to measure phenotypes for the same trait are different in the different populations. For example, the DRP is a standardized DRP in the Nordic population, but the original DRP serves as a deviation from the base population in the Chinese population in our study; as such, the single-trait model would, thus, be implausible.

The reliability of the DRP for each trait is found to be much higher in the Nordic population than in the Chinese population. As such, we, therefore, analyzed whether the Nordic reference population could yield a higher genomic prediction accuracy for the Chinese Holstein population than the Chinese reference population could. The results of the cross-population prediction (Table 5) show that the accuracies are worse when compared with using the own reference population. Likewise, the Chinese reference population does not yield a reasonable prediction accuracy with respect to the Nordic population. However, it is consistent with the report by Ma et al. regarding the prediction of the Chinese Holstein population when using French and USA reference populations [32]. Our results suggest that the foreign reference population, in general, cannot lead to the higher accuracy of prediction that is found in the Chinese reference population. Having said this, however, the LD is found to be similar between the two cattle populations [12], while a joint reference population including the foreign reference population to the Chinese reference populations is found to be helpful.

The genomic Chinese performance index (GCPI) that is currently used in the genomic breeding of the Chinese Holstein cows consists of the seven traits that are investigated in this study. The improvement in the prediction accuracy, regarding the production traits and the SCS traits, is achieved by combining the Nordic population with the Chinese population, which is, thus, found to be helpful for the purposes of Chinese Holstein breeding. Moreover, the longevity traits, reproductive traits, feed conversion traits, etc., will also be gradually added to the GCPI. The information regarding the reference population from the Nordic countries and other countries/organizations would be also helpful in order to improve the genomic prediction efficiency, as indicated in this study. Therefore, it is necessary to carry out this joint genomic selection for China and for other countries.

5. Conclusions

Genomic prediction when using a foreign reference population may not obtain high accuracy. However, the joint reference population can improve the prediction accuracy and prediction unbiasedness. In particular, this is true for the traits that possess moderate-to-high genetic correlations between the populations. As for traits that possess low genetic correlations between the populations, a joint reference population may not improve the prediction accuracy and prediction unbiasedness. However, the difference in prediction accuracy when using own, foreign, and joint reference populations is also dependent on the composition of the reference populations.

Author Contributions: X.D. and G.S. designed the experiments. Z.Z. and S.S. performed statistical analysis and wrote the manuscript. M.S.L., Q.Z., and G.P.A. contributed to the joint reference population. X.D. and G.S. revised the manuscript. All authors have read and agreed to the published version of the manuscript.

Funding: This work was supported by the National Key Research and Development Project (2019YFE0106800). We acknowledge Vikinggenetics and Nordic Cattle Genetic Evaluation for providing data of Nordic reference population.

Institutional Review Board Statement: Not applicable.

Informed Consent Statement: Not applicable.

Data Availability Statement: Not applicable.

Conflicts of Interest: The authors declare no conflict of interest.

References

1. Meuwissen, T.H.; Hayes, B.J.; Goddard, M.E. Prediction of total genetic value using genome-wide dense marker maps. *Genetics* **2001**, *157*, 1819–1829. [CrossRef]
2. Varona, L.; Legarra, A.; Toro, M.A.; Vitezica, Z.G. Non-additive Effects in Genomic Selection. *Front. Genet.* **2018**, *9*, 78. [CrossRef]
3. Naserkheil, M.; Mehrban, H.; Lee, D.; Park, M.N. Evaluation of Genome-Enabled Prediction for Carcass Primal Cut Yields Using Single-Step Genomic Best Linear Unbiased Prediction in Hanwoo Cattle. *Genes* **2021**, *12*, 1886. [CrossRef] [PubMed]
4. König, S.; Simianer, H.; Willam, A. Economic evaluation of genomic breeding programs. *J. Dairy Sci.* **2009**, *92*, 382–391. [CrossRef]
5. García-Ruiz, A.; Cole, J.B.; Vanraden, P.M.; Wiggans, G.R.; Ruiz-López, F.J.; Van Tassell, C.P. Changes in genetic selection differentials and generation intervals in US Holstein dairy cattle as a result of genomic selection. *Proc. Natl. Acad. Sci. USA* **2016**, *113*, 28. [CrossRef] [PubMed]
6. Daetwyler, H.D.; Villanueva, B.; Woolliams, J.A. Accuracy of predicting the genetic risk of disease using a genome-wide approach. *PLoS ONE* **2008**, *3*, e3395. [CrossRef]
7. Goddard, M. Genomic selection: Prediction of accuracy and maximisation of long term response. *Genetica* **2009**, *136*, 245–257. [CrossRef]
8. Li, X.; Lund, M.S.; Zhang, Q.; Costa, C.N.; Ducrocq, V.; Su, G. Short communication: Improving accuracy of predicting breeding values in Brazilian Holstein population by adding data from Nordic and French Holstein populations. *J. Dairy Sci.* **2016**, *99*, 4574–4579. [CrossRef]

9. Lund, M.S.; Roos, A.P.; Vries, A.G.; Druet, T.; Ducrocq, V.; Fritz, S.; Guillaume, F.; Guldbrandtsen, B.; Liu, Z.; Reents, R.; et al. A common reference population from four European Holstein populations increases reliability of genomic predictions. *Genet. Sel. Evol.* **2011**, *43*, 43. [CrossRef]
10. Song, H.; Zhang, J.; Jiang, Y.; Gao, H.; Tang, S.; Mi, S.; Yu, F.; Meng, Q.; Xiao, W.; Zhang, Q.; et al. Genomic prediction for growth and reproduction traits in pig using an admixed reference population1. *J. Anim. Sci.* **2017**, *95*, 3415–3424. [CrossRef] [PubMed]
11. Vanraden, P.M.; Van Tassell, C.P.; Wiggans, G.R.; Sonstegard, T.S.; Schnabel, R.D.; Taylor, J.F.; Schenkel, F.S. Invited Review: Reliability of genomic predictions for North American Holstein bulls. *J. Dairy Sci.* **2009**, *92*, 16–24. [CrossRef] [PubMed]
12. Zhou, L.; Ding, X.; Zhang, Q.; Wang, Y.; Lund, M.S.; Su, G. Consistency of linkage disequilibrium between Chinese and Nordic Holsteins and genomic prediction for Chinese Holsteins using a joint reference population. *Genet. Sel. Evol.* **2013**, *45*, 7. [CrossRef]
13. Calus, M.P.L.; Goddard, M.E.; Wientjes, Y.C.J.; Bowman, P.J.; Hayes, B.J. Multibreed genomic prediction using multitrait genomic residual maximum likelihood and multitask Bayesian variable selection. *J. Dairy Sci.* **2018**, *101*, 4279–4294. [CrossRef]
14. Olson, K.M.; Vanraden, P.M.; Tooker, M.E. Multibreed genomic evaluations using purebred Holsteins, Jerseys, and Brown Swiss. *J. Dairy Sci.* **2012**, *95*, 5378–5383. [CrossRef]
15. Ben, J.; Hayes, P.J.B.A. Accuracy of genomic breeding values in multi-breed dairy cattle populations. *Genet. Sel. Evol.* **2009**, *41*, 1–9.
16. Haile-Mariam, M.; Pryce, J.E.; Schrooten, C.; Hayes, B.J. Including overseas performance information in genomic evaluations of Australian dairy cattle. *J. Dairy Sci.* **2015**, *98*, 3443–3459. [CrossRef]
17. Ding, X.; Zhang, Z.; Li, X.; Wang, S.; Wu, X.; Sun, D.; Yu, Y.; Liu, J.; Wang, Y.; Zhang, Y.; et al. Accuracy of genomic prediction for milk production traits in the Chinese Holstein population using a reference population consisting of cows. *J. Dairy Sci.* **2013**, *96*, 5315–5323. [CrossRef]
18. Browning, S.R.; Browning, B.L. Rapid and Accurate Haplotype Phasing and Missing-Data Inference for Whole-Genome Association Studies by Use of Localized Haplotype Clustering. *Am. J. Hum. Genet.* **2007**, *81*, 1084–1097. [CrossRef] [PubMed]
19. Madsen, P.; Jensen, J.; Labouriau, R.; Christensen, O.F.; Sahana, G. DMU—A Package for Analyzing Multivariate Mixed Models in quantitative Genetics and Genomics. In Proceedings of the 10th World Congress of Genetics Applied to Livestock Productio, Vancouver, BC, Canada, 17–22 August 2014.
20. Meyer, K. Maximum Likelihood Estimation of Variance Components for a Multivariate Mixed Model with Equal Design Matrices. *Biometrics* **1985**, *41*, 153–165. [CrossRef] [PubMed]
21. Schaeffer, L.R.; Wilton, J.W.; Thompson, R. Simultaneous Estimation of Variance and Covariance Components from Multitrait Mixed Model Equations. *Int. Biom. Soc.* **1978**, *34*, 199–208. [CrossRef]
22. Vanraden, P.M. Efficient Methods to Compute Genomic Predictions. *J. Dairy Sci.* **2008**, *91*, 4414–4423. [CrossRef]
23. Su, G.; Madsen, P.; Nielsen, U.S.; Mäntysaari, E.A.; Aamand, G.P.; Christensen, O.F.; Lund, M.S. Genomic prediction for Nordic Red Cattle using one-step and selection index blending. *J. Dairy Sci.* **2012**, *95*, 909–917. [CrossRef]
24. Robertson, A. The Sampling Variance of the Genetic Correlation Coefficient. *Biometrics* **1959**, *15*, 469–485. [CrossRef]
25. Su, G.; Ma, P.; Nielsen, U.S.; Aamand, G.P.; Wiggans, G.; Guldbrandtsen, B.; Lund, M.S. Sharing reference data and including cows in the reference population improve genomic predictions in Danish Jersey. *Animal* **2016**, *10*, 1067–1075. [CrossRef]
26. Mc Hugh, N.; Meuwissen, T.H.E.; Cromie, A.R.; Sonesson, A.K. Use of female information in dairy cattle genomic breeding programs. *J. Dairy Sci.* **2011**, *94*, 4109–4118. [CrossRef]
27. Jiang, Y.; Song, H.; Gao, H.; Zhang, Q.; Ding, X. Exploring the optimal strategy of imputation from SNP array to whole-genome sequencing data in farm animals. *Front. Genet.* **2022**, *13*. [CrossRef] [PubMed]
28. Olson, K.M.; Null, D.J.; Sargolzaei, M.; Winters, M.; van Kaam, J.B.C.H. Reliability Increases from Combining 50,000-and 777,000-Marker Genotypes from Four Countries. In Proceedings of the 2012 Interbull Meeting, Cork, Ireland, 28–31 May 2012; pp. 75–79.
29. Meyer, K. Factor-analytic models for genotype x environment type problems and structured covariance matrices. *Genet. Sel. Evol.* **2009**, *41*, 21. [CrossRef]
30. Song, H.; Zhang, Q.; Ding, X. The superiority of multi-trait models with genotype-by-environment interactions in a limited number of environments for genomic prediction in pigs. *J. Anim. Sci. Biotechnol.* **2020**, *11*, 88. [CrossRef] [PubMed]
31. Song, H.; Zhang, Q.; Misztal, I.; Ding, X. Genomic prediction of growth traits for pigs in the presence of genotype by environment interactions using single-step genomic reaction norm model. *J. Anim. Breed. Genet.* **2020**, *137*, 523–534. [CrossRef] [PubMed]
32. Ma, P.; Huang, J.; Gong, W.; Li, X.; Gao, H.; Zhang, Q.; Ding, X.; Wang, C. The impact of genomic relatedness between populations on the genomic estimated breeding values. *J. Anim. Sci. Biotechnol.* **2018**, *9*, 64. [CrossRef]

animals MDPI

Article

Estimation of the Genetic Components of (Co)variance and Preliminary Genome-Wide Association Study for Reproductive Efficiency in Retinta Beef Cattle

José María Jiménez [1], Rosa María Morales [2,*], Alberto Menéndez-Buxadera [2], Sebastián Demyda-Peyrás [3,4], Nora Laseca [2] and Antonio Molina [2]

1 CEAG The Council of Cadiz, 11400 Jerez de la Frontera, Spain
2 Department of Genetics, Veterinary School, Campus de Rabanales, University of Córdoba, Edificio Gregor Mendel, Ctra. Madrid-Cádiz, km 396, 14014 Córdoba, Spain
3 Departamento de Producción Animal, Facultad de Ciencias Veterinarias, Universidad Nacional de La Plata, La Plata 1900, Argentina
4 Consejo Nacional de Investigaciones Científicas y Técnicas (CONICET), La Plata 1900, Argentina
* Correspondence: v22mocir@uco.es; Tel.: +34-957-21-10-70

Simple Summary: Fertility is one of the most important traits for productivity in extensive beef production systems, since it has a major effect on the number of calves born and, as a result, the quantity of weaned calves produced per year. However, it is difficult to improve this trait under extensive conditions, mainly due to the lack of reliable and easy-to-obtain selection criteria. In this study, fertility was analyzed using reproductive efficiency, which was calculated as the deviation between the optimal and real parity number of females at each age. We demonstrated a high h^2 value (0.30) using a classic repeatability model and a random regression model (ranging from 0.24 to 0.51), which suggests that the latter model can be recommended to improve fertility in beef breeds raised under extensive environmental conditions such as the Retinta. In addition, we performed the first GWAS analysis looking for SNP genetic markers associated with this character in cattle, which showed five markers significantly associated with the trait located on BTA4 and BTA28. Finally, the functional analysis revealed the presence of five candidate genes located within these regions, which were previously shown to be related to fertility in cattle and mice models.

Abstract: In this study, we analyzed the variation of reproductive efficiency, estimated as the deviation between the optimal and real parity number of females at each stage of the cow's life, in 12,554 cows belonging to the Retinta Spanish cattle breed, using classical repeatability and random regression models. The results of the analyses using repeatability model and the random regression model suggest that reproductive efficiency is not homogeneous throughout the cow's life. The h^2 estimate for this model was 0.30, while for the random regression model it increased across the parities, from 0.24 at the first calving to 0.51 at calving number 9. Additionally, we performed a preliminary genome-wide association study for this trait in a population of 252 Retinta cows genotyped using the Axiom Bovine Genotyping v3 Array. The results showed 5 SNPs significantly associated with reproductive efficiency, located in two genomic regions (BTA4 and BTA28). The functional analysis revealed the presence of 5 candidate genes located within these regions, which were previously involved in different aspects related to fertility in cattle and mice models. This new information could give us a better understanding of the genetic architecture of reproductive traits in this species, as well as allow us to accurately select more fertile cows.

Keywords: Retinta beef cattle; age at first calving; calving interval; reproductive efficiency; random regression model; GWAS

1. Introduction

The productivity of extensive beef production systems relies on the sum of several traits. Of these, fertility is one of the most crucial, since it has a major effect on the number of calves born and, therefore, the quantity of weaned calves produced per year [1]. This has been demonstrated by several studies, which went so far as to quantify reproductive traits as three times more important than production traits in semi-intensive cattle production systems [2–5]. Despite the existence of a sizeable environmental effect, it was also demonstrated that there is a considerable genetic component affecting fertility in cows, thus enabling us to select more fertile individuals in order to increase the reproductive efficiency of the whole system [6]. In this context, there is a general consensus that the ideal beef cow, in terms of fertility, should be precocious (i.e., first calving at 24 months in Retinta) and produce one calf per year [7,8]. Therefore, the possibility of selecting individuals that have a genetic predisposition to meet such objectives could represent a key advantage from a breeding point of view.

An increase in the calving interval is one of the most common causes of culling in beef cattle [9]. However, this reproductive trait, closely associated with overall fertility, is characterized by low heritability and slow genetic progress in cattle [10]. For this reason, it is rarely included as a selection criterion in breeding programs of beef breeds raised under extensive environmental conditions. Instead, breeding programs focus more on increasing the live weight of the individuals, using growth traits such as weight and average daily weight gain, which correlate negatively with fertility [11]. However, the inclusion of reproductive traits, either individually or in combination through indexes, has been considered in certain breeds, although it is not yet a very widespread practice [12]. Cammack, et al. [10] summarized average values and heritabilities in thirteen different reproduction traits in beef cattle raised under extensive systems reported over 25 years, showing that most of them focused on the effects of the bull and precocity. However, few studies have looked into the individual fertility of cows, most likely due to the difficulty in obtaining a large and reliable dataset in extensive or grazing production systems.

During the last few years, the use of an indirect fertility criterion, including precocity and the calving interval, which can be estimated indirectly based on reproductive records, has been demonstrated as an interesting option for selecting more fertile females in extensively-bred livestock species [13,14]. This parameter, called reproductive efficiency (Re), is estimated as the percentage deviation of the number of calvings that a cow has at each age, from the number of calvings that this cow could have had in optimal conditions. In the Retinta breed, the optimum age at first calving was considered two years and one year for the optimal calving interval [15]. It has demonstrated increased heritability and reliability in comparison with other reproductive traits in beef cattle, such as calving interval or age at first calving [16]. However, to our knowledge, it has only been analyzed in a few studies in cattle [17,18]. In addition, reproductive traits have mostly been evaluated from a genetic point of view, using repeatability models. This methodology assumes that the Re values at the different calvings are manifestations of the same trait, and therefore the (co)variance across the trajectory of the parities of the cow is constant. Despite the fact that this methodology is widely used, Wilson and Réale [19] concluded that this assumption is not correct since the genetic parameters for this kind of effect vary at different ages. However, there is still a shortage of reports evaluating the changes in such parameters over the lifetime of the cows.

In recent years, the genomic revolution has allowed us to identify single nucleotide polymorphisms (SNPs) associated with phenotypic traits, leading to a better understanding of complex traits. One of the most powerful tools for analyzing such associations is the genome-wide association study (GWAS). In beef, a large number of GWAS analyses have been reported for different traits such as longitudinal, carcass, fatty acid profiles, meat tenderness and quality, and growth traits [20–26]. However, GWAS studies on reproductive traits in beef cows are still scarce [27–30].

Retinta is an autochthonous breed with unique adaptive characteristics linked to an extensive regime in the Dehesa ecosystem. Nowadays, it is widely bred in the south of Spain, due to its excellent adaptation to the harsh environment characterized by marginal pasturelands and the hot, dry climate [31]. In 2018, the breed reported more than 42,000 breeding cows, of which 16,850 were enrolled in the Genealogical Book [32]. Of these, 25% were raised as purebreds, with the rest used in crossbreeding as a maternal line with other continental breeds, such as Charolais and Limousin, with the aim of increasing profitability by using crossbred individuals [33]. For over 40 years now, the Retinta breeders' association has not only gathered pedigree records but also a large phenotypic and environmental dataset, including birth and culling records from all the individuals enrolled in the breed [15]. For this reason, the Retinta breed is an interesting model by which to evaluate quantitative traits accurately, including those related to fertility.

In this study, we aimed to determine the genetic influence on the fertility of Retinta breeding cows by measuring the variance component patterns of reproductive efficiency across the calvings in the Retinta beef breed using random regression models (RRM). Additionally, we performed a preliminary genome-wide association study for reproductive efficiency traits to identify genetic variants, genomic regions, and candidate genes associated with fertility.

2. Material and Methods

2.1. Animal Dataset

In this study, we analyzed 63,421 calving records collected by the National Retinta Breeders' Association. The dataset comprises information on 13,888 cows (from 1356 sires and 9154 dams, of which 5968 are in the data vector), which produced offspring with 3922 different sires. The pedigree was extended to include all the available information in the breed database, with a total of 20,178 animals. The inbreeding coefficients of the cow (Fc) and the bull (Fs) were determined according to the methodology described by Meuwissen and Luo [34] using the optiSel package [35] from the R statistical environment [36] and clustered into 15 classes of 2 percent intervals each. In addition, we estimated the calving number (Cn) and the herd-year-breeding season combination (HYS) in each observation using self-made R-scripts together with the Tidyverse [37] and data.table [38] packages. We estimated the fertility of the cows using the Re parameter, calculated as the deviation between the optimal and real parity number of females at each age, as described by Perdomo-Gonzalez, et al. [39]. After filtering and pruning incomplete and outlying data, 57,018 records from 12,554 cows were retained for the analysis.

2.2. Quantitative Genetic Analysis

Genetic analysis was performed using 2 animal models implemented in ASreml3 [40], without including Fs since it was not significant ($p > 0.05$). First, we tested the classical repeatability model (Rep), in which it is assumed that the Re effect and residual variance are homogeneous across the Cn scale, as follows:

$$y_{ijkl} \approx \mu + X_1Cn_i + X_2Fc_j + Z_1a_m + Z_2p_n + Z_3HYSN_k + e_{ijkmn} \rightarrow \text{Rep}$$

where

y is the dependent variable Re.

μ is the average, Cn is the calving number (fixed effect), HYSN is the herd-year-season of birth of the cow combination (random effect), and Fc is the inbreeding value of the cow (fixed effect). X_1, X_2, Z_1, Z_2, and Z_3 are incidence matrixes (0 or 1). e is the residual effect.

In addition, we tested a random regression methodology (RA), in which (co)variance components for Re varied across the Cn, assuming the existence of 6 classes of residual variance according to the calving number (e_{res} = 1,2,3,4,5, and 6 or more), using the following model:

$$y_{ijkl} \approx \mu + \sum_{r=0}^{n} \Phi_r b_1 Cn_i + \sum_{r=0}^{n} \Phi_r b_2 Fc_j + Z_3 HYS_k + \sum_{r=0}^{n} \Phi_r a_{m:Cn} + I_{Pn} + e_{:res} \rightarrow RA$$

where b_1 and b_2 are the fixed regression coefficients modeled by a Legendre polynome (Φ) of order r = 2 representing the (co)variance variations for Re in Cn (9 classes) and Fc (16 classes), respectively. Z_3 is an incidence matrix (0 or 1).

In both models, $a_{m:Cn}$; p_n and HYS_k (herd-year-breeding season combination in each observation) are random effect vectors for the animal and their ancestors; and the repetition of records of the same trait in the animal and contemporary groups, respectively. Residual effects were considered homogenous (e_{ijkmn}) in Rep, while they were estimated in 6 classes ($e_{:res}$) according to Cn in RA. In Rep, X_1, X_2, Z_1, Z_2, and Z_3 are incidence matrixes (0 or 1) connecting fixed and random effects, which were replaced (with the exception of Z_3) by Legendre polynomial coefficients (Φ) of r = n order to estimate the results in a longitudinal way along the calving trajectory of the cows in RA. The parameters by RA model were estimated using two different approaches with r = 1 and r = 2.

The variance components were estimated per model as follows:

Rep model

$$Var(y) = A\sigma_a^2 + I_p\sigma_p^2 + I_{HYS}\sigma_{HYS}^2 + I_n\sigma_e^2$$

RA model

$$ar(y) = N[0, (\sigma_y^2 = [G_0 = (A \otimes K_a)] + I_p\sigma_p^2 + I_{HYS}\sigma_{HYS}^2 + I_n\sigma_{e:res}^2$$

In both cases, A is the additive genetic relationship matrix; I is the identity matrix (number of cows) of the p order. In Rep, σ_a^2, σ_p^2, σ_{HYS}^2, and σ_e^2 are the additive, permanent environmental, HYS and residual variances, respectively. The genetic parameters and the animal Expected Genetic Values (EGV) for Re were estimated through the direct solution of the Rep model, assuming that they do not vary across Cn.

In the RA model, the same parameters were estimated along the Cn, using the K_a matrix

$$K_a = \Phi_i \begin{bmatrix} \sigma_{a_o}^2 & \sigma_{a_{os}} & \sigma_{a_{oq}} \\ \sigma_{a_{so}} & \sigma_{a_s}^2 & \sigma_{a_{sq}} \\ \sigma_{a_{qo}} & \sigma_{a_{qs}} & \sigma_{a_q}^2 \end{bmatrix} \Phi_i'$$

containing the elements related to the intercept, the slope, and the quadratic term for the additive genetic effects with variances $\sigma_{a_o}^2$, $\sigma_{a_s}^2$ y $\sigma_{a_q}^2$ and their respective covariances $\sigma_{a_{so}}$; $\sigma_{a_{sq}}$ y $\sigma_{a_{oq}}$; and $I_p\sigma_p^2 + I_{HYS}\sigma_{HYS}^2 + I_n\sigma_e^2$ represent the same indicators for the variances and covariances.

(Co)variance components were estimated along the Cn trajectory following the methodology proposed by Jamrozik, et al. [41], using the formula $\sigma_{a_i}^2 = \Phi_i K_a \Phi_i'$ for both genetic variances and covariances.

Heritability (h^2) and repeatability (r) were estimated in Rep by the classical procedure proposed by Falconer and MCKay [42], while h^2 and genetic correlations (r_g) were estimated in each ith Cn point included in the respective Legendre polynomials (Φ_i) in the RA model as follows:

$$h_i^2 = \frac{\Phi_i K_a \Phi_i'}{\Phi_i K_a \Phi_i' + I_p\sigma_p^2 + I_{HYS}\sigma_{HYS}^2 + I_n\sigma_{e:res}^2} \text{ y } r_{g_{ij}} = \frac{\Phi_i K_a \Phi_j'}{\sqrt{\Phi_i K_a \Phi_i' * \Phi_j K_a \Phi_j'}}$$

In addition, the RA model allows us to determine the EGV per individual (a_m) on all the ith points of the Cn trajectory as follows:

$$EGV_m^i = \Phi_i a_m'$$

where Φ_i are the polynomial coefficients of the i point across the Cn trajectory, and a_m = $[a_o\ a_s\ a_q\ ]$ is the animal genetic function estimated by the model described above.

Finally, the Re genetic values obtained using RA in each of the 9 calvings were summarized in a subjacent index (Ipc) estimated using a PCA analysis, according to the procedure described by Menendez-Buxadera, et al. [43].

In the second step, we arrived at a new EGV value based on the PCA results (IpcT) by applying the methodology described by Togashi and Lin [44], as follows:

$$\mathrm{Ipc}_1 = \sum_{i=1}^{9} \mathrm{PC1evi} * \mathrm{EGVi}$$

$$\mathrm{Ipc}_2 = \sum_{i=1}^{9} \mathrm{PC2evi} * \mathrm{EGVi}$$

$$\mathrm{IpcT} = \mathrm{Ipc}_1 + \mathrm{Ipc}_2$$

where EGV_i are the genetic values of the ith calving, expressed as standardized values, and $PC1ev_i$ and $PC2ev_i$ are the estimations of the PC1 and PC2 for the ith calving, respectively.

2.3. *DNA Samples and Genotyping of Samples*

In total, 252 Retinta cows were selected for genotyping by their EGV index based on the PCA results (IpcT), applying the methodology described by Togashi and Lin [44] (described in the Quantitative Genetic Analysis section). Of these, 40 cows had a negative IpcT, and 212 cows had an IpcT in the upper quartile. Genomic DNA was isolated from blood samples using a DNeasy Blood & Tissue Extraction Kit (Qiagen, Hilden, Germany). The cows were genotyped using the Axiom Bovine Genotyping v3 Array, which includes 63,000 markers uniformly distributed across the genome. The raw genotype data were processed following the "Best Practices Workflow" procedure in the Axiom Analysis Suite package v5.0 and were filtered using PLINK v1.9 software [45]. SNP markers with a call-rate $\leq$ 0.95 and with a minor allele frequency (MAF) < 0.05 were excluded. In the final step, the dataset was pruned by linkage disequilibrium in PLINK using the *–indep-pairwise 50 5 0.5* command. The final genomic dataset included 29,037 SNPs from the autosomal and X chromosomes.

2.4. *Genome-Wide Association Study (GWAS)*

GWAS analysis was performed using GEMMA software [46], employing the following univariate linear mixed model:

$$y = W\alpha + X\beta + \mu + \mathcal{E}$$

where y is an n-vector with pseudo-phenotypes (IpcT index); W is an incidence matrix of covariates (fixed effects) including a column of 1s; α is a vector of the corresponding coefficients including the intercept; X is an n-vector of marker genotypes; β is the effect size of the marker; μ is an n-vector of random effects $\mu \sim N(0, \lambda\tau^{-1} K)$, where τ^{-1} is the variance of the residual errors; λ is the ratio between two variance components; K is the genomic relationship matrix (estimated from the markers); and $\mathcal{E}$ is the n-vector of errors. In addition, the model included a correction for population stratification based on the first 10 components of a principal component analysis performed in PLINK v.1.9 as covariates. The statistical significance of the SNP effect was calculated using the Wald test statistic, which determined a *p*-value for each SNP.

2.5. *Functional Analysis*

Potential candidate genes associated within ± 500 Kb of the significant SNPs were annotated using the Ensembl BioMart database with the last available cow reference genome (ARS-UCD1.2. https://www.ensembl.org/Bos_taurus/Info/Index (accessed on 28 September 2022). Finally, the function of these genes and their putative relationship with fertility processes was established by performing an extensive review of the available literature in public databases, as well as in the DAVID V6.8 and Uniprot online resources.

3. Results and Discussion

Fertility is one of the key factors in any extensive livestock production system. However, at the same time, it is one of the most neglected from a genetic point of view, since it is difficult to study models in such production conditions. For this reason, the use of indirect traits obtained from pedigree records is becoming an interesting alternative to evaluating fertility at the population level since it allows us to obtain large phenotypic datasets, which are crucial for making reliable estimations [13,47]. Using such an approach, we were able to determine the genetic and environmental effects in a large cohort of beef cows bred in extensive conditions. To our knowledge, no previous studies have employed this methodology in cattle.

Re is a novel trait that estimates the deviation between the optimal and real parity number of females at each age of the cows (considering precocity and calving interval) throughout their entire productive lives. Our results showed a wide variability in average Re (across calving) among Retinta cows (Table 1).

Table 1. Descriptive statistics for input data.

Parameter	Mean	Minimum	Maximum	Coef.Var.
Reproductive efficiency (%)	72.05 ± 17.33	33	100	24.05
Inbreeding coefficient of the cow (Fc)	0.06 ± 0.08	0	0.46	136.00
Inbreeding coefficient of the bull (Fs)	0.08 ± 0.07	0	0.45	90.93
1st calving age (months)	33.99 ± 6.18	20	48	18.18

Furthermore, Re increased in value almost linearly with the calving number (Figure 1). This positive correlation (0.97) agrees with the results obtained by Mercadante, et al. [48], and Swanepoel and Hoogenboezem [49], who concluded that the cows that remain longer in the herd present better reproductive behavior during their productive lives. In this context, Balieiro, et al. [50] demonstrated that the most fertile cows that do not "miss" any breeding seasons are the ones most likely to be allowed to remain in the herd in extensively-raised beef breeds. Interestingly, those "regular and reliable" individuals show high values of Re, demonstrating the validity of this trait as an indirect estimator of fertility.

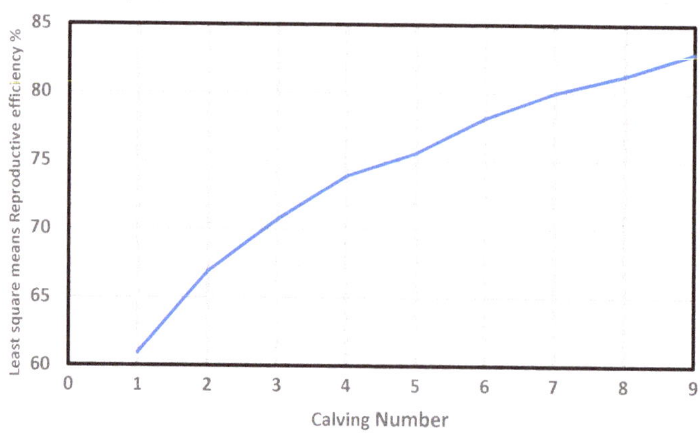

Figure 1. Effects of calving number on reproductive efficiency in Retinta breed.

The same analysis revealed lower variability in the age of first calving (33.9 ± 6.18 months, on average), which can be associated with differences in precocity, as has been convincingly proven in beef breeds [51,52]. This fact is particularly important since precocity is negatively related to longevity, and cows starting their reproductive lives at later stages are less likely to be culled for reproductive reasons [53]. Since precocity is not commonly included in models of calving intervals in beef cattle, we hypothesize that Re may be an interesting estimator of fertility in cows bred in extensive conditions, reducing a potential cause of bias within the evaluations.

3.1. Modeling the Genetic Influence on Re Using a Repeatability Model

In this study, we employed two different approaches to estimate the genetic component affecting Re: the classical Rep model and the RRM. The former is the most common methodology employed, having been used for over 20 years [16]. However, the latter has been increasingly employed during recent years since it allows us to estimate changes in the variance components of Re (or any reproductive trait) across calvings and, therefore, to select individuals with increased fertility values at any particular age [54]. However, regardless of which model is employed, fertility traits in livestock bred in extensive conditions must be analyzed using large reproductive datasets, such as the dataset used with Re as the trait of interest.

The variance components obtained using the Rep model are shown in Table 2. Heritability was particularly high (near 0.3) for a reproductive character in comparison with previous reports on beef cows, which are closer to 0.1 (reviewed by Cammack, et al. [10]). However, none of these employed Re as a fertility estimator for the cows. Likewise, similar h^2 values (near 0.25) were recently reported in goats and horses [13,47]. In addition, it was also noticeable that the permanent environmental effect of the cow accounted for nearly 35% of the total variance and the contemporary group (HYS) for only 9%. This result for the permanent environmental effect is expected in any extensive system in which environmental factors (such as metritis or retained placentas) with poor treatment can permanently delay the cow's fertility within the season [55]. Regarding HYS, the lower value fits in with the homogeneity of the production conditions across herds and years in the Spanish Dehesa in which the Retinta cows are bred [33].

Table 2. Variance components and parameter estimates for Reproductive efficiency.

Effect	Variance Components	% Relative Importance
Additive genetic effect	91.82	30.18
Permanent environmental effect	105.55	34.69
Herd-year-season effect	27.38	9.00
Residual effect	79.52	26.13
Total	304.27	

Figure 2 shows the evolution of Re over the years under study, where a clear positive trend in the breeding values can be seen over recent decades. This could be partly due to selection against carriers of the Robertsonian translocation 1/29 (rob(1;29)). Despite the incidence of carrier individuals in the whole population decresing during the last 30 years from 15.73% in 1992 to 1.06% in 2020, an increase was observed in Re phenotypic values in the same period. Although this increase in Re can be partially related to the decrease in the translocation rate, it may also be explained partly by other effects, such as improvements in animal management [15].

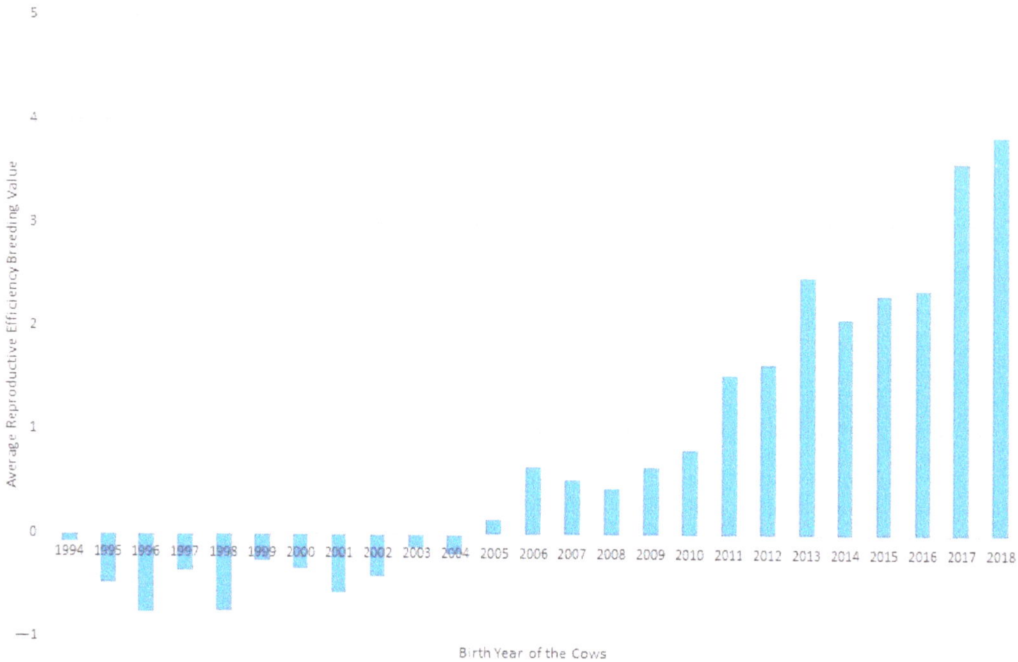

Figure 2. Evolution of the effect of the year of birth on the reproductive efficiency of the cow.

3.2. Modeling Fertility across the Lifespan of the Cow

One of the drawbacks of the Rep model is the assumption of homogeneity of the variance components of Re across calvings. This hypothesis can be problematic for traits with low-moderate heritability, such as fertility-related traits. In contrast, RA models take into account the possible h^2 variations across the stage of production of the individual (the calving number of the cow, in our case) [54]. In our study, both RA models showed better adjustments than those obtained using the Rep model (lower logL, AIC, and BIC, Table 3), which supports the hypothesis that Re behavior is not homogeneous for each calving of the cow. Among the RA models, the second-order RA (RA2) showed the best adjustment to the data, suggesting that it can produce the most unbiased estimation of the (co)variance components.

Table 3. Comparison among repeatability (REP) and random regression models (RA).

Model	N of Parameters	logL	AIC	BIC
REP	4	−183,253	366,514	366,550
RA r = 1	11	−173,721	347,464	347,563
RA r = 2	14	−169,279	338,587	338,714

In the RA model, heritabilities rose gradually with increased calving number, from 0.246 in the 1st calving to 0.583 in the 9th, mostly due to a slower pace of reduction observed in genetic variance in comparison with phenotypic variance (Table 4). In this context, only the estimated heritability at the first calving in RA was lower than the estimated value using the Rep model. Therefore, the use of RA models allows us to give a better explanation of the intrinsic variation of the trait, which is not possible using Re. However, the reduction in the total variance observed for the later calvings can limit the selection possibilities of the trait, despite the higher heritabilities observed. This can be seen clearly in Figure 3, which shows the range of Re variation per calving. The size of this variation is close to 40% for the first calving but falls to 29% and 24% in the fourth and rest of the calvings, respectively, which implies that a better response to the selection based on EGVs is expected in the first calving. However, it is also worth mentioning that the correlated responses for the rest of the calvings will be lower, due to the decreasing pattern observed in the genetic correlations between the first and other calvings (Figure 4), despite the fact that all the correlations estimated among the genetic values for different calvings were high (none were lower than 0.6, Figure 4), and even higher (>0.83) when three or more calvings were considered. For this reason, our results suggest that the Re estimation for the third calving could be the best trait for selecting cows for fertility since it combines precocity in the selection practices, genetic variability, and reliability. In addition, it demonstrates that the order of merit for Re will vary among calvings, and therefore, selective practices should take this into account.

Table 4. Genetic parameters for reproductive efficiency of the Retinta breed throughout the different calvings were estimated by repeatability (REP) and random regression models (RA).

	Random Regression Models of Second-Order		
Calving	**Genetic Var**	**Total Var**	**h^2**
1	123.18	506.83	0.24
2	85.02	241.00	0.35
3	62.21	163.82	0.38
4	49.64	128.36	0.39
5	43.28	109.33	0.40
6	40.22	94.96	0.43
7	38.60	87.35	0.44
8	37.69	79.86	0.47
9	37.83	74.39	0.51
Repeatability model	91.82	304.27	0.30 ± 0.003

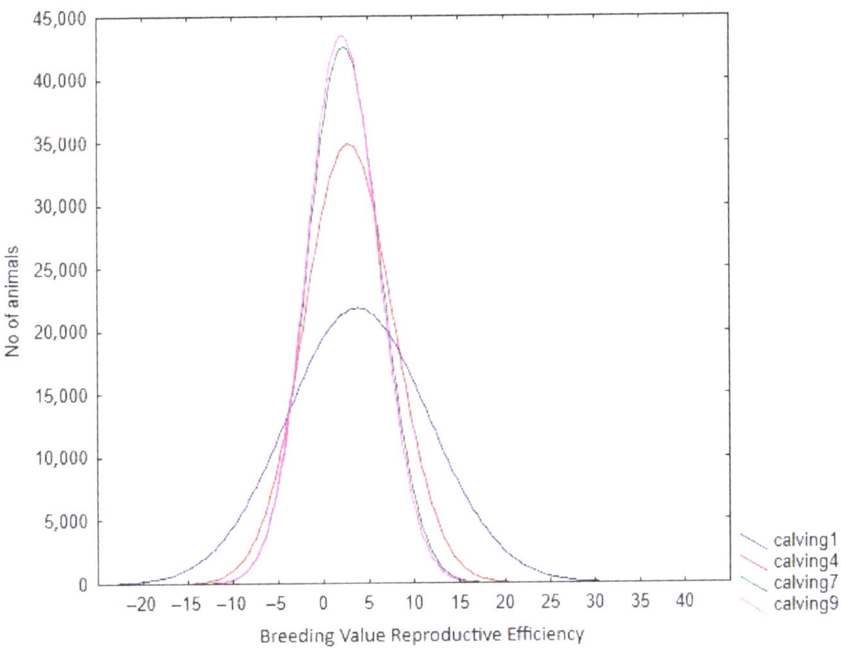

Figure 3. Variation in breeding value for reproductive efficiency per calving in the Retinta breed, estimated by random regression models.

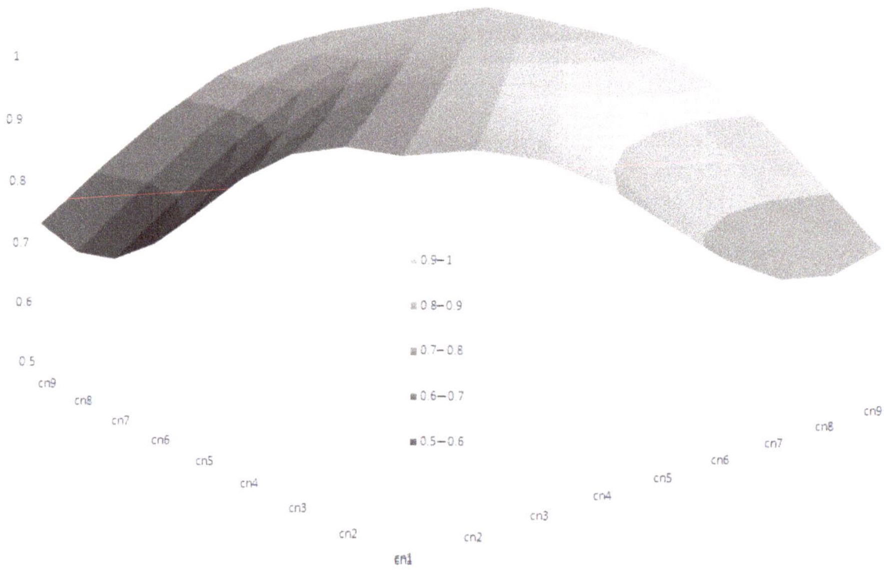

Figure 4. Genetic correlations of breeding value for reproductive efficiency between calvings, estimated by random regression genetic models.

Finally, the dataset employed in this study allows us to analyze the evolution of EGV for Re during the last 3 decades, since the beginning of the improvement program (from 1992 to 2018, Figure 5), which showed a positive trend towards increased Re genetic values in both sexes (7% on average). However, this value was higher in the EGV of the 1st and 3rd calving, which implies that breeding decisions based on fertility were made by breeders at the early stages of the lives of these cows and bulls.

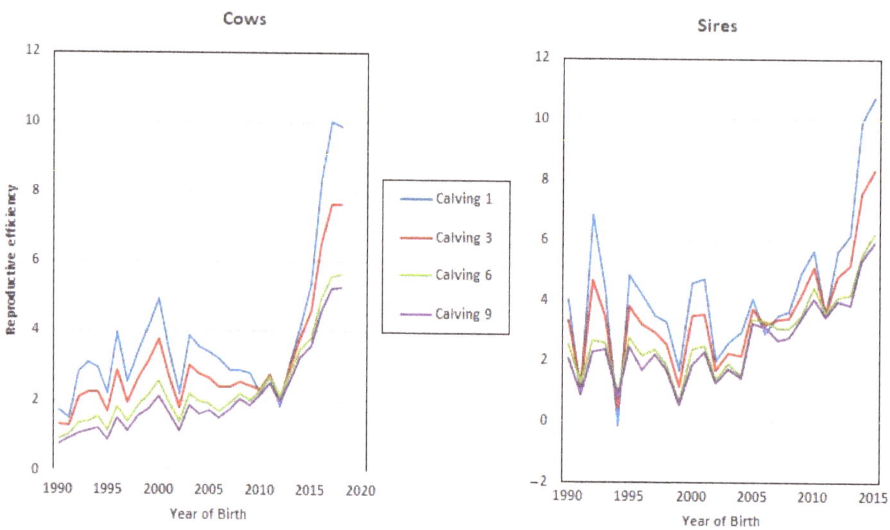

Figure 5. Genetic progress achieved in reproductive efficiency (Re) over a 30-year period.

3.3. Practical Implications of the RA Results

One of the problems of RA models is the vast amount of information produced in each analysis. In our case, we estimated 9 EGVs for 38,058 individuals, making it difficult to obtain reliable and clear conclusions. To cope with this situation, we performed a multivariate principal component analysis (PCA) in which all the available information was transformed into vectors (principal components, PC), which account for a certain % of the (co)variances existing among the variables studied. In this procedure, originally developed by Togashi and Lin [44], PCs are ordered by the percentage of the total variance explained across all the response surfaces (all the calvings analyzed, in our case), in order to summarize the information produced by the RA analysis in a simple and useful way.

Our analysis revealed that 99.7% of EGV (co)variance was explained by two major λ_i (PC1 and PC2), which demonstrates that all the variability determined by the RA analysis can be interpreted in a reliable way using this methodology (Table 5). In our case, all the eigenvectors (EVs) in PC1 showed a very similar positive trend (close to 0.3), suggesting that any variation observed will be linear across the calvings (size vector). In contrast, PC2 (known as the vector of shape) showed positive values until the 4th calving, when it became negative.

Table 5. Estimated eigenvalues (λ_i) and eigenvectors (evi) for the principal component analysis results with the genetic values of reproductive efficiency for each calving, from a total of 38,058 animals.

Variable	PC1	PC2
ev1	0.29	0.58
ev2	0.31	0.45
ev3	0.32	0.29
ev4	0.33	0.11
ev5	0.33	−0.07
ev6	0.32	−0.21
ev7	0.32	−0.30
ev8	0.32	−0.33
ev9	0.32	−0.23
λ_i	9.37	0.59
Variance %	93.6	6.1

Finally, the estimated IpcT (calculated as the combination of Ipc values for PC1 and 2) showed a much more accurate fit to the breeding value distribution observed in all the population in comparison to that obtained by estimating the average EGV values for each individual using the RA model (Figure 6). This value has an average of 0 ± 0.014 varying from −6.812 to 9.70. However, IpcT also allows us to summarize in a single value all the Re variability per individual across the calvings, thus allowing us to increase the accuracy of the breeding decisions. The correlation between the breeding values for Rep and IpcT shows a significant relationship of medium magnitude (0.45), which supports the use of this index as a pseudophenotype instead of the global EGV of the animal for this trait.

In addition, the use of IpcT permits a significant reduction in the risk of the possible interaction genotype x environment interaction problems (a different genetic potential in each calving number, as shown the Table 5) by adding the eingvectors, which by definition are uncorrelated. The use of PCA in different species and breed selection [43,56–58] and the results of the correlations between the EGV of each calving and the IpcT demonstrate it (Table 6).

Table 6. Correlations between the estimated EGV1 and the estimated EGV for each calving number and Correlations between the IpcT and the estimated EGV (first 9 calvings only).

	EGV1	EGV2	EGV3	EGV4	EGV5	EGV6	EGV7	EGV8	EGV9
EGV1	1.00	0.99	0.97	0.92	0.85	0.78	0.74	0.72	0.74
IpcT	0.97	0.99	0.99	0.99	0.96	0.92	0.90	0.89	0.91

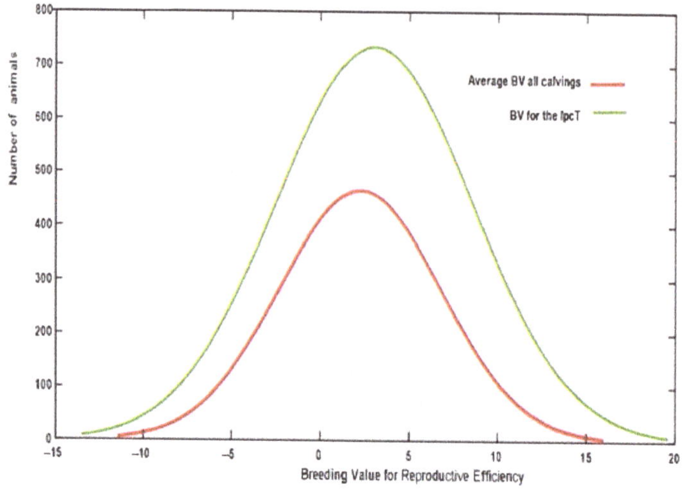

Figure 6. Variation in reproductive efficiency in the whole population estimated using PCA (IpcT) or the average RE_{EGV} for all the calvings.

3.4. Differences among Individuals in Re Development and Variation in Calvings

One interesting finding is the fact that individuals with similar EGV values at early ages can show opposite behaviors in terms of fertility at later stages. This pattern is observed in Figure 7A, in which 100 individuals with the same genetic estimation for Re at the first calving diverge profoundly at the fifth calving, in which some individuals showed a positive EGV while others showed the opposite. In the same way, Figure 7B showed the Re estimation per calving of 100 individuals with the same value at the 5th calving, in which some of them showed a highly positive estimation at the 1st calving (with values ranging around 20%), whereas another group of individuals showed negative values at the same calving. From a biological point of view, the differences observed among individuals can be explained by the phenomenon known as "plasticity", described by de Jong and Bijma [59] and largely employed in modeling production traits in livestock systems [60,61]. However, it is worth mentioning that 82% of the best 100 individuals ranked using EGV1 and IpcT were coincident, and the divergent behaviors were shown by less of the population, which supports the use of PCA analysis as a robust methodology to perform better matings.

In our case, this parameter, defined as the difference between the genetic values obtained in the same character in the 1st and 9th calving, showed wide variability over the whole population. In addition, we determined an asymmetrical distribution with an increased proportion of negative values, which suggests increased reproductive efficiency at the first calving. We hypothesize that this deviation could be caused by the increased environmental variability affecting the Re values at the first calving caused by the age of the cow, since the age at first calving is influenced not only by the age at puberty but also by when each breeder decides to include the heifer in the mating lot. This breeder's influence is reduced when the calving number increases (in addition to reducing the variability of the parameter with the elimination of less fertile animals).

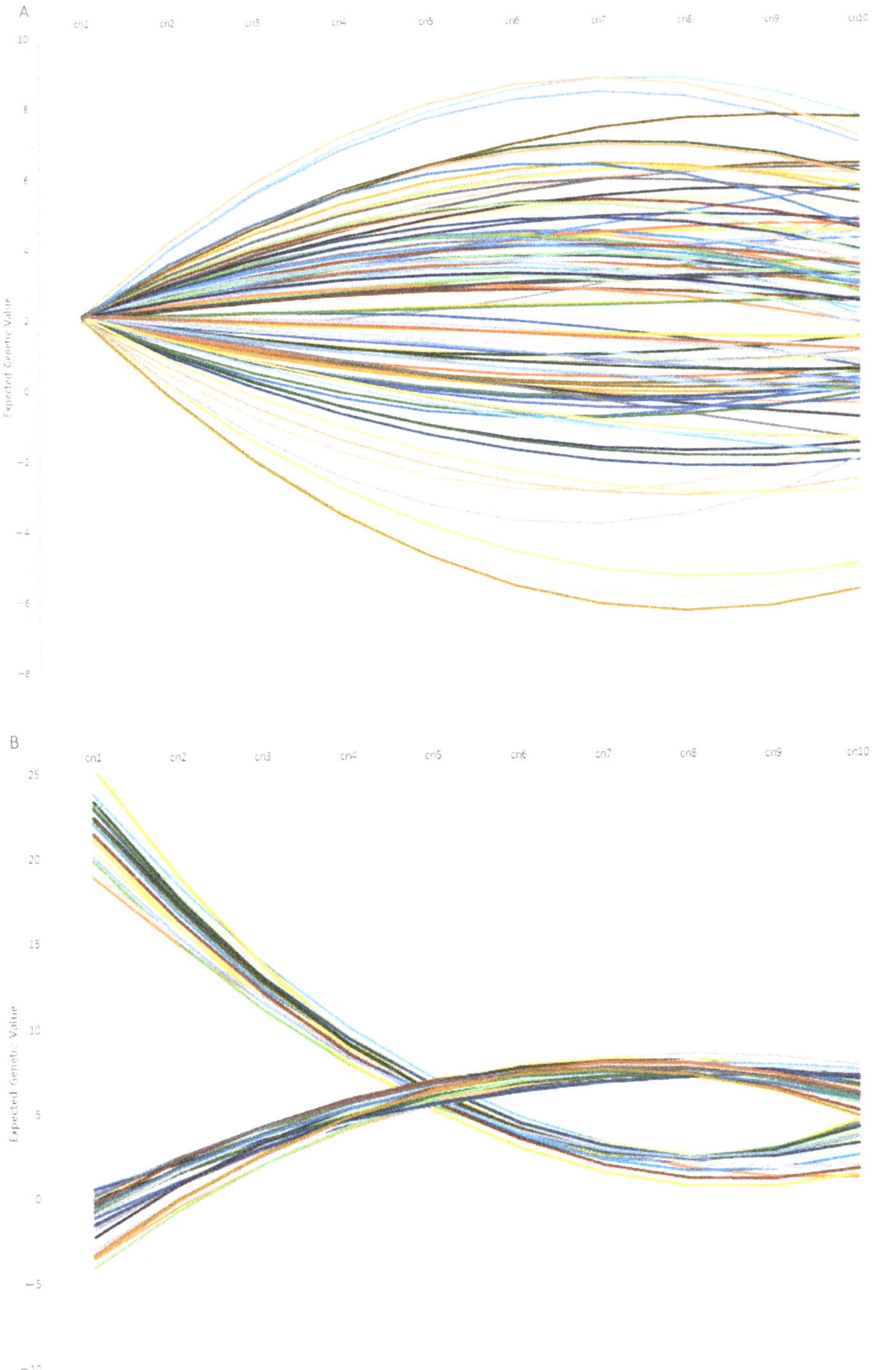

Figure 7. Variation on the Expected Genetic Values (EGV) across calvings (cn) in 100 individuals with similar EGV at 1st (**A**) or 5th calving (**B**).

3.5. Genome-Wide Association Study (GWAS) and Functional Analysis

Genome-wide association analyses detected 5 SNPs significantly associated with reproductive efficiency (p-value $< 10^{-4}$, Table 7), located in 2 genomic regions on chromosomes BTA4 and BTA28 (Figure 8). The estimated average effect of the significant SNPs in absolute values was 4.91. The in-silico functional analysis revealed the presence of 5 candidate genes (*NRF1, SSMEM1, CPA5, RYR2,* and *ZP4*) previously linked with known biological processes, molecular functions, and pathways related to fertility located within ± 500 Kb of the significant SNPs (Table 7), among which there were processes related to steroidogenesis, embryonic development, and spermatogenesis in cattle, goat, rabbit, mouse, and human models.

Table 7. List of SNPs associated with reproductive efficiency and candidate genes related to fertility in Retinta beef cattle.

Chr	SNP	Position (bp)	β	se	p-Wald	Candidate Gene
4	AX-124382279	93,739,544	−5.414	1.334	7.199054×10^{-5}	*NRF1, SSMEM1; CPA5*
4	AX-106723907	97,052,195	−4.382	0.964	9.616885×10^{-6}	
4	AX-185120444	97,052,759	−4.290	0.979	1.926352×10^{-5}	
4	AX-115113656	97,128,968	−5.936	1.229	2.755483×10^{-6}	
28	AX-169372743	10,126,454	−4.530	1.086	4.607665×10^{-5}	*RYR2; ZP4*

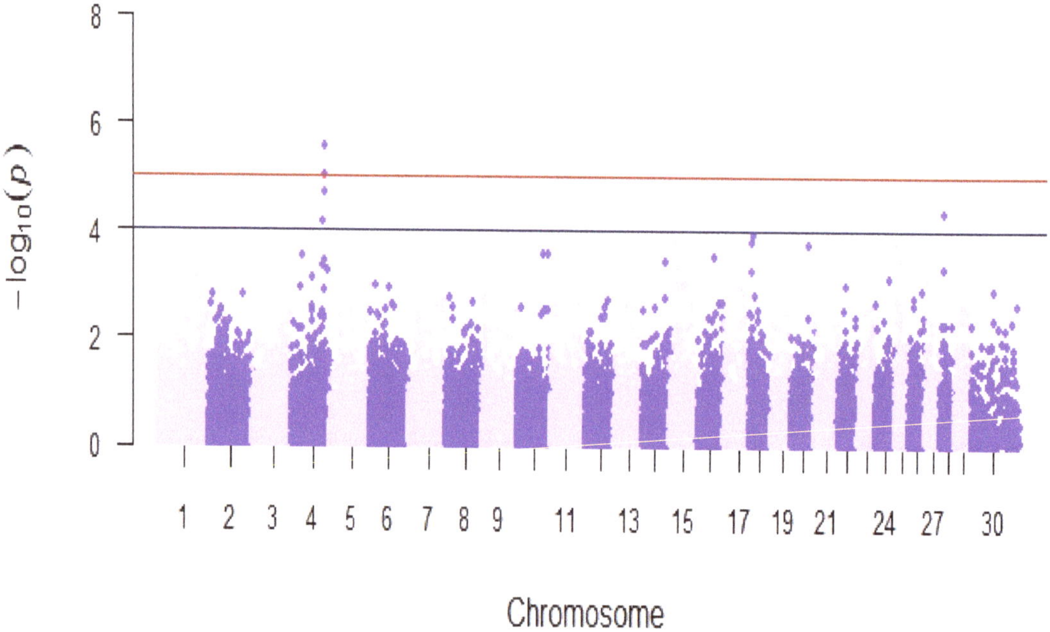

Figure 8. Manhattan plot of genome-wide association analyses for reproductive efficiency in the Retinta breed. The red line shows a genome-wide significance threshold ($-\log10(p) = 5$) and the blue line a genome-wide suggestive threshold ($-\log10(p) = 4$).

The candidate genes located within the significant regions of the SNP AX-124382279 have previously been described in important biological pathways. One of these genes was nuclear respiratory factor 1 (*NRF1*), which is involved in mitochondrial biogenesis, signal transduction, and protein synthesis. In mice, this gene has been related to late gestational embryonic lethality when a loss of function occurs [62]. Recently, Zhang, et al. [63] demonstrated the effects of *NRF1* on steroidogenesis and cell apoptosis in goat luteinized granulosa cells. An attenuated expression of *NRF1* led to mitochondrial dysfunction, disrupted the cellular redox balance, impaired steroid synthesis, and finally resulted in granulosa cell apoptosis through the mitochondria-dependent pathway. The other two candidate genes found in this region were *SSMEM1* and *CPA5*, both related to male fertility. Serine-rich single-pass membrane protein 1 (*SSMEM1*) is a conserved testis-specific gene in mammals. Nozawa, et al. [64] demonstrated that *SSMEM1* is essential for male fertility in mice and found that the *SSMEM1* protein is expressed during spermatogenesis but not in mature sperm. The sterility of the *SSMEM1* KO mice was associated with globozoospermia and a loss of sperm motility, which is crucial for fertilization. Similarly, *CPA5* (carboxypeptidase A5) is involved in spermiogenesis and in the regulation of sperm function, as it is exclusively or highly expressed in spermatogenetic cells, especially from the secondary spermatocyte to elongated spermatid stages. However, Zhou, et al. [65] reported that there were no perceptible changes in reproductive phenotypes in *CPA5*-KO mice, suggesting that it is a dispensable factor for spermatogenesis and male fertility in mice.

On BTA28, we found the *RYR2* and *ZP4* genes located very close to the significant AX-169372743 marker. *RYR2* encodes an intracellular calcium-release channel and is expressed in many tissues, including the ovaries. In humans, some variants of *RYR2* were significantly associated with weight loss in early pregnancy [66], providing evidence that it may play a role in fertility. However, its expression was also associated with amphiregulin, a key mediator of the effect of LH/hCG and a marker for oocyte competence [67]. Finally, the *ZP4* gene codified a glycoprotein that formed the zona pellucida (ZP), the extracellular matrix that houses mammalian oocytes and embryos. In rabbits, Lamas-Toranzo, et al. [68] reported that *ZP4* has a structural role in the zona pellucida and that rabbits without *ZP4* were subfertile. In vitro, the loss of *ZP4* did not affect ovulation, fertilization, or the early stages of the development of embryos. However, in vivo development was severely impaired in embryos covered by a ZP4-devoid zone. In addition, two recent studies in rats showed that *ZP4* is not responsible for gamete-specific interaction [69,70].

4. Conclusions

In this study, we were able to model the fertility of cows bred in extensive conditions using a novel fertility trait, reproductive efficiency (Re), which can be estimated through the analysis of reproductive records, using two different genetic methodologies. Our results showed that Re has a significant genetic component, but also that the genetic influence on Re is not homogeneous during the length of life of the cows, and individuals were observed showing very different patterns of variability in the breeding values across parities, which cannot be detected by the Rep model. Therefore, the use of the RA model would be an option to enhance fertility in the Retinta breeding program. We also determined that the use of a composite index, based on a principal component analysis could satisfactorily integrate all the information produced by RA models in an efficient and user-friendly way for the breeding program. Finally, we demonstrated that evaluating reproductive efficiency using the RA model makes it possible to differentiate the cows' genetic potential throughout the calving trajectory and enjoy greater flexibility in selecting female breeders. Finally, our preliminary GWAS analysis shows the existence of specific genomic regions influencing the fertility of Retinta cows. This new information could help us understand the genetic architecture of reproductive traits in the species better as well as allow us to select more fertile cows with greater accuracy. However, further analyses including large populations and different breeds would be required to validate our genomic findings.

Author Contributions: Conceptualization, J.M.J., R.M.M., A.M.-B., and A.M.; methodology, A.M.-B., S.D.-P., and A.M.; formal analysis, J.M.J., R.M.M., A.M.-B., and N.L.; data curation, J.M.J., R.M.M., A.M.-B., and N.L.; writing—original draft preparation, J.M.J., R.M.M., A.M.-B., S.D.-P., and A.M.; writing—review and editing, R.M.M., A.M.-B., N.L., S.D.-P., and A.M.; supervision A.M.-B. and A.M. All authors have read and agreed to the published version of the manuscript.

Funding: This research received funding from the Ministry of Agriculture, Fisheries, and Food (MAPA) (SelGenBeef (AGL2010-15903) and GENE2FARM (KBBE.2011.1.3-06), the National Association of Breeders of Selected Retinta Cattle (REF.OTRI:12021081), and the MERAGEM—PAI AGR-158 Research Group (DU.24.5A.00.03).

Institutional Review Board Statement: Not applicable.

Informed Consent Statement: All the individuals analyzed in this study are involved in the Retinta breeding program which acknowledges the use of samples and data for scientific purposes.

Data Availability Statement: The data sets employed in this study are property of the ACRE (Asociación Nacional de Criadores de Ganado Vacuno Selecto de Raza Retinta—National Association of Breeders of Selected Retinta Cattles) and were provided for scientific purposes under a specific collaboration arrangement. The data set can be made available for scientific purposes to other authors by the ACRE technical department, upon reasonable request.

Acknowledgments: The authors would like to thank the ACRE (Asociación Nacional de Criadores de Ganado Vacuno Selecto de Raza Retinta—National Association of Breeders of Selected Retinta Cattle) for providing the data for this study. We would also like to thank the Commission for the Genomic Evaluation of Native Beef Bovine Breeds, the National Institute of Agricultural and Food Research and Technology (INIA), the Ministry of Agriculture, Fisheries, and Food (MAPA), the Cádiz Provincial Council, and Córdoba University for their collaboration.

Conflicts of Interest: The authors declare no conflict of interest.

References

1. Eler, J.; Ferraz, J.; Balieiro, J.; Mattos, E. Genetic analysis of average annual productivity of Nellore breeding cows (COWPROD). *Genet. Mol. Res.* **2008**, *7*, 234–242. [CrossRef] [PubMed]
2. Åby, B.A.; Aass, L.; Sehested, E.; Vangen, O. A bio-economic model for calculating economic values of traits for intensive and extensive beef cattle breeds. *Livest. Sci.* **2012**, *143*, 259–269. [CrossRef]
3. Fernández-Perea, M.; Jiménez, R.A. Economic weights for a selection index in Avileña purebred beef cattle. *Livest. Prod. Sci.* **2004**, *89*, 223–233. [CrossRef]
4. Koots, K.R.; Gibson, J.P. Economic values for beef production traits from a herd level bioeconomic model. *Can. J. Anim. Sci.* **1998**, *78*, 29–45. [CrossRef]
5. Albera, A.; Carnier, P.; Groen, A. Definition of a breeding goal for the Piemontese breed: Economic and biological values and their sensitivity to production circumstances. *Livest. Prod. Sci.* **2004**, *89*, 66–77. [CrossRef]
6. Berry, D.P.; Wall, E.; Pryce, J.E. Genetics and genomics of reproductive performance in dairy and beef cattle. *Animal* **2014**, *8*, 105–121. [CrossRef] [PubMed]
7. Ponzoni, R.W.; Newman, S. Developing breeding objectives for australian beef cattle production. *Anim. Sci.* **1989**, *49*, 35–47. [CrossRef]
8. Amer, P.; Simm, G.; Keane, M.; Diskin, M.; Wickham, B. Breeding objectives for beef cattle in Ireland. *Livest. Prod. Sci.* **2001**, *67*, 223–239. [CrossRef]
9. Snelling, W.M.; Golden, B.L.; Bourdon, R.M. Within-herd genetic analyses of stayability of beef females. *J. Anim. Sci.* **1995**, *73*, 993–1001. [CrossRef]
10. Cammack, K.; Thomas, M.; Enns, R. Reproductive Traits and Their Heritabilities in Beef Cattle. *Prof. Anim. Sci.* **2009**, *25*, 517–528. [CrossRef]
11. Grossi, D.; Frizzas, O.; Paz, C.; Bezerra, L.; Lôbo, R.; Oliveira, J.; Munari, D. Genetic associations between accumulated productivity, and reproductive and growth traits in Nelore cattle. *Livest. Sci.* **2008**, *117*, 139–146. [CrossRef]
12. Berry, D.P.; Evans, R.D. Genetics of reproductive performance in seasonal calving beef cows and its association with performance traits. *J. Anim. Sci.* **2014**, *92*, 1412–1422. [CrossRef]
13. Ziadi, C.; Molina, A.; Muñoz-Mejías, E.; Sánchez, M.; López, M.; González-Casquet, O. Selection Criteria for Improving Fertility in Spanish Goat Breeds: Estimation of Genetic Parameters and Designing Selection Indices for Optimal Genetic Responses. *Animals* **2021**, *11*, 409. [CrossRef]
14. Laseca, N.; Molina, A.; Ramón, M.; Valera, M.; Azcona, F.; Encina, A.; Demyda-Peyrás, S. Fine-Scale Analysis of Runs of Homozygosity Islands Affecting Fertility in Mares. *Front. Veter. Sci.* **2022**, *9*, 754028. [CrossRef] [PubMed]
15. Jiménez, J.M.; Morales, R.; Molina, A.; Moreno-Millán, M.; Demyda-Peyrás, S. Effect of the rob(1;29) translocation on the fertility of beef cattle reared under extensive conditions: A 30-year retrospective study. *Reprod. Domest. Anim.* **2021**, *57*, 349–356. [CrossRef] [PubMed]
16. Rust, T.; Groeneveld, E. Variance component estimation on female fertility traits in beef cattle. *S. Afr. J. Anim. Sci.* **2001**, *31*, 131–141. [CrossRef]
17. Mercadante, M.E.Z.; Lôbo, R.B.; Oliveira, H.N.D. Estimativas de (co) variâncias entre características de reprodução e de crescimento em fêmeas de um rebanho Nelore. *Rev. Bras. Zootec.* **2000**, *29*, 997–1004. [CrossRef]

18. Bertazzo, R.P.; Freitas, R.T.F.D.; Gonçalves, T.D.M.; Pereira, I.G.; Eler, J.P.; Ferraz, J.B.S.; Oliveira, A.I.G.D.; Andrade, I.F.D. Parâmetros genéticos de longevidade e produtividade de fêmeas da raça Nelore. *Rev. Bras. Zootec.* **2004**, *33*, 1118–1127. [CrossRef]
19. Wilson, A.J.; Réale, D. Ontogeny of Additive and Maternal Genetic Effects: Lessons from Domestic Mammals. *Am. Nat.* **2006**, *167*, E23–E38. [CrossRef]
20. Du, L.; Duan, X.; An, B.; Chang, T.; Liang, M.; Xu, L.; Zhang, L.; Li, J.; E, G.; Gao, H. Genome-Wide Association Study Based on Random Regression Model Reveals Candidate Genes Associated with Longitudinal Data in Chinese Simmental Beef Cattle. *Animals* **2021**, *11*, 2524. [CrossRef]
21. Niu, Q.; Zhang, T.; Xu, L.; Wang, T.; Wang, Z.; Zhu, B.; Zhang, L.; Gao, H.; Song, J.; Li, J.; et al. Integration of selection signatures and multi-trait GWAS reveals polygenic genetic architecture of carcass traits in beef cattle. *Genomics* **2021**, *113*, 3325–3336. [CrossRef]
22. Amorim, S.T.; Stafuzza, N.B.; Kluska, S.; Peripolli, E.; Pereira, A.S.C.; da Silveira, L.F.M.; de Albuquerque, L.G.; Baldi, F. Genome-wide interaction study reveals epistatic interactions for beef lipid-related traits in Nellore cattle. *Anim. Genet.* **2021**, *53*, 35–48. [CrossRef] [PubMed]
23. Berton, M.P.; de Lemos, M.V.A.; Stafuzza, N.B.; Fonseca, L.F.S.; Silva, D.B.D.S.; Peripolli, E.; Pereira, A.S.C.; Magalhães, A.F.B.; Albuquerque, L.G.; Baldi, F. Integration analyses of structural variations and differential gene expression associated with beef fatty acid profile in Nellore cattle. *Anim. Genet.* **2022**, *53*, 570–582. [CrossRef]
24. de Souza Fonseca, P.A.; Caldwell, T.; Mandell, I.; Wood, K.; Cánovas, A. Genome-wide association study for meat tenderness in beef cattle identifies patterns of the genetic contribution in different post-mortem stages. *Meat Sci.* **2022**, *186*, 108733. [CrossRef] [PubMed]
25. Li, J.; Wang, Y.; Mukiibi, R.; Karisa, B.; Plastow, G.S.; Li, C. Integrative analyses of genomic and metabolomic data reveal genetic mechanisms associated with carcass merit traits in beef cattle. *Sci. Rep.* **2022**, *12*, 3389. [CrossRef] [PubMed]
26. Smith, J.L.; Wilson, M.L.; Nilson, S.M.; Rowan, T.N.; Schnabel, R.D.; Decker, J.E.; Seabury, C.M. Genome-wide association and genotype by environment interactions for growth traits in U.S. Red Angus cattle. *BMC Genom.* **2022**, *23*, 517. [CrossRef]
27. Purfield, D.C.; Evans, R.D.; Carthy, T.R.; Berry, D.P. Genomic Regions Associated with Gestation Length Detected Using Whole-Genome Sequence Data Differ between Dairy and Beef Cattle. *Front. Genet.* **2019**, *10*, 1068. [CrossRef]
28. Keogh, K.; Carthy, T.; McClure, M.; Waters, S.; Kenny, D. Genome-wide association study of economically important traits in Charolais and Limousin beef cows. *Animal* **2020**, *15*, 100011. [CrossRef]
29. SStegemiller, M.R.; Murdoch, G.K.; Rowan, T.N.; Davenport, K.M.; Becker, G.M.; Hall, J.B.; Murdoch, B.M. Genome-Wide Association Analyses of Fertility Traits in Beef Heifers. *Genes* **2021**, *12*, 217. [CrossRef]
30. Butler, M.L.; Hartman, A.R.; Bormann, J.M.; Weaber, R.L.; Grieger, D.M.; Rolf, M.M. Genome-wide association study of beef bull semen attributes. *BMC Genom.* **2022**, *23*, 74. [CrossRef]
31. Morales, R.M.; Menéndez-Buxadera, A.; Demyda-Peyrás, S.; Molina, A. Genetic effects of season on the preweaning growth of beef cattle: A first approach to Retinta calves. *Rev. Colomb. Cienc. Pecu.* **2019**, *33*, 134–143. [CrossRef]
32. MAPAMA, M.o.A. Fisheries and Feeding. In *Study of the Spanish Sector of Suckler Cows;* Catalog of Publications of the General State Administration; General Sub Directorate of Livestock Products, General Directorate of Agricultural Productions and Markets: Beijing, China, 2018.
33. Morales, R.; Phocas, F.; Solé, M.; Demyda-Peyrás, S.; Menéndez-Buxadera, A.; Molina, A. Breeding beef cattle for an extended productive life: Evaluation of selection criteria in the Retinta breed. *Livest. Sci.* **2017**, *204*, 115–121. [CrossRef]
34. Meuwissen, T.H.E.; Luo, Z. Computing inbreeding coefficients in large populations. *Genet. Sel. Evol.* **1992**, *24*, 305–313. [CrossRef]
35. Wellmann, R. Optimum contribution selection for animal breeding and conservation: The R package optiSel. *BMC Bioinform.* **2019**, *20*, 25. [CrossRef]
36. R Core Team. *R: A Language and Environment for Statistical Computing V4.1.3 "One Push Up"*; R Foundation for Statistical Computing: Vienna, Austria, 2022.
37. Wickham, H.; Averick, M.; Bryan, J.; Chang, W.; McGowan, L.D.A.; François, R.; Grolemund, G.; Hayes, A.; Henry, L.; Hester, J.; et al. Welcome to the Tidyverse. *J. Open Source Softw.* **2019**, *4*, 1686. [CrossRef]
38. Dowle, M.; Srinivasan, A. Data.table: Extension of data.frame. R Package Version, 2019.
39. Perdomo-González, D.; Molina, A.; Sánchez-Guerrero, M.J.; Bartolomé, E.; Varona, L.; Valera, M. Genetic inbreeding depression load for fertility traits in Pura Raza Española mares. *J. Anim. Sci.* **2021**, *99*, skab316. [CrossRef]
40. Gilmour, A. *ASReml User Guide Release 3.0*; VSN International Ltd.: Hemel Hempstead, UK, 2009.
41. Jamrozik, J.; Kistemaker, G.; Dekkers, J.; Schaeffer, L. Comparison of Possible Covariates for Use in a Random Regression Model for Analyses of Test Day Yields. *J. Dairy Sci.* **1997**, *80*, 2550–2556. [CrossRef]
42. Falconer, D.; MCKay, T. *Introduction to Quantitative Genetics*, 4th ed.; Longman Group: Harlow, UK, 1996.
43. Menéndez-Buxadera, A.; Muñoz-Mejías, E.; Sánchez, M.; Serradilla, J.; Molina, A. Use of Principal Component Analysis to Combine Genetic Merit for Heat Stress and for Fat and Protein Yield in Spanish Autochthonous Dairy Goat Breeds. *Animals* **2021**, *11*, 736. [CrossRef]
44. Togashi, K.; Lin, C. Selection for Milk Production and Persistency Using Eigenvectors of the Random Regression Coefficient Matrix. *J. Dairy Sci.* **2006**, *89*, 4866–4873. [CrossRef]

45. Purcell, S.; Neale, B.; Todd-Brown, K.; Thomas, L.; Ferreira, M.A.R.; Bender, D.; Maller, J.; Sklar, P.; de Bakker, P.I.W.; Daly, M.J. et al. PLINK: A Tool Set for Whole-Genome Association and Population-Based Linkage Analyses. *Am. J. Hum. Genet.* **2007**, *81* 559–575. [CrossRef] [PubMed]
46. Zhou, X.; Stephens, M. Genome-wide efficient mixed-model analysis for association studies. *Nat. Genet.* **2012**, *44*, 821–824 [CrossRef] [PubMed]
47. Laseca, N.; Demyda-Peyrás, S.; Valera, M.; Ramón, M.; Escribano, B.; Perdomo-González, D.; Molina, A. A genome-wide association study of mare fertility in the Pura Raza Español horse. *Animal* **2022**, *16*, 100476. [CrossRef] [PubMed]
48. Mercadante, M.E.Z.; Lôbo, R.B.; De Oliveira, H.N. Estimates of (Co)Variances among Reproductive and Growth Traits in Female Nellore Cattle. *Rev. Bras. Zootec.* **2000**, *29*, 997–1004. [CrossRef]
49. Swanepoel, F.J.C.; Hoogenboezem, J. *Interrelationship among Cow Size, Lifetime Cow Fertility, Milk Production and Pre-Weaning Calf Growth in Sub-Tropically Adapted Beef Cattle*; Australian Society of Animal Production: Sydney, Australia, 1994.
50. Balieiro, J.; Eler, J.; Ferraz, J.; Mattos, E.; Balieiro, C. Genetic parameters for productive life traits and reproductive efficiency traits at 6 years in Nellore cattle. *Genet. Mol. Res.* **2008**, *7*, 1312–1318. [CrossRef]
51. Bourdon, R.M.; Brinks, J.S. Genetic, Environmental and Phenotypic Relationships among Gestation Length, Birth Weight, Growth Traits and Age at First Calving in Beef Cattle. *J. Anim. Sci.* **1982**, *55*, 543–553. [CrossRef]
52. Gutiérrez, J.; Alvarez, I.; Fernández, I.; Royo, L.; Díez, J.; Goyache, F. Genetic relationships between calving date, calving interval, age at first calving and type traits in beef cattle. *Livest. Prod. Sci.* **2002**, *78*, 215–222. [CrossRef]
53. Damiran, D.; A Larson, K.; Pearce, L.T.; E Erickson, N.; A Lardner, B.H. Effect of calving period on beef cow longevity and lifetime productivity in western Canada. *Transl. Anim. Sci.* **2018**, *2*, S61–S65. [CrossRef]
54. Oliveira, T.V.V.B.; Silva, H.T.; Filho, I.C.; Ramos, P.V.B.; Santana, T.E.Z.; Guimarães, J.D.; Guimarães, S.E.F.; Lopes, P.S.; e Silva, F.F.; Veroneze, R. Alternative bayesian models for genetic evaluation of biometrical, physical, and morphological reproductive traits in nelore bulls. *Livest. Sci.* **2020**, *244*, 104313. [CrossRef]
55. Goyache, F.; Gutiérrez, J.; Fernández, I.; Royo, L.; Álvarez, I. Genetic analysis of days open in beef cattle. *Livest. Prod. Sci.* **2005**, *93*, 283–289. [CrossRef]
56. Togashi, K.; Lin, C.Y. Improvement of lactation milk and persistency using the eigenvectors of the genetic covariance matrix between lactation stages. *Livest. Sci.* **2007**, *110*, 64–72. [CrossRef]
57. Boligon, A.A.; Vicente, I.S.; Vaz, R.Z.; Campos, G.S.; Souza, F.R.; Carvalheiro, R.; Albuquerque, L.G. Principal component analysis of breeding values for growth and reproductive traits and genetic association with adult size in beef cattle. *J. Anim. Sci.* **2016**, *94*, 5014–5022. [CrossRef] [PubMed]
58. Carabaño, M.J.; Logar, B.; Bormann, J.; Minet, J.; Vanrobays, M.L.; Díaz, C.; Tychon, B.; Gengler, N.; Hammami, H. Modeling heat stress under different environmental conditions. *J. Dairy Sci.* **2016**, *99*, 3798–3814. [CrossRef]
59. de Jong, G.; Bijma, P. Selection and phenotypic plasticity in evolutionary biology and animal breeding. *Livest. Prod. Sci.* **2002**, *78*, 195–214. [CrossRef]
60. Menendez-Buxadera, A.; Carabaño, M.; Gonzalez-Recio, O.; Cue, R.; Ugarte, E.; Alenda, R. Reaction norm of fertility traits adjusted for protein and fat production level across lactations in Holstein cattle. *J. Dairy Sci.* **2013**, *96*, 4653–4665. [CrossRef]
61. Santana, M.L.; Eler, J.; Cardoso, F.; Albuquerque, L.; Ferraz, J.B.S. Phenotypic plasticity of composite beef cattle performance using reaction norms model with unknown covariate. *Animal* **2013**, *7*, 202–210. [CrossRef]
62. Leung, L.; Kwong, M.; Hou, S.; Lee, C.; Chan, J.Y.; Jenkins-Kruchten, A.E.; Bennaars-Eiden, A.; Ross, J.R.; Shen, W.-J.; Kraemer, F.B.; et al. Deficiency of the Nrf1 and Nrf2 Transcription Factors Results in Early Embryonic Lethality and Severe Oxidative Stress. *J. Biol. Chem.* **2003**, *278*, 48021–48029. [CrossRef]
63. Zhang, G.-M.; Deng, M.-T.; Lei, Z.-H.; Wan, Y.-J.; Nie, H.-T.; Wang, Z.-Y.; Fan, Y.-X.; Wang, F.; Zhang, Y.-L. Effects of NRF1 on steroidogenesis and apoptosis in goat luteinized granulosa cells. *Reproduction* **2017**, *154*, 111–122. [CrossRef]
64. Nozawa, K.; Zhang, Q.; Miyata, H.; Devlin, D.J.; Yu, Z.; Oura, S.; Koyano, T.; Matsuyama, M.; Ikawa, M.; Matzuk, M.M. Knockout of serine-rich single-pass membrane protein 1 (Ssmem1) causes globozoospermia and sterility in male mice†. *Biol. Reprod.* **2020**, *103*, 244–253. [CrossRef]
65. Zhou, Y.; Zhang, X.; Xiong, S.; Zeng, X.; Zhang, X. Predicted gene 31453 (Gm31453) and the gene encoding carboxypeptidase A5 (Cpa5) are not essential for spermatogenesis and male fertility in the mouse. *Reprod. Fertil. Dev.* **2021**, *33*, 401–409. [CrossRef] [PubMed]
66. Fejzo, M.S.; Myhre, R.; Conde, L.C.; MacGibbon, K.; Sinsheimer, J.S.; Reddy, M.P.L.; Pajukanta, P.; Nyholt, D.; Wright, M.; Martin, N.G.; et al. Genetic analysis of hyperemesis gravidarum reveals association with intracellular calcium release channel (RYR2). *Mol. Cell. Endocrinol.* **2016**, *439*, 308–316. [CrossRef] [PubMed]
67. Grøndahl, M.L.; Andersen, C.Y.; Bogstad, J.; Borgbo, T.; Boujida, V.H.; Borup, R. Specific genes are selectively expressed between cumulus and granulosa cells from individual human pre-ovulatory follicles. *Mol. Hum. Reprod.* **2012**, *18*, 572–584. [CrossRef] [PubMed]
68. Lamas-Toranzo, I.; Balvís, N.F.; Querejeta-Fernández, A.; Izquierdo-Rico, M.J.; González-Brusi, L.; Lorenzo, P.L.; García-Rebollar, P.; Avilés, M.; Bermejo-Álvarez, P. ZP4 confers structural properties to the zona pellucida essential for embryo development. *Elife* **2019**, *8*, e48904. [CrossRef]

69. Izquierdo-Rico, M.J.; Moros-Nicolás, C.; Pérez-Crespo, M.; Laguna-Barraza, R.; Gutiérrez-Adán, A.; Veyrunes, F.; Ballesta, J.; Laudet, V.; Chevret, P.; Avilés, M. ZP4 Is Present in Murine Zona Pellucida and Is Not Responsible for the Specific Gamete Interaction. *Front. Cell Dev. Biol.* **2021**, *8*, 626679. [CrossRef]
70. Zeng, M.-H.; Wang, Y.; Huang, H.-L.; Quan, R.-P.; Yang, J.-T.; Guo, D.; Sun, Y.; Lv, C.; Li, T.-Y.; Wang, L.; et al. *Zp4* is completely dispensable for fertility in female rats. *Biol. Reprod.* **2021**, *104*, 1282–1291. [CrossRef]

animals

MDPI

Article

Reverse Genetic Screen for Deleterious Recessive Variants in the Local Simmental Cattle Population of Switzerland

Irene M. Häfliger [1], Franz R. Seefried [2] and Cord Drögemüller [1,*]

[1] Institute of Genetics, Vetsuisse Faculty, University of Bern, 3012 Bern, Switzerland; irene.haefliger@vetsuisse.unibe.ch
[2] Qualitas AG, Chamerstrasse 56, 6300 Zug, Switzerland; franz.seefried@qualitasag.ch
* Correspondence: cord.droegemueller@vetsuisse.unibe.ch; Tel.: +41-316842529

Simple Summary: Today's Swiss Simmental represents a local dual-purpose breed of cattle. Within closed populations, deleterious variants can reach problematic frequencies, explaining substantial proportions of inbreeding depression. Depletions in homozygous genotypes for certain haplotypes among large cohorts of animals genotyped for the purpose of genomic selection is a widely used approach to pinpoint undesired recessive alleles. In the course of a reverse genetic screen, we aimed to identify single recessive Mendelian variants that potentially affect fertility and rearing success without any phenotypic information available. We detected eleven genome regions showing obvious depletion of homozygosity based on genome-wide SNP data. Furthermore, after performing whole-genome sequencing of selected animals, we propose three candidate causative variants affecting different genes with possibly detrimental effects for embryonic development. The established haplotypes, as well as the identified protein-changing variants, can be directly implemented into breeding practice to avoid the risk of mating carriers and thereby increase breeding success.

Abstract: We herein report the result of a large-scale reverse genetic screen in the Swiss Simmental population, a local dual-purpose cattle breed. We aimed to detect possible recessively inherited variants affecting protein-coding genes, as such deleterious variants can impair fertility and rearing success significantly. We used 115,000 phased SNP data of almost 10 thousand cattle with pedigree data. This revealed evidence for 11 genomic regions of 1.17 Mb on average, with haplotypes (SH1 to SH11) showing a significant depletion in homozygosity and an allele frequency between 3.2 and 10.6%. For the proposed haplotypes, it was unfortunately not possible to evaluate associations with fertility traits as no corresponding data were available. For each haplotype region, possible candidate genes were listed based on their known function in development and disease. Subsequent mining of single-nucleotide variants and short indels in the genomes of 23 sequenced haplotype carriers allowed us to identify three perfectly linked candidate causative protein-changing variants: a SH5-related *DIS3*:p.Ile678fs loss-of-function variant, a SH8-related *CYP2B6*:p.Ile313Asn missense variant, and a SH9-related *NUBPL*:p.Ser143Tyr missense variant. None of these variants occurred in homozygous state in any of more than 5200 sequenced cattle of various breeds. Selection against these alleles in order to reduce reproductive failure and animal loss is recommended.

Keywords: cattle; *Bos taurus*; reproduction; breeding; fertility; embryonic lethality; loss-of-function variants; whole-genome sequencing; SNP genotyping

1. Introduction

Simmental is a globally recognized cattle breed, originating in the Simmental valley in the canton Bern in Switzerland. Autochthonous of Switzerland, today up to 40 million cattle worldwide are designated as members of the Simmental breed. However, Simmental are characterized by very different local breeding objectives (http://wsff.info; accessed on 24 October 2021). In the 19th century, crossbreeding of local cattle with

Simmental cattle exported from Switzerland led to Fleckvieh populations in neighboring countries, including Austrian and German Fleckvieh, French Montbeliarde, and Italian Pezzata Rossa cattle. In the 20th century, introduction of animals of the Holstein breed into the Swiss Simmental population led to today's Fleckvieh population in Switzerland (Swiss Fleckvieh). A recent analysis of the population structure of Swiss cattle showed a clear differentiation between today's Swiss Simmental cattle and all other Swiss cattle populations, including the modern Swiss Fleckvieh [1,2]. Various historically younger Central European Simmental populations that descended from the Swiss Simmental can also be clearly distinguished from today's Swiss Simmental animals, which showed the highest inbreeding level [3]. In contrast, recent studies found a comparatively low degree of genomic inbreeding in purebred Swiss Simmental that might be explained by the continued use of natural service sires, which is likely the major reason for their remarkably high level of genetic diversity although this population has been closed for a long time [1,2]. Therefore, the current so-called Original Simmental breed of Switzerland represents a unique purebred population. At the end of 2020, more than 23 thousand dual-purpose animals in the Simmental population were registered in the Swiss herdbook (https://www.swissherdbook.ch/fileadmin/Domain1/PDF_Dokumente/05-Statistiken-Formulare/53-Jahresstatistik/1_Wichtigste_Zahlen/D_sh_JS_2020_HBZahlen_web.pdf; accessed on 24 October 2021).

A universal problem in cattle breeding is reproductive failure. It was shown that the reproduction success is negatively associated with production traits [4,5]. These effects have been thoroughly studied and possible reasons are selection programs focusing on production traits coupled with negative correlation with reproductive traits [6]. Regarding early pregnancy loss in Simmental cattle, e.g., a Croatian study described the body condition score, parity, and milk yield as important influencing factors [7]. Lower reproductive success is an economic problem for farmers and a major reason for cattle slaughter [8,9]. To tackle fertility issues, one approach is to apply a so-called reverse genetic screen, where only genomic data is used rather than phenotypic information. A whole-genome sequencing (WGS)-based approach was suggested by Charlier et al. (2016), where sequencing data of entire genomes is mined for variants affecting fertility by causing embryonic lethality [10]. A similar approach proposes scanning the cumulative population-wide SNP genotyping data from genomic selection programs to identify haplotype regions indicative of a depletion in homozygosity [11]. The analysis includes the statistical evaluation of segregating haplotypes regarding the Hardy–Weinberg equilibrium (HWE) [11]. Such genomic regions harboring deficient homozygous haplotypes indicate potential causal variants for embryonic death or unwanted phenotypes in newborns, leading to the exclusion of homozygous animals from breeding programs [11]. This SNP-based reverse genetics approach has been used successfully in diverse breeds and species (e.g., [11–31]). Haplotypes never observed in homozygous state indicate the presence of recessive, predominantly embryonic lethal variants. Potential causative variants were detected by analyzing whole-genome or whole-exome sequencing data linked to the identified haplotypes (e.g., [10,19–30,32–35]). In general, these approaches are especially interesting for small populations where phenotypic information is sparse.

In Simmental-derived breeds in Central Europe, eight causal protein-changing variants for recessively inherited disorders are known so far. In German Fleckvieh cattle, a SNP-based reverse genetic study identified four deficient homozygous haplotypes located on chromosomes 1, 10, and 12 and proposed two candidate causal variants in two genes: the FH2-related frameshift variant in *SLC2A2* and the FH4-related missense variant in *SUGT1* [21]. In addition, a fifth haplotype is described, affecting the calf survival due to a congenital heart failure and severe liver damage (https://www.lfl.bayern.de/itz/rind/122227/index.php; accessed on 24 October 2021). Based on forward genetic studies, where affected individuals were examined, several causal variants for Mendelian disorders were found. A frameshift variant in *GON4L* was associated with the autosomal recessively inherited disorder dwarfism (OMIA 001985-

9913) [36]. Furthermore, a nonsense variant in *PLD4* causing a recessive genodermatosis observed in German Fleckvieh (OMIA 001935-9913) and a missense variant in *OPA3* associated with a form of dilated cardiomyopathy predominantly affecting Swiss Fleckvieh (OMIA 000162-9913) were described [37,38]. In German Fleckvieh, a *MOCS1*-related form of arachnomelia was identified to be due to a frameshift variant (OMIA 001541-9913) [39]. Furthermore, the BH2-related *TUBD1* missense variant, known to cause juvenile mortality in Braunvieh cattle (OMIA 001939-9913), was also observed in German and Austrian Fleckvieh [23]. To the best of our knowledge, so far, no carriers for any of these eight deleterious alleles have been found in the purebred Swiss Simmental population.

In purebred Swiss Simmental cattle, no genomic analysis has yet been carried out to systematically identify recessively inherited harmful variants that affect female reproduction or calf rearing. Based on a reverse genetic screen, the present study aimed at a comprehensive analysis of available SNP and WGS data to identify genomic regions containing deficient homozygous haplotypes as well as linked candidate causal variants for hidden phenotypes.

2. Materials and Methods

The national breeding association provided us with the available SNP data, including all genotyped purebred Swiss Simmental cattle born after 2009 and their ancestors. The genomic positions of the markers relate to the latest cattle reference sequence ARS-UCD1.2 [40,41]. Due to the application of several routinely available SNP arrays ranging from 9000 to 150,000 SNPs, the data had to be imputed. Therefore, the software Fimpute v2.2 [42] was used with default parameters to increase the number of markers and correct for wrongly called markers. In order to assure the quality, SNPs with a minor allele frequency <0.01 were excluded from the dataset. Furthermore, SNPs that could not pass the quality measures for the call rate per SNP >0.99 were excluded and animals with call rates <0.8 were excluded too. Quality control was applied before and after imputation. This resulted in a final SNP data set of 114,890 markers for 9965 animals.

As a first step, haplotypes showing a deviation from the HWE, indicated by depletion of homozygosity, were identified. The first subset of data analyzed included only fully genotyped trios where the complete trio (sire, dam, and offspring) were genotyped (n = 2626), further called "trio" approach. The second dataset analyzed included genotyped trios where an offspring and two paternal animals (sire and maternal grandfather) were genotyped (n = 3969), subsequently called parent–grandparent "pgp" approach. Both data sets, trio and pgp, were used in screening for window size of 50 markers within the software snp1101 [43]. The analyzed haplotypes overlap, as the windows were continuously moved marker by marker. The snp1101 software used the Fisher exact test of HWE [44] to analyze the resulting haplotypes. Furthermore, the p-values were corrected for a false discovery rate with the Benjamini–Yekultieli method and a significance level of 5% was applied [45]. For each significant haplotype region, we selected the most significant haplotype (lowest p-value). Regarding the identification of candidate variants, haplotype regions were defined by the selected haplotypes and increased on both sides by 2 Mb in order to make sure that the regions of overlapping significant haplotypes were included.

Within the 11 regions detected, between 14 and 262 genes were annotated, giving a total of 669 genes. To these genes, information regarding associated phenotypes were collected from the various online databases: HUGO Gene Nomenclature Committee (https://www.genenames.org; accessed on 7 October 2021), Mouse Genome Informatics (http://www.informatics.jax.org; accessed on 7 October 2021), International Mouse Phenotyping Consortium (https://www.mousephenotype.org; accessed on 7 October 2021), Online Mendelian Inheritance in Man (https://www.omim.org; accessed on 7 October 2021), Online Mendelian Inheritance in Animals (https://omia.org/home/; accessed on 7 October 2021), and Decipher (https://www.deciphergenomics.org; accessed on 7 October 2021).

With the selected haplotypes, we predicted individual diplotypes that represent if an animal carries one, two, or no copies of the haplotype. Based on these diplotypes, we selected three carrier animals for whole-genome sequencing (WGS) for each haplotype region. Therefore, 23 Simmental cattle (1 female and 22 male) carrying one or more significant haplotypes were selected for this study. WGS data were prepared as previously described [46]. However, recalibration was performed with the variant catalogue of the 1000 Bull Genomes Project run 7 (BQSR file version 2) [47,48]. The 23 genomes sequenced in this study were submitted to the 1000 Bull Genomes project [47,48] and are therefore part of the 5116 animals in the recent variant catalogue (run9). This international dataset was used to evaluate candidate causative variants in a larger cohort and across breeds, to evaluate breed specificity as well as reduced homozygosity. Furthermore, we had access to an additional 115 publicly available genomes from the Swiss Comparative Bovine Resequencing project, deposited in the European Nucleotide Archive under project accession PRJEB18113, that were yet not added to the 1000 Bull Genomes project. The combined WGS dataset of 5231 genomes includes 62 purebred Swiss Simmental cattle.

In order to identify haplotype-associated candidate causal variants, linkage disequilibrium (LD) analysis was performed using plink v1.9 [49]. LD (r^2) was calculated between the diplotype state and the WGS-derived genotypes for the 23 animals that were selected for WGS. Analysis was restricted to protein-changing variants, including variants annotated by SnpEff (version 4.3; [50]) with the following sequence ontology terms: missense_variant, start_lost, stop_gained (nonsense), stop_lost, stop_retained_variant, splice_acceptor_variant, splice_donor_variant, conservative_inframe_deletion, conservative_inframe_insertion, disruptive_inframe_deletion, disruptive_inframe_insertion, exon_loss_variant, frameshift_variant, and gene_fusion.

Furthermore, to improve the understanding of candidate causal variants, their base conservation scores from UCSC database called PhyloP and PhastCons were applied [51,52]. To apply these values, firstly, the variants of the bovine genome needed to be mapped to the human genome 38 [53] with the tool LiftOver of the UCSC tools. Secondly, the conservation scores of 99 vertebrates for these human positions were obtained. Additionally, the effects of protein-changing variant were estimated using PROVEAN [54] and PredictSNP [55].

3. Results

In this reverse genetic study, we used SNP data to identify haplotypes showing a depletion in homozygosity and applied WGS data to pinpoint candidate causal variants in purebred Swiss Simmental animals. Reduced homozygosity due to hidden recessive variants in cattle could, so far, only be causally explained by coding variants. Therefore, we focused on variants having a moderate impact, such as missense variants, conservative inframe insertions and deletions up to a size of 50 bp, as well as on all other protein-changing with high impact including loss-of-function variants, such as stop-gains (nonsense), splice site-disrupting SNVs, frameshift indels in a coding sequence, or deletions that remove coding exons.

3.1. Identification of Deficient Homozygous Haplotypes in Swiss Simmental Cattle

We detected seven haplotype regions with the trio approach and nine haplotype regions applying the pgp approach (Table 1, Figure 1, Table S1). Five of these haplotype regions appear in both analyses. We named the haplotype regions, in accordance to previous studies performed in other cattle breeds, as Simmental Haplotypes (SH) 1 to 11 [22] (Table 1, Table S1). All 11 identified haplotypes show a deficiency of at least 85 percent of the expected homozygous animals within the studied Swiss Simmental population (Table 1, Table S1). Four selected haplotypes (SH5, SH7, SH8, and SH10) presented a complete deficit of observed homozygous animals, whereas the others showed a partial deficiency ranging from 85 to 96% of the expected homozygotes (Table 1; Table S1). The average length of the eleven haplotypes is 1.17 Mb and ranges from 0.73 to 1.94 Mb (Table 1).

Table 1. List of deficient homozygous haplotypes in Swiss Simmental cattle.

Haplotype [a]	Analysis	Chr	Position (Mb) [b]	Length (Mb)	Number of Homozygotes Observed	Expected	Deficiency (%)	Allele Frequency
SH1	trio and pgp	1	65.793–66.891	1.10	3	17	85	0.042
SH2	trio	2	1.191–1.920	0.73	1	16	94	0.040
SH3	pgp	7	104.784–105.640	0.86	1	17	94	0.041
SH4	pgp	11	29.389–30.353	0.96	4	22	85	0.047
SH5	trio	12	46.828–47.837	1.01	0	38	100	0.062
	pgp		46.831–47.843	1.01				
SH6	pgp	14	11.304–12.373	1.07	2	15	88	0.039
SH7	trio	16	6.120–8.011	1.89	0	30	100	0.055
	pgp		6.122–8.060	1.94	5	111	96	0.106
SH8	pgp	18	48.763–50.005	1.24	0	11	100	0.0337
	trio		48.806–50.017	1.21				0.0338
SH9	trio and pgp	21	41.855–42.851	1.00	2	19	96	0.043
SH10	pgp	24	41.945–43.140	1.20	0	10	100	0.032
SH11	trio	28	38.937–40.069	1.13	1	12	92	0.035

[a] SH meaning Simmental Haplotype. [b] in accordance with the reference sequence ARS-UCD1.2.

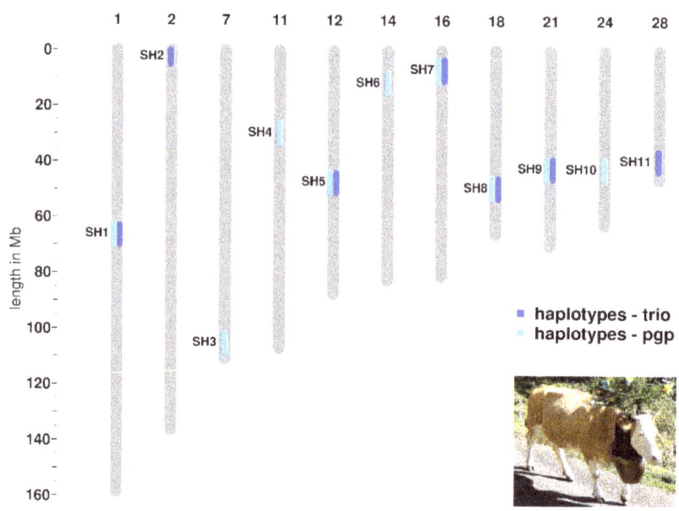

Figure 1. Genomic co-localization of the deficient homozygous haplotype regions detected in Swiss Simmental cattle using two different approaches. A typical breed-specific cow is shown in the lower right corner.

3.2. Identification of Candidate Genes in Haplotype Regions

The intensive analysis of all annotated protein-coding genes in the defined haplotype regions, extended by 2 Mb on each side, led to a comprehensive list of candidate genes possibly affecting either prenatal or postnatal lethality or associated sub-lethal phenotypes. We extracted 145 positional candidate genes of special interest, as they are associated with mammalian autosomal recessive disorders in human, mice, or other animals and listed in the consulted databases (Table S2). Loss-of-function mouse models of many of these genes have revealed defects that affect embryonic or perinatal to pre-weaning survival and therefore represent suitable functional candidates for this study. The presented short list of the 43 most probable candidate genes includes all genes that are associated with sub-lethal or lethal phenotypes (Table 2).

Table 2. List of candidate genes known to cause recessive disorders with lethal and sub-lethal phenotypes in human.

Haplotype	Gene	Gene Description	Recessive Disorder
SH1	*IQCB1* §	IQ motif containing B1	Senior-Loken syndrome 5
	MYLK §	myosin light chain kinase	Megacystis-microcolon-intestinal hypoperistalsis syndrome 1
	CASR *	calcium sensing receptor	Hyperparathyroidism, neonatal
	POGLUT1 *	protein O-glucosyltransferase 1	Limb-girdle muscular dystrophy-dystroglycanopathy
	TIMMDC1 *	translocase of inner mitochondrial membrane domain containing 1	Mitochondrial complex I deficiency, nuclear type 31
SH2	*HERC2* *,§	HECT and RLD domain containing E3 ubiquitin protein ligase 2	Mental retardation, MRT 38
	OCA2 *,§	OCA2 melanosomal transmembrane protein	Albinism, brown oculocutaneous
SH4	*CRIPT* *	CXXC repeat containing interactor of PDZ3 domain	Short stature with microcephaly and distinctive facies
	EPCAM *	epithelial cell adhesion molecule	Congenital diarrhea with tufting enteropathy
	FSHR *,§	follicle stimulating hormone receptor	Ovarian dysgenesis 1
	MSH2 §	mutS homolog 2	Mismatch repair cancer syndrome 2
	MSH6	mutS homolog 6	Mismatch repair cancer syndrome 3
	PIGF *	phosphatidylinositol glycan anchor biosynthesis class F	Onychodystrophy, osteodystrophy, impaired intellectual development, and seizures syndrome
	PPP1R21 *	protein phosphatase 1 regulatory subunit 21	Neurodevelopmental disorder with hypotonia, facial dysmorphism, and brain abnormalities
	TTC7A *	tetratricopeptide repeat domain 7A	Gastrointestinal defects and immunodeficiency syndrome
SH5	*PIBF1* *	progesterone immunomodulatory binding factor 1	Joubert syndrome 33
SH6	*MYC* *	MYC proto-oncogene, bHLH transcription factor	Burkitt lymphoma, somatic
SH7	*CFH*	complement factor H	Basal laminar drusen, complement factor H deficiency
	CR2 §	complement C3d receptor 2	Immunodeficiency, CVID7
	IL10 §	interleukin 10	Critical role in the control of immune responses
SH8	*ARHGEF1* §	Rho guanine nucleotide exchange factor 1	Immunodeficiency 62
	BCKDHA *	branched chain keto acid dehydrogenase E1 subunit alpha	Maple syrup urine disease, type Ia @
	B9D2 *	B9 domain containing 2	Meckel syndrome 10, Joubert syndrome 34
	COQ8B *	coenzyme Q8B	Nephrotic syndrome, type 9
	DLL3 §	delta-like canonical Notch ligand 3	Spondylocostal dysostosis 1
	ERF *	ETS2 repressor factor	Spondylocostal dysostosis 1
	ETHE1 §	ETHE1 persulfide dioxygenase	Ethylmalonic encephalopathy

Table 2. *Cont.*

Haplotype	Gene	Gene Description	Recessive Disorder
	LTBP4 §	latent transforming growth factor beta binding protein 4	Cutis laxa, autosomal recessive, type IC
	MEGF8 *	multiple EGF-like domains 8	Carpenter syndrome 2
	PLEKHG2	pleckstrin homology and RhoGEF domain containing G2	Leukodystrophy and acquired microcephaly
	RYR1 *	ryanodine receptor 1	Neuromuscular disease
	SMG9 *	SMG9 nonsense mediated mRNA decay factor	Heart and brain malformation syndrome
	SPINT2 *	serine peptidase inhibitor, Kunitz type 2	Diarrhea 3, secretory sodium, congenital, syndromic
	SPTBN4	spectrin beta, non-erythrocytic 4	Neurodevelopmental disorder with hypotonia, neuropathy, and deafness
	TGFB1 *	transforming growth factor beta 1	Inflammatory bowel disease, immunodeficiency, and encephalopathy
	TIMM50 *	translocase of inner mitochondrial membrane 50	3-methylglutaconic aciduria, type IX
	XRCC1 *	X-ray repair cross complementing 1	Spinocerebellar ataxia, SCAR26
SH9	*NUBPL* *	nucleotide binding protein-like	Mitochondrial complex I deficiency, nuclear type 21
SH10	*AFG3L2* *	AFG3-like matrix AAA peptidase subunit 2	Spastic ataxia 5
	LAMA1 *	laminin subunit alpha 1	Poretti–Boltshauser syndrome
	MC2R §	melanocortin 2 receptor	Glucocorticoid deficiency due to ACTH unresponsiveness
	NDUFV2 *	NADH:ubiquinone oxidoreductase core subunit V2	Mitochondrial complex I deficiency, nuclear type 7
	PIEZO2 *	piezo type mechanosensitive ion channel component 2	Arthrogryposis, distal, with impaired proprioception and touch

* Homozygous lethal with complete penetrance, § homozygous lethal with incomplete penetrance, @ disorder described in cattle (OMIA-ID: 000627-9913).

3.3. Identification of Candidate Causal Variants

For three deficient homozygous haplotypes (SH5, SH8, and SH9), by linkage disequilibrium analysis, we found perfectly linked ($r^2 = 1$) candidate causal variants. These three haplotypes were detected with both the pgp and the trio approach. For each of these haplotypes, we propose a protein changing SNV (Table 3; Table S3). These three variants never occur in homozygous state in the analyzed 5231 bovine genomes of various cattle breeds (Table S3). Interestingly, the SH8-associated variant is apparently specific for Swiss Simmental; however, the variants associated with SH5 and SH9 occur sporadically in single animals of some other breeds (Table S3).

Table 3. Candidate causal variants for deficiency of homozygotes in Swiss Simmental cattle.

Haplotype	Gene	OMIM	Variant	Transcript [a]	Coding DNA Change	Predicted Protein Change
SH5	*DIS3*	607533	1 bp insertion (frameshift)	XM_025000110.1	c.2032dupA	p.Ile678AsnTer2
SH8	*CYP2B6*	123930	SNV (missense)	NM_001075173.1	c.938T > A	p.Ile313Asn
SH9	*NUBPL*	613621	SNV (missense)	NM_001193042.1	c.428C > A	p.Ser143Tyr

[a] According to the NCBI Annotation Release 106 (National Center for Biotechnology Information, 2018b).

Among the three proposed non-synonymous variants are two missense variants altering evolutionary conserved residues and a frameshift variant that significantly truncates the encoded protein. The SH8-related SNV in exon 4 of the bovine *cytochrome P450 family 2 subfamily B member 6* (*CYP2B6*) gene on chromosome 18 at position 50296371 is a missense variant (NM_001075173.1: p.Ile313Asn) that was predicted by PROVEAN to have a deleterious effect (Table 3, Table S3). The SH9-related SNV located in exon 6 of the bovine *nucleotide binding protein-like* (*NUBPL*) gene on chromosome 21 at position 42154344 represents a missense variant (NM_001193042.1: p.Ser143Tyr) predicted to be deleterious by PROVEAN and PredictSNP (Figure 2, Table 3, Table S3). Both presented missense mutations altering evolutionary conserved amino acids (Figure 2). The SH5-associated 1 bp insertion located in exon 16 of the bovine *DIS3 homolog, exosome endoribonuclease and 3′-5′ exoribonuclease* (*DIS3*) gene on chromosome 12 at position 47511687 represents a loss-of-function variant. It was predicted to result in a frameshift after isoleucine 678 with a premature stop codon (NP_025000110.1: p.Ile678AsnTer2), resulting in a significantly shortened amino acid sequence, if expressed, when compared with the wild-type protein (Table 3; Table S3). In addition, the comparative DNA sequence approach (PhyloP and phastCons) showed a high conservation across species for all three variant positions (Table S3).

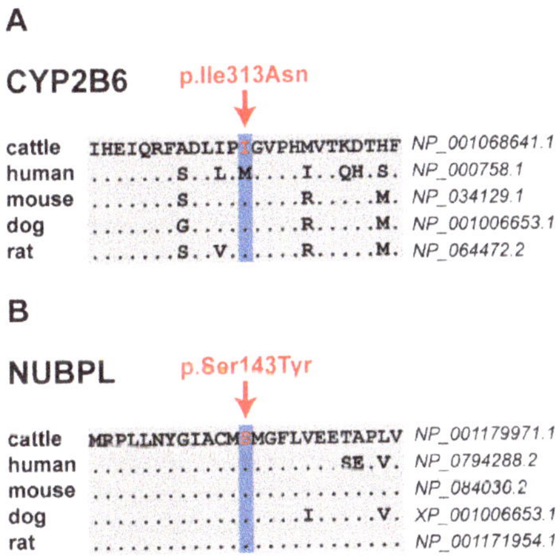

Figure 2. Multiple sequence alignment of the amino acid sequences of CYP2B6 (**A**) and NUBPL (**B**) around the position of the missense variants (shown in red), indicating evolutionary conservation across all species.

4. Discussion

For the first time, the genomic data of the current Swiss Simmental dual-purpose cattle population were analyzed for reduced homozygosity due to hidden recessive monogenic variants and validated using the international variant catalogue of the 1000 Bull Genomes Project. In addition to environmental factors, inherited deleterious variants lead to natural or artificial selection against homozygous individuals, which also explains embryonic lethality, reduce rearing success, or the exclusion from the breeding population due to poor development. Unfortunately, these phenomena are not systematically monitored and are therefore difficult phenotypes to study. Although reduced reproductive success can theoretically be detected in the sires estimated breeding values for certain fertility traits, these effects are only noticeable for deleterious alleles that have reached a high frequency in the population [56]. To overcome this issue, we performed a genome-wide missing homozygosity scan, revealing eleven haplotype regions with considerable homozygous depletion. After subsequent mining of genome sequence data for candidate causal variants, we propose three non-synonymous variants that probably cause the obvious deficiency of homozygous animals.

We applied two different approaches, a trio and pgp approach, that include genotyped trios and applies a Fisher exact test. Therefore, we were able to detect haplotypes that segregate at a lower allele frequency. As expected, we found more haplotype regions with the pgp approach in comparison to the trio approach, most likely because more genotyped groups were available. Nevertheless, the trio approach, detecting two regions not found with pgp, appears very powerful, probably because it directly traces the inheritance of the haplotypes. Previous approaches used the assumption of random mating and the deviation from HWE based on allele frequencies or used the deviation from expected number of homozygous offspring based on the haplotype state of the sire and maternal grandsire [14,20]. In contrast, the herein applied trio approach allows performing such an analysis with a small population such as Simmental. The five haplotype regions that were found in both analyses are the most probable genome regions harboring hidden harmful variants. Especially the haplotypes that never occur in homozygous state led to the suspicion of embryonic lethal variants segregating in the population. It is suspected that some haplotypes arose due to imputation errors introduced due to genotyping bias, SNP density/panel, sample size, and a bias introduced by the chosen software [57]. This would explain the haplotype regions SH3 and SH11 that arose in regions for which we could not identify any plausible functional candidate genes. Otherwise, we were able to identify candidate genes within all missing homozygous regions. Unfortunately, effect estimations of the haplotypes towards traits of female fertility and rearing success were not reasonable, as the available phenotypic data is currently limited. To avoid the detection of mostly sporadic associations rather than actual effects, it is planned to conduct such haplotype association studies in the future.

Candidate causal variants are proposed for the haplotypes SH5, SH8, and SH9 in the genes *DIS3*, *CYP2B6*, and *NUBPL*, respectively. These variants all show complete depletion in homozygosity, perfect LD to the associated haplotype, and high conservation scores when compared across 99 genomes, indicating their importance in basic biological functions. As in other mammals, it is expected that ~100 harmful recessive variants will be found per individual in cattle, of which up to five of these impact essential genes and cause embryonic lethality or severe disease when homozygous [10]. Nevertheless, it is recognized that it is very difficult to clarify the actual deleterious functions of these variants, although given the genes involved, it is assumed that these variants influence fitness. Reverse genetic screens to identify genes with major effects, as used in the current study, are therefore helpful to assign function to variants in candidate genes and/or so far less characterized genes such as *DIS3* and *CYP2B6*.

The SH5-related loss-of-function variant found in bovine *DIS3*, most likely leading to missing homozygosity, represents the first time a pathogenic variant that most likely causes embryonic lethality has been identified. If the mutant mRNA transcript were to escape nonsense-mediated decay, even if this truncated protein was expressed, it would lack roughly 30 percent of the C-terminal part, and therefore it is not expected to contribute any function. *DIS3*, also known as ribosomal RNA-processing protein 44 (RRP44), is a RNase II/R-like enzyme located primarily in the nucleus (Table S3) [58]. The protein has catalytic function in the RNA exosome complex, which is responsible for 3′-end processing and RNA degradation of a broad variety of RNAs [58,59]. Biological functions are associated with RNA metabolism, mitotic control, spindle-fiber formation, antibody diversification, microtubule production, and growth and development [60]. Recently, pathogenic variants in genes encoding both structural and catalytic subunits of the RNA exosome have been linked to human disease, such as *EXOSC3* and *EXOSC8* related forms of pontocerebellar hypoplasia, representing recessive neurodegenerative diseases [61]. To our knowledge, the *DIS3* gene has not yet been associated with Mendelian diseases, but variants are reported to be associated with various types of cancers and multiple myeloma [60–62]. Interestingly, the book of Fasken et al. (2020)

provides a summary of the most common variants in *DIS3* which all occur in heterozygous state, are associated with multiple myeloma, and seem to have mild effects only, while Tomecki et al. (2014) suggests the potential lethality of mutations in the PIN domain of *DIS3*. In *Drosophila*, a knock-down model led to wingless animals implicating an important role in development [63]. Nevertheless, the inactivation of *DIS3* in B cells was shown to lead to an increase in unbalanced DNA translocations [64]. Lastly, public databases for mouse phenotypes indicate a complete pre-weaning lethality of *DIS3* knock-out mice (https://www.mousephenotype.org/data/genes/MGI:1919912; http://www.informatics.jax.org/marker/MGI:1919912; accessed on 7 October 2021).

Despite a long list of candidate genes located within the SH8-region, we propose a missense variant in the *CYP2B6* gene as a candidate causal variant. The main reason for this is the perfect LD to the haplotype, the complete absence of homozygous animals, and the prediction of the DNA position to be highly conserved and the amino acid exchange to be deleterious. Regarding the gene function, CYP2B6 is a protein of the cytochromes P450 subfamily 2B (HGNC: 20604). This enzyme is known to be of importance for drug metabolism, as well as endogenous compounds, environmental toxins, and other substances [65,66]. For example, the susceptibility to Efavirenz depends on the individual *CYP2B6* genotype (OMIM: 123930). Several SNV were detected in human that are associated with the expression level and activity (increased and decreased) of *CYP2B6* with a population-wide importance [65] (https://www.pharmvar.org/gene/CYP2B6; accessed on 26 October 2021). In monkeys and human, it was shown that *CYP2B6* is expressed in the brain and affected by nicotine and alcohol consumption; however, its neurological function remains unclear [65–67]. The protein is also expressed in the placenta [68] and it was shown that the pregnancy hormone estradiol induces the expression of *CYP2B6* [69]. Nevertheless, what the function and importance of *CYP2B6* is in maintaining pregnancy is unclear. As the herein identified bovine variant is predicted to be deleterious, we speculate that function of CYP2B6 might be impaired during development.

Lastly, we propose the missense variant in the bovine *NUBPL* gene, exchanging a strongly conserved residue predicted to be deleterious and affecting a highly conserved nucleotide, to be causal for the deficit of homozygosity of SH9. As we observe few haplotype carriers in our data, we speculate that the effect of the variant is not fully penetrant or that signs of poor development appear later in life, after initial genotyping of young animals. NUBPL, also known as *iron-sulfur protein required for NADH dehydrogenase* (*IND1*), is a protein that is vital for the assembly of the respiratory complex I [70]. More precisely, the NUBPL supplies Fe/S clusters to the respiratory complex I and thereby ensures that important subunits are delivered to build the whole complex [70]. In human, pathogenic variants affecting *NUBPL* are associated with the autosomal recessive mitochondrial complex I deficiency disorder (OMIM: 613621, 618242) [71–73]. These variants include missense, frameshift and splice site variants, as well as small and large insertions and deletions. Clinical symptoms of mitochondrial complex I deficiency include, among others, ataxia, dysarthria, hypotonia, nystagmus, spasticity, and tremor [73]. Furthermore, variants in *NUBPL* were hypothesized as risk factors for Parkinson's disease [74]. A mouse model identified the necessity of the protein as knock-out alleles led to homozygous lethality (MGI: 1924076) [75].

However, if the proposed variants are indeed depleted in a number of homozygous animals, it needs to be evaluated by further genotyping of larger cohorts before implementation into selection schemes. This is planned by adding the variants to a custom array for genotyping larger cohorts of further animals, both Swiss Simmental as well as other local cattle populations. An alternative approach to confirm the absence of homozygous animals, particularly in the offspring of carrier-by-carrier mating's, would strongly support the deleterious nature of the variants as shown before [10]. In particular, for the *NUBPL*-associated variant, the observed homozygous animals should be examined in detail.

The chosen approach requires that the haplotype and the causative variant are in near perfect linkage disequilibrium and, obviously, this is not always the case. This could be an explanation for the fact that we could not identify any potentially causal variants in the other haplotype regions. An alternative approach, therefore, is to mine the genome sequence data for candidate variants, e.g., loss-of-function in the listed essential candidate genes, and to genotype these directly in large cohorts of Swiss Simmental cattle. Using this approach, nine causal variants were uncovered in cattle that would not have been detected using SNP-based haplotype approaches [10].

Finally, another drawback of our study was the restriction to protein-changing variants. Moreover, we had no evidence for perfectly linked non-coding regulatory variants, as well as the limitation to consider only SNV and short indels, overlooking possible larger structural variants.

5. Conclusions

In the presented project, we mined SNP and WGS data by applying a reverse genetic approach. Without any phenotypic evidence but mining the data of almost 10 thousand SNP genotyped Swiss Simmental cattle and more than 5200 WGS animals from a variety of breeds, we propose three candidate variants in the genes *DIS3*, *CYP2B6*, and *NUBPL* causing embryonic lethality and/or yet unknown recessive developmental disorders. After phenotypic validation of these variants, selection against these variants is recommended.

Supplementary Materials: The following are available online at https://www.mdpi.com/article/10.3390/ani11123535/s1, Table S1: Haplotype information of SH1 to SH11, Table S2: Comprehensive list of candidate genes located in the haplotype regions, including gene information from MGI, IMPC, OMIM, and OMIA, Table S3: Comprehensive list of potential candidate causal variants.

Author Contributions: Conceptualization, C.D. and F.R.S.; formal analysis, I.M.H. and F.R.S.; investigation, I.M.H. and F.R.S.; data curation, I.M.H.; writing—original draft preparation, I.M.H.; writing—review and editing, C.D., F.R.S.; visualization, I.M.H.; supervision, C.D. and F.R.S.; project administration and funding acquisition, C.D. All authors have read and agreed to the published version of the manuscript.

Funding: This research was funded by the Swiss National Science Foundation, grant number 172911.

Institutional Review Board Statement: This study did not require official or institutional ethical approval as it was not an experimental study, but an analysis of existing genotype data.

Data Availability Statement: The SNP data of Swiss Simmental cattle are owned by the breeding association swissherdbook. Therefore, we ask interested people to contact the authors or the breeding association directly in order to gain access to the SNP data. The WGS data are publicly available at the European Nucleotide Archive under project accession PRJEB18113.

Acknowledgments: We are very thankful for the genomic data provided by the Swiss cattle breeding association swissherdbook. The authors also wish to thank Swissgenetics and Selectstar for providing semen. The authors are thankful to the University of Bern institutions, namely the Next Generation Sequencing Platform and the Interfaculty Bioinformatics Unit, for performing the WGS experiments and providing high performance computing infrastructure, respectively. Finally, we acknowledge the 1000 Bull Genomes Project consortium, led by Christy Vander Jagt, Amanda Chemberlain, and Hans Daetwyler, for providing the variant catalogue.

Conflicts of Interest: The authors declare no conflict of interest.

Ethics Approval and Consent to Participate: According to the Swiss laws, no ethics approval was needed. In this study, available data were reused and mined using novel approaches.

References

1. Signer-Hasler, H.; Burren, A.; Neuditschko, M.; Frischknecht, M.; Garrick, D.; Stricker, C.; Gredler, B.; Bapst, B.; Flury, C. Population structure and genomic inbreeding in nine Swiss dairy cattle populations. *Genet. Sel. Evol.* **2017**, *49*, 83. [CrossRef]
2. Stranzinger, G.F.; Steiger, D.; Kneubühler, J.; Hagger, C. Y chromosome polymorphism in various breeds of cattle (*Bos taurus*) in Switzerland. *J. Appl. Genet.* **2007**, *48*, 241–245. [CrossRef]
3. Cesarani, A.; Gaspa, G.; Pauciullo, A.; Degano, L.; Vicario, D.; Macciotta, N.P.P. Genome-wide analysis of homozygosity regions in european simmental bulls. *J. Anim. Breed. Genet.* **2021**, *138*, 69–79. [CrossRef]
4. Pryce, J.E.; Woolaston, R.; Berry, D.P.; Wall, E.; Winters, M.; Butler, R.; Shaffer, M. World Trends in Dairy Cow Fertility. In Proceedings of the 10th World Congress on Genetics Applied to Livestock Production, Vancouver, BC, Canada, 17–22 August 2014; Volume 10, p. 6.
5. Walsh, S.W.; Williams, E.J.; Evans, A.C.O. A review of the causes of poor fertility in high milk producing dairy cows. *Anim. Reprod. Sci.* **2011**, *123*, 127–138. [CrossRef] [PubMed]
6. Ma, L.; Sonstegard, T.S.; Cole, J.B.; Vantassell, C.P.; Wiggans, G.R.; Crooker, B.A.; Tan, C.; Prakapenka, D.; Liu, G.E.; Da, Y. Genome changes due to artificial selection in U.S. Holstein cattle. *BMC Genom.* **2019**, *20*, 128. [CrossRef] [PubMed]
7. Zobel, R.; Tkalčić, S.; Pipal, I.; Buić, V. Incidence and factors associated with early pregnancy losses in Simmental dairy cows. *Anim. Reprod. Sci.* **2011**, *127*, 121–125. [CrossRef]
8. Seegers, H.; Beaudeau, F.; Fourichon, C.; Bareille, N. Reasons for culling in French Holstein cows. *Prev. Vet. Med.* **1998**, *36*, 257–271. [CrossRef]
9. Rilanto, T.; Reimus, K.; Orro, T.; Emanuelson, U.; Viltrop, A.; Mõtus, K. Culling reasons and risk factors in Estonian dairy cows. *BMC Vet. Res.* **2020**, *16*, 173. [CrossRef]
10. Charlier, C.; Li, W.; Harland, C.; Littlejohn, M.; Coppieters, W.; Creagh, F.; Davis, S.; Druet, T.; Faux, P.; Guillaume, F.; et al. NGS-based reverse genetic screen for common embryonic lethal mutations compromising fertility in livestock. *Genome Res.* **2016**, *26*, 1333–1341. [CrossRef] [PubMed]
11. VanRaden, P.M.; Olson, K.M.; Null, D.J.; Hutchison, J.L. Harmful recessive effects on fertility detected by absence of homozygous haplotypes. *J. Dairy Sci.* **2011**, *94*, 6153–6161. [CrossRef]
12. Casellas, J.; Canas-Alvarez, J.J.; Gonzalez-Rodriguez, A.; Puig-Oliveras, A.; Fina, M.; Piedrafita, J.; Molina, A.; Diaz, C.; Baro, J.A.; Varona, L. Bayesian analysis of parent-specific transmission ratio distortion in seven Spanish beef cattle breeds. *Anim. Genet.* **2017**, *48*, 93–96. [CrossRef] [PubMed]
13. Casellas, J.; Gularte, R.J.; Farber, C.R.; Varona, L.; Mehrabian, M.; Schadt, E.E.; Lusis, A.J.; Attie, A.D.; Yandell, B.S.; Medrano, J.F. Genome Scans for Transmission Ratio Distortion Regions in Mice. *Genetics* **2012**, *191*, 247–259. [CrossRef] [PubMed]
14. Derks, M.F.L.; Megens, H.J.; Bosse, M.; Lopes, M.S.; Harlizius, B.; Groenen, M.A.M. A systematic survey to identify lethal recessive variation in highly managed pig populations. *BMC Genom.* **2017**, *18*, 858. [CrossRef]
15. Derks, M.F.L.; Megens, H.J.; Bosse, M.; Visscher, J.; Peeters, K.; Bink, M.; Vereijken, A.; Gross, C.; de Ridder, D.; Reinders, M.J.T.; et al. A survey of functional genomic variation in domesticated chickens. *Genet. Sel. Evol.* **2018**, *50*, 17. [CrossRef]
16. Georges, M.; Charlier, C.; Hayes, B. Harnessing genomic information for livestock improvement. *Nat. Rev. Genet.* **2019**, *20*, 135–156. [CrossRef]
17. Abdalla, E.A.; Id-Lahoucine, S.; Cánovas, A.; Casellas, J.; Schenkel, F.S.; Wood, B.J.; Baes, C.F. Discovering lethal alleles across the turkey genome using a transmission ratio distortion approach. *Anim. Genet.* **2020**, *51*, 876–889. [CrossRef] [PubMed]
18. Hoff, J.L.; Decker, J.E.; Schnabel, R.D.; Taylor, J.F. Candidate lethal haplotypes and causal mutations in Angus cattle. *BMC Genom.* **2017**, *18*, 799. [CrossRef]
19. Mesbah-Uddin, M.; Hoze, C.; Michot, P.; Barbat, A.; Lefebvre, R.; Boussaha, M.; Sahana, G.; Fritz, S.; Boichard, D.; Capitan, A. A missense mutation (p.Tyr452Cys) in the CAD gene compromises reproductive success in French Normande cattle. *J. Dairy Sci.* **2019**, *102*, 6340–6356. [CrossRef] [PubMed]
20. Fritz, S.; Capitan, A.; Djari, A.; Rodriguez, S.C.; Barbat, A.; Baur, A.; Grohs, C.; Weiss, B.; Boussaha, M.; Esquerre, D.; et al. Detection of haplotypes associated with prenatal death in dairy cattle and identification of deleterious mutations in GART, SHBG and SLC37A2. *PLoS ONE* **2013**, *8*, e65550. [CrossRef]
21. Pausch, H.; Schwarzenbacher, H.; Burgstaller, J.; Flisikowski, K.; Wurmser, C.; Jansen, S.; Jung, S.; Schnieke, A.; Wittek, T.; Fries, R. Homozygous haplotype deficiency reveals deleterious mutations compromising reproductive and rearing success in cattle. *BMC Genom.* **2015**, *16*, 312. [CrossRef]
22. Cole, J.B.; VanRaden, P.M.; Null, D.J.; Hutchison, J.L.; Cooper, T.A.; Hubbard, S.M. Haplotype Tests for Recessive Disorders that Affect Fertility and Other Traits. 2021. Available online: https://www.aipl.arsusda.gov/reference/recessive_haplotypes_ARR-G3.html (accessed on 15 April 2021).
23. Schwarzenbacher, H.; Burgstaller, J.; Seefried, F.R.; Wurmser, C.; Hilbe, M.; Jung, S.; Fuerst, C.; Dinhopl, N.; Weissenböck, H.; Fuerst-Waltl, B.; et al. A missense mutation in TUBD1 is associated with high juvenile mortality in Braunvieh and Fleckvieh cattle. *BMC Genom.* **2016**, *17*, 400. [CrossRef] [PubMed]
24. Sonstegard, T.S.; Cole, J.B.; VanRaden, P.M.; Van Tassell, C.P.; Null, D.J.; Schroeder, S.G.; Bickhart, D.; McClure, M.C. Identification of a nonsense mutation in CWC15 associated with decreased reproductive efficiency in Jersey cattle. *PLoS ONE* **2013**, *8*, e54872. [CrossRef] [PubMed]

25. Charlier, C.; Agerholm, J.S.; Coppieters, W.; Karlskov-Mortensen, P.; Li, W.; de Jong, G.; Fasquelle, C.; Karim, L.; Cirera, S.; Cambisano, N.; et al. A Deletion in the Bovine FANCI Gene Compromises Fertility by Causing Fetal Death and Brachyspina. *PLoS ONE* **2012**, *7*, e43085. [CrossRef]
26. McClure, M.; Kim, E.; Bickhart, D.; Null, D.; Cooper, T.; Cole, J.; Wiggans, G.; Ajmone-Marsan, P.; Colli, L.; Santus, E.; et al. Fine Mapping for Weaver Syndrome in Brown Swiss Cattle and the Identification of 41 Concordant Mutations across NRCAM, PNPLA8 and CTTNBP2. *PLoS ONE* **2013**, *8*, e59251. [CrossRef]
27. Schütz, E.; Wehrhahn, C.; Wanjek, M.; Bortfeld, R.; Wemheuer, W.E.; Beck, J.; Brenig, B. The Holstein Friesian lethal haplotype 5 (HH5) results from a complete deletion of TBF1M and cholesterol deficiency (CDH) from an ERV-(LTR) insertion into the coding region of APOB. *PLoS ONE* **2016**, *11*, e0154602. [CrossRef]
28. Null, D.J.; Huthcins, J.L.; Bickhart, D.M.; VanRaden, P.M.; Cole, J.B. Discovery of a haplotype affecting fertility in Ayrshire dairy cattle and identification of a putative causal variant. In Proceedings of the American Dairy Science Association Annual Meeting, Pittsburgh, PA, USA, 25–29 June 2017; Volume 100, p. 199.
29. Guarini, A.R.; Sargolzaei, M.; Brito, L.F.; Kroezen, V.; Lourenco, D.A.L.; Baes, C.F.; Miglior, F.; Cole, J.B.; Schenkel, F.S. Estimating the effect of the deleterious recessive haplotypes AH1 and AH2 on reproduction performance of Ayrshire cattle. *J. Dairy Sci.* **2019**, *102*, 5315–5322. [CrossRef]
30. Cooper, T.A.; Wiggans, G.R.; Null, D.J.; Hutchison, J.L.; Cole, J.B. Genomic evaluation, breed identification, and discovery of a haplotype affecting fertility for Ayrshire dairy cattle. *J. Dairy Sci.* **2014**, *97*, 3878–3882. [CrossRef]
31. Ben Braiek, M.; Fabre, S.; Hozé, C.; Astruc, J.-M.; Moreno-Romieux, C. Identification of homozygous haplotypes carrying putative recessive lethal mutations that compromise fertility traits in French Lacaune dairy sheep. *Genet. Sel. Evol.* **2021**, *53*, 41. [CrossRef]
32. Michot, P.; Fritz, S.; Barbat, A.; Boussaha, M.; Deloche, M.C.; Grohs, C.; Hoze, C.; Le Berre, L.; Le Bourhis, D.; Desnoes, O.; et al. A missense mutation in PFAS (phosphoribosylformylglycinamidine synthase) is likely causal for embryonic lethality associated with the MH1 haplotype in Montbeliarde dairy cattle. *J. Dairy Sci.* **2017**, *100*, 8176–8187. [CrossRef]
33. Fritz, S.; Hoze, C.; Rebours, E.; Barbat, A.; Bizard, M.; Chamberlain, A.; Escouflaire, C.; Vander Jagt, C.; Boussaha, M.; Grohs, C.; et al. An initiator codon mutation in SDE2 causes recessive embryonic lethality in Holstein cattle. *J. Dairy Sci.* **2018**, *101*, 6220–6231. [CrossRef]
34. Hozé, C.; Escouflaire, C.; Mesbah-Uddin, M.; Barbat, A.; Boussaha, M.; Deloche, M.C.; Boichard, D.; Fritz, S.; Capitan, A. Short communication: A splice site mutation in CENPU is associated with recessive embryonic lethality in Holstein cattle. *J. Dairy Sci.* **2020**, *103*, 607–612. [CrossRef]
35. Kunz, E.; Rothammer, S.; Pausch, H.; Schwarzenbacher, H.; Seefried, F.R.; Matiasek, K.; Seichter, D.; Russ, I.; Fries, R.; Medugorac, I. Confirmation of a non-synonymous SNP in PNPLA8 as a candidate causal mutation for Weaver syndrome in Brown Swiss cattle. *Genet. Sel. Evol.* **2016**, *48*, 21. [CrossRef] [PubMed]
36. Schwarzenbacher, H.; Wurmser, C.; Flisikowski, K.; Misurova, L.; Jung, S.; Langenmayer, M.C.; Schnieke, A.; Knubben-Schweizer, G.; Fries, R.; Pausch, H. A frameshift mutation in GON4L is associated with proportionate dwarfism in Fleckvieh cattle. *Genet. Sel. Evol.* **2016**, *48*, 25. [CrossRef] [PubMed]
37. Jung, S.; Pausch, H.; Langenmayer, M.C.; Schwarzenbacher, H.; Majzoub-Altweck, M.; Gollnick, N.S.; Fries, R. A nonsense mutation in PLD4 is associated with a zinc deficiency-like syndrome in Fleckvieh cattle. *BMC Genom.* **2014**, *15*, 623. [CrossRef] [PubMed]
38. Owczarek-Lipska, M.; Plattet, P.; Zipperle, L.; Drögemüller, C.; Posthaus, H.; Dolf, G.; Braunschweig, M.H. A nonsense mutation in the optic atrophy 3 gene (OPA3) causes dilated cardiomyopathy in Red Holstein cattle. *Genomics* **2011**, *97*, 51–57. [CrossRef]
39. Seichter, D.; Russ, I.; Förster, M.; Medugorac, I. SNP-based association mapping of Arachnomelia in Fleckvieh cattle. *Anim. Genet.* **2011**, *42*, 544–547. [CrossRef]
40. National Center for Biotechnology Information ARS-UCD1.2 Assembly. 2018. Available online: https://www.ncbi.nlm.nih.gov/assembly/GCF_002263795.1 (accessed on 31 May 2018).
41. Rosen, B.D.; Bickhart, D.M.; Schnabel, R.D.; Koren, S.; Elsik, C.G.; Tseng, E.; Rowan, T.N.; Low, W.Y.; Zimin, A.; Couldrey, C.; et al. De novo assembly of the cattle reference genome with single-molecule sequencing. *Gigascience* **2020**, *9*, giaa021. [CrossRef]
42. Sargolzaei, M.; Chesnais, J.P.; Schenkel, F.S. A new approach for efficient genotype imputation using information from relatives. *BMC Genom.* **2014**, *15*, 478. [CrossRef]
43. Sargolzaei, M. *SNP1101 User's Guide. Version 1.0*; HiggsGene Solut. Inc.: Guelph, ON, Canada, 2014.
44. Wigginton, J.E.; Cutler, D.J.; Abecasis, G.R. A note on exact tests of Hardy-Weinberg equilibrium. *Am. J. Hum. Genet.* **2005**, *76*, 887–893. [CrossRef]
45. Benjamini, Y.; Yekutieli, D. The control of the false discovery rate in multiple testing under dependency. *Ann. Stat.* **2001**, *29*, 1165–1188. [CrossRef]
46. Häfliger, I.M.; Wiedemar, N.; Švara, T.; Starič, J.; Cociancich, V.; Šest, K.; Gombač, M.; Paller, T.; Agerholm, J.S.; Drögemüller, C. Identification of small and large genomic candidate variants in bovine pulmonary hypoplasia and anasarca syndrome. *Anim. Genet.* **2020**, *51*, 382–390. [CrossRef]
47. Bovine Research Community 1000 Bull Genomes Project. 2018. Available online: http://www.1000bullgenomes.com/ (accessed on 15 June 2018).
48. Hayes, B.J.; Daetwyler, H.D. 1000 Bull Genomes Project to Map Simple and Complex Genetic Traits in Cattle: Applications and Outcomes. *Annu. Rev. Anim. Biosci.* **2019**, *7*, 89–102. [CrossRef] [PubMed]

49. Purcell, S.; Neale, B.; Todd-Brown, K.; Thomas, L.; Ferreira, M.A.; Bender, D.; Maller, J.; Sklar, P.; de Bakker, P.I.; Daly, M.J.; et al. PLINK: A tool set for whole-genome association and population-based linkage analyses. *Am. J. Hum. Genet.* **2007**, *81*, 559–575. [CrossRef]
50. Cingolani, P.; Platts, A.; Wang, L.L.; Coon, M.; Nguyen, T.; Wang, L.; Land, S.J.; Lu, X.; Ruden, D.M. A program for annotating and predicting the effects of single nucleotide polymorphisms, SnpEff. *Fly* **2012**, *6*, 80–92. [CrossRef] [PubMed]
51. Pollard, K.S.; Hubisz, M.J.; Rosenbloom, K.R.; Siepel, A. Detection of nonneutral substitution rates on mammalian phylogenies. *Genome Res.* **2010**, *20*, 110–121. [CrossRef]
52. Siepel, A.; Bejerano, G.; Pedersen, J.S.; Hinrichs, A.S.; Hou, M.; Rosenbloom, K.; Clawson, H.; Spieth, J.; Hillier, L.D.W.; Richards, S.; et al. Evolutionarily conserved elements in vertebrate, insect, worm, and yeast genomes. *Genome Res.* **2005**, *15*, 1034–1050. [CrossRef] [PubMed]
53. Lander, E.S.; Linton, L.M.; Birren, B.; Nusbaum, C.; Zody, M.C.; Baldwin, J.; Devon, K.; Dewar, K.; Doyle, M.; Fitzhugh, W.; et al. Initial sequencing and analysis of the human genome. *Nature* **2001**, *409*, 860–921. [CrossRef]
54. Choi, Y.; Sims, G.E.; Murphy, S.; Miller, J.R.; Chan, A.P. Predicting the Functional Effect of Amino Acid Substitutions and Indels. *PLoS ONE* **2012**, *7*, e46688. [CrossRef]
55. Bendl, J.; Stourac, J.; Salanda, O.; Pavelka, A.; Wieben, E.D.; Zendulka, J.; Brezovsky, J.; Damborsky, J. PredictSNP: Robust and accurate consensus classifier for prediction of disease-related mutations. *PLoS Comput. Biol.* **2014**, *10*, e1003440. [CrossRef] [PubMed]
56. Kadri, N.K.; Sahana, G.; Charlier, C.; Iso-Touru, T.; Guldbrandtsen, B.; Karim, L.; Nielsen, U.S.; Panitz, F.; Aamand, G.P.; Schulman, N.; et al. A 660-Kb Deletion with Antagonistic Effects on Fertility and Milk Production Segregates at High Frequency in Nordic Red Cattle: Additional Evidence for the Common Occurrence of Balancing Selection in Livestock. *PLoS Genet.* **2014**, *10*, e1004049. [CrossRef]
57. Shi, S.; Yuan, N.; Yang, M.; Du, Z.; Wang, J.; Sheng, X.; Wu, J.; Xiao, J. Comprehensive Assessment of Genotype Imputation Performance. *Hum. Hered.* **2018**, *83*, 107–116. [CrossRef]
58. Tomecki, R.; Kristiansen, M.S.; Lykke-Andersen, S.; Chlebowski, A.; Larsen, K.M.; Szczesny, R.J.; Drazkowska, K.; Pastula, A.; Andersen, J.S.; Stepien, P.P.; et al. The human core exosome interacts with differentially localized processive RNases: hDIS3 and hDIS3L. *EMBO J.* **2010**, *29*, 2342–2357. [CrossRef]
59. Schmid, M.; Jensen, T.H. *The Nuclear RNA Exosome and Its Cofactors BT—The Biology of mRNA: Structure and Function*; Oeffinger, M., Zenklusen, D., Eds.; Springer International Publishing: Cham, Switzerland, 2019; pp. 113–132. ISBN 978-3-030-31434-7.
60. Robinson, S.R.; Oliver, A.W.; Chevassut, T.J.; Newbury, S.F. The 3′ to 5′ Exoribonuclease DIS3: From Structure and Mechanisms to Biological Functions and Role in Human Disease. *Biomolecules* **2015**, *5*, 1515–1539. [CrossRef]
61. Fasken, M.B.; Morton, D.J.; Kuiper, E.G.; Jones, S.K.; Leung, S.W.; Corbett, A.H. *The RNA Exosome and Human Disease BT—The Eukaryotic RNA Exosome: Methods and Protocols*; LaCava, J., Vaňáčová, Š., Eds.; Springer New York: New York, NY, USA, 2020; pp. 3–33. ISBN 978-1-4939-9822-7.
62. Tomecki, R.; Drazkowska, K.; Kucinski, I.; Stodus, K.; Szczesny, R.J.; Gruchota, J.; Owczarek, E.P.; Kalisiak, K.; Dziembowski, A. Multiple myeloma-associated hDIS3 mutations cause perturbations in cellular RNA metabolism and suggest hDIS3 PIN domain as a potential drug target. *Nucleic Acids Res.* **2014**, *42*, 1270–1290. [CrossRef]
63. Towler, B.P.; Jones, C.I.; Viegas, S.C.; Apura, P.; Waldron, J.A.; Smalley, S.K.; Arraiano, C.M.; Newbury, S.F. The 3′-5′ exoribonuclease Dis3 regulates the expression of specific microRNAs in Drosophila wing imaginal discs. *RNA Biol.* **2015**, *12*, 728–741. [CrossRef] [PubMed]
64. Laffleur, B.; Lim, J.; Zhang, W.; Chen, Y.; Pefanis, E.; Bizarro, J.; Batista, C.R.; Wu, L.; Economides, A.N.; Wang, J.; et al. Noncoding RNA processing by DIS3 regulates chromosomal architecture and somatic hypermutation in B cells. *Nat. Genet.* **2021**, *53*, 230–242. [CrossRef]
65. Turpeinen, M.; Zanger, U.M. Cytochrome P450 2B6: Function, genetics, and clinical relevance. *Drug Metabol. Drug Interact.* **2012**, *27*, 185–197. [CrossRef] [PubMed]
66. Lee, A.M.; Miksys, S.; Palmour, R.; Tyndale, R.F. CYP2B6 is expressed in African Green monkey brain and is induced by chronic nicotine treatment. *Neuropharmacology* **2006**, *50*, 441–450. [CrossRef]
67. Miksys, S.; Lerman, C.; Shields, P.G.; Mash, D.C.; Tyndale, R.F. Smoking, alcoholism and genetic polymorphisms alter CYP2B6 levels in human brain. *Neuropharmacology* **2003**, *45*, 122–132. [CrossRef]
68. Li, L.; Zhang, Q.-Y.; Ding, X. A CYP2B6-humanized mouse model and its potential applications. *Drug Metab. Pharmacokinet.* **2018**, *33*, 2–8. [CrossRef]
69. Koh, K.H.; Jurkovic, S.; Yang, K.; Choi, S.-Y.; Jung, J.W.; Kim, K.P.; Zhang, W.; Jeong, H. Estradiol induces cytochrome P450 2B6 expression at high concentrations: Implication in estrogen-mediated gene regulation in pregnancy. *Biochem. Pharmacol.* **2012**, *84*, 93–103. [CrossRef] [PubMed]
70. Sheftel, A.D.; Stehling, O.; Pierik, A.J.; Netz, D.J.A.; Kerscher, S.; Elsässer, H.-P.; Wittig, I.; Balk, J.; Brandt, U.; Lill, R. Human ind1, an iron-sulfur cluster assembly factor for respiratory complex I. *Mol. Cell. Biol.* **2009**, *29*, 6059–6073. [CrossRef] [PubMed]
71. Calvo, S.E.; Tucker, E.J.; Compton, A.G.; Kirby, D.M.; Crawford, G.; Burtt, N.P.; Rivas, M.; Guiducci, C.; Bruno, D.L.; Goldberger, O.A.; et al. High-throughput, pooled sequencing identifies mutations in NUBPL and FOXRED1 in human complex I deficiency. *Nat. Genet.* **2010**, *42*, 851–858. [CrossRef]

72. Kevelam, S.H.; Rodenburg, R.J.; Wolf, N.I.; Ferreira, P.; Lunsing, R.J.; Nijtmans, L.G.; Mitchell, A.; Arroyo, H.A.; Rating, D.; Vanderver, A.; et al. NUBPL mutations in patients with complex I deficiency and a distinct MRI pattern. *Neurology* **2013**, *80*, 1577–1583. [CrossRef] [PubMed]
73. Kimonis, V.; Al Dubaisi, R.; Maclean, A.E.; Hall, K.; Weiss, L.; Stover, A.E.; Schwartz, P.H.; Berg, B.; Cheng, C.; Parikh, S. NUBPL mitochondrial disease: New patients and review of the genetic and clinical spectrum. *J. Med. Genet.* **2021**, *58*, 314–325. [CrossRef]
74. Eis, P.S.; Huang, N.; Langston, J.W.; Hatchwell, E.; Schüle, B. Loss-of-Function NUBPL Mutation May Link Parkinson's Disease to Recessive Complex I Deficiency. *Front. Neurol.* **2020**, *11*, 1332. [CrossRef]
75. Dickinson, M.E.; Flenniken, A.M.; Ji, X.; Teboul, L.; Wong, M.D.; White, J.K.; Meehan, T.F.; Weninger, W.J.; Westerberg, H.; Adissu, H.; et al. High-throughput discovery of novel developmental phenotypes. *Nature* **2016**, *537*, 508–514. [CrossRef]

animals

MDPI

Article

Association between Polymorphism in the Janus Kinase 2 (*JAK2*) Gene and Selected Performance Traits in Cattle and Sheep

Nicola Oster [1], Małgorzata Anna Szewczuk [1,*], Sławomir Zych [2], Tomasz Stankiewicz [3], Barbara Błaszczyk [3] and Marta Wieczorek-Dąbrowska [4]

1 Department of Monogastric Animal Science, Faculty of Biotechnology and Animal Husbandry, West Pommeranian University of Technology in Szczecin, 29 Klemensa Janickiego, 71-270 Szczecin, Poland; nicola.oster@zut.edu.pl

2 Laboratory of Chromatography and Mass Spectroscopy, Faculty of Biotechnology and Animal Husbandry, West Pomeranian University of Technology in Szczecin, 29 Klemensa Janickiego, 71-270 Szczecin, Poland

3 Department of Animal Reproduction Biotechnology and Environmental Hygiene, Faculty of Biotechnology and Animal Husbandry, West Pomeranian University of Technology in Szczecin, 29 Klemensa Janickiego, 71-270 Szczecin, Poland; tomasz.stankiewicz@zut.edu.pl (T.S.); bblaszczyk@zut.edu.pl (B.B.)

4 National Research Institute of Animal Production, Kraków, Experimental Department, Kołbacz, 1 Warcisława Street, 74-106 Stare Czarnowo, Poland

* Correspondence: malgorzata.szewczuk@zut.edu.pl

Simple Summary: An important element of livestock breeding is the selection that leads to the consolidation and improvement of the traits with utility value. The use of single-nucleotide polymorphism (SNP) markers offers the opportunity to select the appropriate genotype linked to the phenotype, which manifests itself in significant breeding benefits. Polymorphic sites within the genes coding the somatotropic axis are directly and indirectly associated with the phenotype, particularly concerning the production properties of growth and carcass as well as reproductive function. This study analyzed the effect of selected polymorphisms in the Janus Kinase 2 (*JAK2*) gene on performance characteristics of cattle and sheep. The impact of the mutations on the development of selected traits, which are important from the economic standpoint, was evaluated. Such studies can be utilized to enhance the growth characteristics of cattle and sheep, thereby facilitating the development of more profitable and sustainable breeding methods.

Abstract: The Janus Kinase 2 (JAK2) tyrosine kinase is an essential component of signal transduction of the class II cytokine receptors, including the growth hormone receptor. Therefore, it may play a crucial role in the signaling pathway of the somatotropic axis, which influences growth, development, and reproductive traits in ruminants. For this purpose, for three breeds of cattle (Hereford, Angus, and Limousin; a total of 781 individuals), two polymorphic sites located in exon 16 (rs210148032; p.Ile704Val, within pseudokinase (JH2)) and exon 23 (silent mutation rs211067160, within JH1 kinase domain) were analyzed. For two breeds of sheep (Pomeranian and Suffolk; 333 individuals in total), two polymorphic sites in exon 6 (rs160146162 and rs160146160; encoding the FERM domain) and one polymorphic site in exon 24 of the *JAK2* gene (rs160146116; JH1 kinase domain) were genotyped. In our study, the associations examined for cattle were inconclusive. However, Hereford and Limousin cattle with genotypes *AA* (e16/*Rsa*I) and *AA* (e23/*Hae*III) tended to have the highest body weight and better daily gains ($p \leq 0.05$). No clear tendency was observed in the selected reproductive traits. In the case of sheep, regardless of breed, individuals with the *AA* (e6/*Ear*I), *GG* (e6/seq), and *AA* (e24/*Hpy*188III) genotypes had the highest body weights and daily gains in the study periods ($p \leq 0.01$). The same individuals in the Pomeranian breed also had better fertility and lamb survival ($p \leq 0.01$). To the best of our knowledge, these are the first association studies for all these polymorphic sites. Single-nucleotide polymorphisms in the *JAK2* gene can serve as genetic markers for growth and selected reproductive traits in ruminants given that they are further investigated in subsequent populations and analyzed using haplotype and/or combined genotype systems.

Keywords: ruminants; *JAK2*; tyrosine kinase; reproductive performance; growth performance

1. Introduction

Genomic selection using single-nucleotide polymorphisms (SNP) is revolutionizing livestock breeding. The primary advantage of genomic selection, in comparison to traditional pedigree-based methods, is the ability to obtain individuals with the desired phenotype and performance traits in the initial generation of selection. There are many studies confirming the effectiveness of the above-mentioned selection. Furthermore, a strong correlation between phenotypic data and predicted genomic values has been consistently demonstrated [1,2].

The biological foundation underlying the development of livestock performance characteristics is rooted in the elements comprising the somatotropic axis [3]. The somatotropic axis in the postnatal life stimulates cell hyperplasia and hypertrophy, thus contributing to the further growth of organs and tissues [4]. Proper growth requires the growth hormone (GH) secreted by the pituitary gland to bind to its specific receptor and activate a complex signaling cascade. The GH receptor (GHR) is initially produced as a dimer and then transported unliganded to the cell surface. Binding of GH to the GHR dimer results in a conformational change of the dimer, activation of intracellular Janus Kinase 2 (JAK2), and phosphorylation of the signal transducer and activator of transcription 5B (STAT5B). The phosphorylated STAT5B dimers are then translocated into the nucleus, where they transcriptionally activate various genes, including insulin-like growth factor 1 (*IGF1*), *IGF* binding protein 3, and the acid-labile subunit (*ALS*) [5]. The presence of two JAK2 molecules results in the transphosphorylation and activation of one JAK2 by the other, which in turn leads to the phosphorylation of downstream GHR tyrosine residues. Phosphorylated and dimerized GHR triggers a multiple-signal cascade translocated to the nucleus, where STATs or other non-receptor tyrosine kinases bind to a specific site in the DNA sequence, activating target genes [6].

The primary structure of JAK2 kinase consists of seven homologous domains (Figure 1), designated JH1–JH7, from the carboxyl terminus to the amino terminus [7]. The C-terminal kinase (JH1) domain is responsible for the correct enzymatic activity, while the JH2 domain (pseudokinase) acts as an inhibitor and regulator of the aforementioned JH1 domain. The SH2 homologous region (JH3–JH4 domains) is located in the center of JAK2 proteins, binding directly to JAK-interacting proteins. The N-terminal part of the JAK is built by the FERM domain (JH5–JH7; F for 4.1 protein, E for ezrin, R for radixin, and M for moesin). The FERM domain is highly conserved among many cellular proteins and is involved in the localization of JAKs in relation to the cell membrane and cytoskeleton–including binding to target receptors, e.g., GHR [8]. Constitutive JAK2 activity can occur in various neoplastic processes [9]. Alternatively, JAK2 inhibition may be associated with additional regulatory proteins. A typical JAK kinase is a relatively large protein (more than 1100 amino acids) with a molecular weight of 120–140 kDa. Depending on the species, the genes encoding the individual kinases are located on different chromosomes [10].

The cytokine-activated kinase (JAK) and signal transducer and activator of transcription (STAT) signaling pathway serves as a crucial link between proteins within the cell, directly influencing processes such as cell proliferation, apoptosis, mammary gland development, lactation, and immune responses. This pathway plays a vital role in transmitting information from cell surface receptors to the cell nucleus, ultimately regulating gene expression through transcription [11].

There are many studies confirming the influence of Janus Kinase 2 on the performance characteristics of dairy cows. Among other things, it has been shown that the JAK-STAT pathway regulates lactation and that PI3K/Akt within the JAK-STAT pathway is overexpressed in lactating cows [12]. Analysis of gene deletions in mice has documented the important role of JAK-STAT signaling in lactation and mammary gland development [13].

Moreover, an important role of the JAK-STAT pathway in blood cell differentiation and casein gene regulation during milk production has been documented [14]. Prolactin also uses JAK-STAT signaling and regulates lactation and reproduction in mammals [15]. In the research conducted by Szewczuk [16], a clear relationship was found between *JAK2/e20/Rsa*I polymorphism and milk-yield traits. The SNP was associated with higher milk, protein, and fat yield. With regard to growth and reproductive traits, there is a lack of literature describing the influence of the *JAK2* gene on, among other things, body weight and daily gains at different periods of the animal's life as well as fertility, prolificacy, or calving interval. The only studies in this area are the results presented by Padzik and Szewczuk [17], which also involved the bovine *JAK2*/e20/*Rsa*I polymorphic site. In contrast, with regard to the ovine JAK2 gene, a preliminary study by Padzik et al. [18] only focused on allele and genotype frequencies and did not consider animal performance traits.

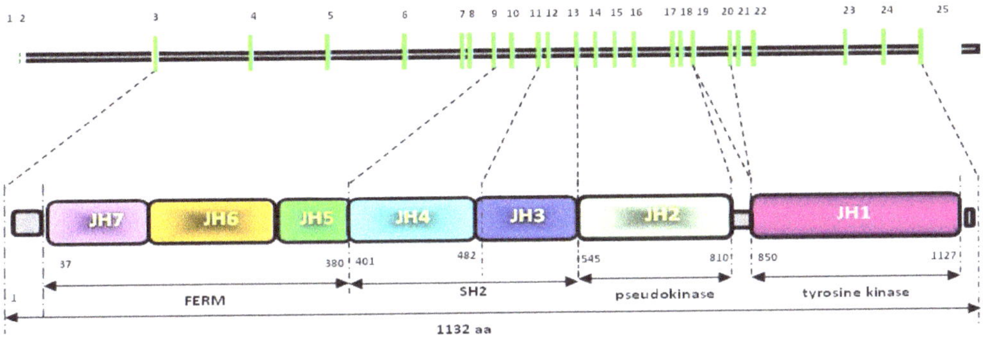

Figure 1. The primary structure of JAK2 kinase (own elaboration).

The aim of the present study was to analyze the possible relationship between polymorphisms in the Janus Kinase 2 (*JAK2*) gene and selected growth and reproductive traits in cattle and sheep. Attention was primarily focused on gene fragments encoding crucial domains that determine JAK2 kinase activity.

2. Material and Methods

2.1. Animals

A total of 781 blood samples were collected from three breeds of cattle, including Hereford (n = 276), Angus (n = 345), and Limousin (n = 160). In regard to sheep, a total of 333 blood samples of two breeds were collected (138 Pomeranian and 195 Suffolk sheep). All animals were kept on a single farm located in West Pomeranian Province, Poland. All animals were kept in a pasture and indoor system. Feeding was carried out in accordance with the standards accepted for these species [19], based on grass and other roughage and concentrates depending on the season. The animals had constant access to water and salt licks.

2.2. DNA Isolation

DNA was extracted using a MasterPure DNA Purification Kit Version II (Epicentre Technologies, Madison, WI, USA) from 3 mL of blood collected into tubes containing EDTA (IMPROVACUTER® K3 EDTA, Guangzhou, China). Briefly, this kit uses a rapid desalting process (red and white cell lysis solutions followed by protein precipitation solution) to remove contaminating macromolecules and avoid toxic organic solvents. Rinsing twice with 70% ethanol and resuspending the DNA in TE buffer (10 mM Tris-HCl (pH 7.5), 1 mM EDTA) allows samples to be stored for many months at <−20 °C.

2.3. *Sequencing*

Three fragments of the bovine *JAK2* gene (covering whole exons 3, 16, and 23) and four fragments of the ovine *JAK2* gene (exons 6, 12, 13, and 24) were selected for sequencing (Table S1 included in the Supplementary Materials). Each of the selected exons potentially contained a minimum of two known polymorphic sites (according to Ensembl–the ARS-UCD1.2 assembly (cow) and the Oar_rambouillet_v1.0 assembly (sheep)). For each exon, 10 pooled samples consisting of 10 individuals were sequenced (Genomed, Warsaw, Poland). Fluorograms were analyzed using Chromas ver. 2.6.6. (Technelysium Pty Ltd., Sydney, Australia). Only the polymorphic sites that had three or more heterozygote-specific results after sequencing of the pooled samples were selected for further genotyping. An additional consideration was the type of mutation detected (order of importance: missense, synonymous, intronic), and all of the detected SNPs had to be present within all breeds under analysis.

2.4. *Bovine JAK2 Gene Polymorphism*

Because the most used commercial enzymes were not available for the native sequence, genotyping of both polymorphisms detected within the three breeds under study was carried out using a polymerase chain reaction–artificially created restriction site (PCR-ACRS) method (Table 1). Primers were designed using Primer3web ver. 4.1.0 online software (https://primer3.ut.ee/) (accessed on 22 July 2021), and correct enzyme digestions after the introduction of a single-nucleotide mismatch were confirmed online by WebCutter v. 2.0 (http://heimanlab.com/cut2.html) (accessed on 22 July 2021) and NEBcutter V2.0 (http://nc2.neb.com/NEBcutter2/) (accessed on 22 July 2021).

Table 1. Characterization of selected single-nucleotide polymorphisms within bovine *JAK2* gene.

Primer Sequence (5′ → 3′)	Polymorphism Position ** and Type	Codon and Amino Acid(s) ***	PCR-ACRS (Restriction Enzyme, Cleavage Site, and Genotypes)
JAK2e16mF gggccctggacatactaagtg JAK2e16mR gtcttttggcaaaactgtaG *	rs210148032 g.39400906A>G Missense	ATT → GTT p.Ile704Val	*Rsa*I (gt^ac) PCR amplicon 211 bp *AA* 190 bp + 21 bp *AG* 211 bp + 190 bp + 21 bp *GG* 211 bp (no cut)
JAK2e23mF catatattgacaagagtaaaagccG * JAK2e23mR tccccacctttcaaaacttc	rs211067160 g.39383281A>G Synonymous	CCA → CCG p.Pro1057	*Hae*III (gg^cc) PCR amplicon 211 bp *AA* 211 bp (no cut) *AG* 211 bp + 185 bp + 26 bp *GG* 185 bp + 26 bp

* Mismatch is underlined (ACRS-PCR); ** position according to: RefSeq NC_037335.1 Chromosome 8 ARS-UCD1.2 Primary Assembly; *** amino acids abbreviations: Ile, isoleucine; Val, valine; Pro, proline.

2.5. *Ovine JAK2 Gene Polymorphism*

After analysis of the sequencing data, two polymorphic sites in the ovine *JAK2* gene were selected and genotyped using a polymerase chain reaction–restriction fragment length polymorphism (PCR-RFLP) method. Unfortunately, it was not possible to select an optimal methodology for the third SNP (rs160146160). Therefore, all samples were individually sequenced (Genomed, Warsaw, Poland) (Table 2).

Table 2. Characterization of selected single-nucleotide polymorphisms within ovine *JAK2* gene.

Primer Sequence (5′ → 3′)	Polymorphism Position * and Type	Codon and Amino Acid **	PCR-RFLP (Restriction Enzyme, Cleavage Site, and Genotypes)
oJAK2e6F ttgaccttgttaaatgtatatgttctg	rs160146162 g.73397955A>G Synonymous	GA<u>A</u> → GA<u>G</u> p.Glu177	*Ear*I (CTCTTC(1/4)^) PCR amplicon 280 bp *AA* 280 bp (no cut) *AG* 280 bp + 170 bp + 110 bp *GG* 170 bp + 110 bp
oJAK2e6R ttgcataagaaaattacctgatagagc	rs160146160 g.73398012G>A Synonymous	AC<u>G</u> → AC<u>A</u> p.Thr196	PCR amplicon 280 bp not applicable (sequencing)
oJAK2e24F tctgcttgaaattaaatgtaccaaa oJAK2e24R tcagtgaactgcataaactgacc	rs160146116 g.73442093A>G Synonymous	CT<u>A</u> → CT<u>G</u> p.Leu1082	*Hpy*188III (TC^NNGA) PCR amplicon 261 bp *AA* 205 bp + 56 bp *AG* 205 bp + 137 bp + 68 bp + 56 bp *GG* 137 bp + 68 bp + 56 bp

* Position according to: RefSeq NC_056055.1 Chromosome 2 ARS-UI_Ramb_v2.0 Primary Assembly; ** amino acids abbreviations: Glu, glutamic acid; Thr, threonine; Leu, leucine.

2.6. PCR Conditions

The PCR reaction was carried out with a total volume of 25 µL of the solution mixture consisting of 12.5 µL 2X PCR Master Mix (Thermo Scientific, ABO, Gdansk, Poland), 0.1 µL of primers (each at concentration of 10 pmol/mL), 2 µL of genomic DNA, and nuclease-free water up to 25 µL. Amplification was performed using Biometra TGradient thermal Cycler (Analytik, Jena, Germany). The first run was at 95 °C for 5 min, followed by subsequent 33 cycles of 95 °C × 30 s, 60 °C × 45 s (all primers), and 72 °C × 45 s and the last run at 72 °C for 5 min (final elongation). The yield and specificity of PCR products were evaluated by electrophoresis in 2% agarose gels (BASICA LE GQT, Prona, ABO, Gdansk, Poland) with ethidium bromide, viewed under UV light and scored in a gel documentation system.

2.7. Digestion

In regard to the first polymorphism within the bovine *JAK2* gene (rs210148032), amplicons (211 bp) were digested with 5U of the *Rsa*I restriction enzyme (37 °C/3 h; MBI Fermentas, ABO, Gdansk, Poland), which recognized one GT↓AC sequence within the PCR product. The second bovine polymorphic site (rs211067160) was also characterized by an amplicon of 211 bp. (However, it relates to a different region of the gene; Table 1) In this case, the *Hae*III restriction enzyme recognized one polymorphic site containing this SNP (5 U at 37 °C/3 h; MBI Fermentas, ABO, Gdansk, Poland). A polymorphism within the 280 bp fragment of exon 6 of the ovine *JAK2* gene was detected using the *Ear*I restrictase (also 5 U, 37 °C/3 h; MBI Fermentas, ABO, Gdansk, Poland). The ovine variant within exon 24 was differentiated using the *Hpy*188III restriction endonuclease (5 U at 37 °C/3 h; New England BioLabs Inc., Ipswich, MA, USA). Electrophoretic separation and visualization were identical to that described above.

2.8. Statistical Analysis

Analyses of the association between genotype and selected growth and reproductive traits were carried out on the basis of data obtained from the breeding records of individual herds maintained by the Polish Union of Meat Cattle Breeders and Producers and the Polish Union of Sheep Farmers. For cattle, the following traits were analyzed: birth weight (BWT), weaning weight adjusted to 210 days of age (WWT_{210}), average daily gain from birth to weaning (ADG), age and body weight at first calving, and calving interval. Traits studied in sheep included body weight at 2, 30, and 56 days of age; average daily gain from 2 to 56 and 30 to 56 days of age; and body weight at mating. Additionally, fertility, prolificacy, and lamb survival were analyzed.

Statistical analysis was performed using the STATISTICA software (13.3 PL software package, Statsoft Inc., Kraków, Poland, 2020). The differences between particular genotypes were evaluated with Duncan's test. Statistical calculations were performed using a general linear model (GLM).

The following statistical models were used:

- (Growth traits—cattle: BWT, ADG, and WWT_{210}; sheep: BW2D, BW30D, BW56D, BWM, $ADG_{2–56}$, and $ADG_{30–56}$)

(1) $Y_{ijklm} = \mu + G_i + s_j + BYS_k + [LS_l] + e_{ijklm}$ where Y_{ijklm} is the analyzed trait; μ the overall mean; G_i the fixed effect of *JAK2* genotype (i = 1, ...3); s_j the random effect of sire (Hereford j = 1, ...31; Angus j = 1, ...39; Limousin j = 1, ...24; Pomeranian j = 1, ...3; Suffolk j = 1, ...5); BYS_k the fixed effect of birth year/season (Hereford k = 1, ...20; Angus k = 1, ...20; Limousin k = 1, ...20; Pomeranian k = 1, ...16; Suffolk k = 1, ...16); $[LS_l]$ the fixed effect of litter size of mother (only in sheep: 1 single, 2 twins), and e_{ijklm} the random error.

- (Cattle: age and body weight at first calving as well as calving interval)

(2) $Y_{ijkl} = \mu + G_i + s_j + CYS_k + e_{ijkl}$ where Y_{ijkl} is the analyzed trait; μ the overall mean; G_i the fixed effect of *JAK2* genotype (i = 1, ...3); s_j the random effect of sire (Hereford j = 1, ...30; Angus j = 1, ...39; Limousin j = 1, ...21); CYS_k the fixed effect of year/season of 1st calving (Hereford k = 1, ...22; Angus k = 1, ...27; Limousin k = 1, ...21); and e_{ijkl} the random error.

- (Only sheep: fertility, prolificacy, and lamb survival)

(3) $Y_{ijkl} = \mu + G_i + LS_j + LY_k + e_{ijkl}$ where Y_{ijkl} is the analyzed trait; μ the overall mean; G_i the fixed effect of *JAK2* genotype (i = 1, ...3); LS_j the fixed effect of litter size (1 single, 2 twins); LY_k the fixed effect of lambing year (Pomeranian k = 1, ...6; Suffolk k = 1, ...8); and e_{ijkl} the random error.

The chi-square test was used to verify whether each population was in Hardy–Weinberg equilibrium.

3. Results

3.1. Sequencing

After a total of 100 individuals from three breeds of cattle were sequenced, two polymorphic sites were identified in the coding region of the bovine *JAK2* gene (Figure 2). In the case of the sheep *JAK2* gene, three polymorphic sites were previously identified and described by Padzik et al. [17].

3.2. Bovine JAK2 Frequencies

A total of 769 individuals from three breeds of cattle were genotyped for the *JAK2*/e16/*Rsa*I polymorphism (Table 3; Figures S1 and S2 included in the Supplementary Materials). Regardless of breed, the largest number of individuals was identified with the *AA* genotype (p.Ile704 variant; 58.3–59.4%), followed by heterozygotes (33–34.4%). In contrast, individuals with the *GG* genotype (p.Val704 variant; 7.25–8.41%) were rarely identified. For this reason, the A allele was most commonly observed (75%, regardless of breed). The genotype frequencies of Hereford and Limousin cows were consistent with HWE (*p*-value 0.25 and 0.24, respectively), while the distribution of genotypes in the herd of Angus cows was inconsistent with HWE (*p*-value = 0.03).

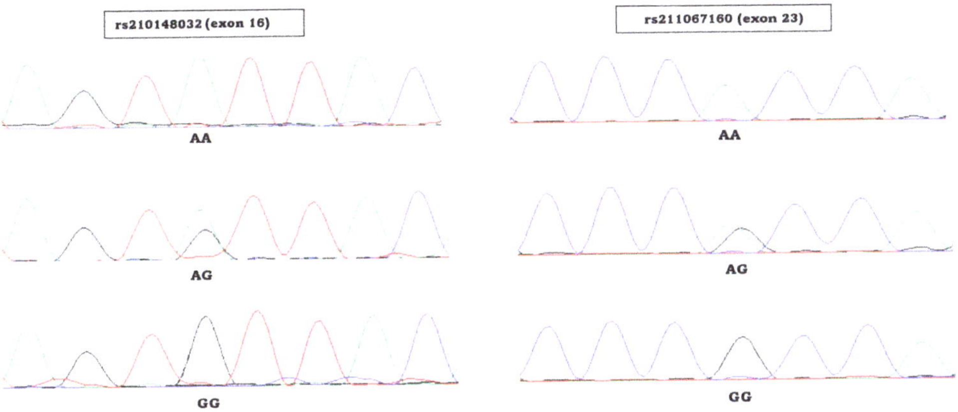

Figure 2. Chromatograms for two polymorphic sites identified within the bovine *JAK2* gene. green—adenine, black—guanine, red—thymine, blue—cystosine.

Table 3. The number and frequency of the *JAK2*/e16/*Rsa*I and the *JAK2*/e23/*Hae*III genotypes and alleles in the three bovine breeds under study.

Breed		*JAK2*/e16/*Rsa*I			Total	allele	
		AA	*AG*	*GG*		*A*	*G*
Hereford	*n* Frequency	161 0.5833	95 0.3442	20 0.0725	276 1.0000	0.7554	0.2446
Angus	*n* Frequency	195 0.5856	110 0.3303	28 0.0841	333 1.0000	0.7508	0.2492
Limousin	*n* Frequency	95 0.5937	53 0.3313	12 0.0750	160 1.0000	0.7594	0.2406
Breed		*JAK2*/e23/*Hae*III			Total	allele	
		AA	*AG*	*GG*		*A*	*G*
Hereford	*n* Frequency	165 0.6088	83 0.3063	23 0.0849	271 1.0000	0.7620	0.2380
Angus	*n* Frequency	61 0.1768	132 0.3826	152 0.4406	345 1.0000	0.3681	0.6319
Limousin	*n* Frequency	28 0.1958	72 0.5035	43 0.3007	143 1.0000	0.4476	0.5524

n, number of individuals.

More diverse results between the three breeds of cattle in terms of genotype and allele frequencies were obtained for the *JAK2*/e23/*Hae*III polymorphism (Table 3) after testing 759 individuals. In the case of the Hereford breed, as in the previous SNP, the *AA* genotype was most frequently identified (60.9%), followed by heterozygotes (30.6%). Hereford cows with the *GG* genotype were rarely identified (8.5%). This distribution was not consistent with HWE (p-value = 0.01). The opposite situation was noted for the Angus breed. The most numerous were cattle with the *GG* genotype (44%) and slightly less often heterozygous individuals (38.3%). Individuals with the *AA* genotype constituted only 17.7% of the population of Angus cows. This distribution was also not consistent with HWE (p-value = 0.00). The distribution of genotypes in the herd of Limousin cows was different. Half of the herd consisted of heterozygous individuals (50.3%). There were also numerous cows with the *GG* genotype (30.1%). The *AA* genotype was the least frequently identified (19.6%). The distribution was consistent with HWE (p-value = 0.83). The above results were reflected in allele frequencies. The A allele clearly dominated only in the

Hereford breed (76.2%) and the G allele in the Angus herd (63.2%). The allele frequency in the Limousin herd was more even (allele A accounted for 44.8% of cases and allele G for 55.2%).

3.3. Bovine *JAK2/e16/RsaI* Associations

In the case of the polymorphism in exon 16 (rs210148032, p.Ile704Val) (Table 4), the birth weight of Hereford and Limousin calves with the *AA* genotype (p.Ile704) was significantly higher than that of the individuals with the *GG* genotype (p.Val704; +1.8 kg and +2.6 kg, respectively; $p \leq 0.05$). However, for the Angus breed, the opposite results were reported, with *GG* individuals born heavier by 2.2 kg compared to *AA* individuals ($p \leq 0.05$). A similar situation was noted for the weaning weight (WWT_{210}). Hereford and Limousin calves with the *AA* genotype were significantly heavier compared to the *GG* genotypes (+11.1 kg ($p \leq 0.05$) and +27.2 kg ($p \leq 0.01$), respectively). For the Angus breed, the differences in WWT_{210} were nonsignificant. In addition, although only for the Hereford breed, there were significant differences in average daily gains (ADG) ($p \leq 0.01$). Individuals with the frequent *AA* genotype had higher ADG compared to both heterozygotes (+132 g) and individuals with the rare *GG* genotype (+165 g). For the other breeds, the differences in ADG were statistically nonsignificant. The observed trends were maintained until the first calving. The body weight of the cow at the first calving differed significantly depending on the identified genotype. In the case of Hereford, individuals with the *AA* genotype were significantly heavier than both heterozygous individuals (+17.9 kg, $p \leq 0.05$) and individuals with the rare *GG* genotype (+28.8 kg, $p \leq 0.01$). In contrast, Limousin cows with the *AA* and *AG* genotypes had an almost identical average body weight at first calving (~610 kg) and were significantly different ($p \leq 0.05$) from the *GG* genotypes, which had an average body weight 29 kg lower than the other cows.

Table 4. The effect of *JAK2/e16/RsaI* genotypes on the growth performance and selected reproductive traits in cows.

Breed	Genotype	Birth Weight (kg)	Average Daily Gain (g)	Weaning Weight (kg)	Age at First Calving (Days)	Body Weight at First Calving (kg)	Calving Interval (Days)
Hereford	*AA*	34.7 [a] (0.36)	1091 [AB] (13.11)	262.2 [a] (2.85)	1039 [A] (28.20)	589.4 [Aa] (2.11)	499 [ab] (16.06)
	AG	34.0 (0.33)	959 [A] (14.84)	254.3 (3.50)	975 [a] (31.00)	571.5 [a] (5.90)	420 [a] (13.57)
	GG	32.9 [a] (0.51)	926 [B] (16.60)	251.1 [a] (6.97)	813 [Aa] (37.69)	560.6 [A] (6.79)	399 [b] (16.36)
Angus	*AA*	36.2 [a] (0.30)	994 (7.76)	247.1 (1.87)	1076 (19.20)	564.6 (3.33)	416 [Aa] (2.02)
	AG	37.3 (0.39)	1011 (8.13)	251.6 (2.50)	1057 (26.91)	562.3 (3.81)	403 [ab] (2.23)
	GG	38.4 [a] (0.59)	1012 (8.58)	255.5 (7.89)	1011 (25.83)	574.5 (8.46)	389 [Ab] (1.22)
Limousin	*AA*	35.2 [a] (0.41)	1001 (23.62)	275.5 [A] (3.22)	1077 (14.55)	610.8 [a] (10.27)	469 (21.51)
	AG	34.2 (0.57)	1001 (35.12)	261.3 (3.61)	1045 (14.83)	610.3 [b] (13.77)	479 (27.96)
	GG	32.6 [a] (0.65)	954 (10.35)	248.3 [A] (6.19)	1090 (15.67)	581.2 [ab] (5.69)	466 (6.69)

The means in the columns marked with the same letters within the same breed differ significantly at: small letters, $p \leq 0.05$; capitals, $p \leq 0.01$. Numbers in parentheses are the standard errors of the means.

Some relationships were also noted for selected reproductive traits of cattle, although they were statistically confirmed only for Hereford and Angus breeds. Hereford individuals with the *GG* genotype, despite having the significantly lowest birth weight and body weight in later periods, as described above, had the lowest average age of first calving (~27 months) compared to the other genotypes, with the difference between the *GG* genotype and the heterozygous *AG* genotype being 162 days (i.e., ~5.5 months difference, $p \leq 0.05$) and between *GG* and the most common *AA* genotype being as much as 226 days (i.e., ~7.5 months difference, $p \leq 0.01$). In the case of the Angus and Limousin breeds, no significant differences were found because the age of the first calving was similar, occurring at around 33–36 months of age for cows of both breeds. Cows with the *GG* genotype also had a shorter calving interval. Within the Hereford breed, the differences were as much as 100 days compared to cows with the *AA* genotype ($p \leq 0.05$). Also, heterozygotes in the *JAK2*/e16/*Rsa*I polymorphism for the Hereford breed had a 79-day shorter calving interval compared to cows with the *AA* genotype ($p \leq 0.05$). The calving interval of Angus cows was even more diverse within the genotypes. As with Hereford, cows with *GG* genotype had a 27-day shorter calving interval compared to cows with *AA* genotype ($p \leq 0.01$) and 14 days shorter than heterozygous cows ($p \leq 0.05$). Heterozygotes also had a shorter calving interval by 13 days compared to *AA* cows ($p \leq 0.05$). In the case of the Limousin breed, there were no significant differences in the calving interval of cows with the different *JAK2*/e16/*Rsa*I genotypes. Irrespective of breed and genotype, all calvings were single and uncomplicated.

3.4. Bovine *JAK2/e23/Hae*III Associations

Regardless of breed, there was no association between the silent mutation in exon 23 (rs211067160) and birth weight of calves (Table 5). In the case of weaning weight in the Hereford and Limousin populations, individuals with *AA* genotype were heavier by 16.4 kg and 14.3 kg, respectively, compared to individuals with *GG* genotype ($p \leq 0.05$) and had better average daily weight gains of 154 g ($p \leq 0.01$) and 91 g, respectively ($p \leq 0.05$). In addition, heterozygous Hereford cows also had higher ADG than *GG* cows (+112 g; $p \leq 0.01$). However, Limousin cows with the previously preferred *AA* genotype had 36.8–38.3 kg lower body weight at first calving compared to cows with the other genotypes ($p \leq 0.05$). For other breeds, no statistical differences in body weight at the first calving were noted.

Similarly to the previously discussed polymorphic site, the age of first calving in Angus and Limousin cows (from 34 to 36 months) did not depend on *JAK2*/e23/*Hae*III genotypes in contrast to Hereford cows, for which, again, individuals with the rare *GG* genotype had significantly earlier age at first calving (~29 months) than *AA* cows (~34 months) ($p \leq 0.05$). At the same time, cows of this breed, also with the *GG* genotype, had a 90-day shorter calving interval compared to the *AA* genotype ($p \leq 0.05$). In the population of Limousin cows, the most numerous heterozygotes had a significantly longer calving interval by 90 days than cows with the rarest *AA* genotype ($p \leq 0.05$).

For the Angus breed, no statistically significant differences were observed between *JAK2*/e23/*Hae*III genotypes for the growth and reproductive traits. Additionally, it was noted that regardless of breed or genotype, all births were single and occurred without complications or human assistance.

Table 5. The effect of *JAK2*/e23/*Hae*III genotypes on the growth performance and selected reproductive traits in cows.

Breed	Genotype	Birth Weight (kg)	Average Daily Gain (g)	Weaning Weight (kg)	Age at First Calving (Days)	Body Weight at First Calving (kg)	Calving Interval (Days)
Hereford	*AA*	34.5 (0.34)	1060 [A] (12.86)	260.8 [a] (2.53)	1020 [a] (27.83)	584.9 (3.26)	487 [a] (15.87)
	AG	34.1 (0.38)	1018 [B] (24.03)	257.9 (4.95)	988 (37.85)	575.6 (4.05)	430 (14.72)
	GG	33.9 (1.03)	906 [AB] (16.51)	244.4 [a] (3.47)	874 [a] (51.07)	571.9 (7.08)	397 [a] (16.73)
Angus	*AA*	37.16 (0.71)	1005 (14.27)	249.6 (3.66)	1065 (25.94)	559.4 (5.42)	423 (4.37)
	AG	37.40 (0.42)	1017 (11.25)	250.8 (2.92)	1021 (25.59)	563.9 (4.72)	400 (1.71)
	GG	36.3 (0.29)	993 (7.37)	248.4 (1.99)	1084 (20.98)	566.0 (3.28)	410 (2.08)
Limousin	*AA*	35.5 (0.77)	1058 [a] (17.42)	274.5 [a] (4.38)	1075 (18.39)	580.5 [ab] (3.13)	413 [a] (28.82)
	AG	34.2 (0.40)	990 (32.90)	271.3 (3.91)	1078 (15.23)	618.8 [a] (12.95)	504 [a] (27.69)
	GG	35.2 (0.73)	967 [a] (36.69)	261.2 [a] (4.13)	1056 (25.25)	617.3 [b] (17.77)	483 (32.79)

The means in the columns marked with the same letters within the same breed differ significantly at: small letters, $p \leq 0.05$; capitals, $p \leq 0.01$. Numbers in parentheses are the standard errors of the means.

3.5. Ovine *JAK2* Frequencies

For three polymorphic sites, 138 Pomeranian and 195 Suffolk individuals were genotyped (Table 6; Figures S3 and S4 included in the Supplementary Materials).

Table 6. The number and frequency of genotypes and alleles for three different *JAK2* polymorphisms among two sheep breeds under study.

Breed		*JAK2*/e6/*Ear*I			Total	allele	
		AA	*AG*	*GG*		*A*	*G*
Pomeranian	*n*	32	58	48	138	0.4420	0.5580
	Frequency	0.2319	0.4203	0.3478	1.0000		
Suffolk	*n*	29	43	123	195	0.2590	0.7410
	Frequency	0.1487	0.2205	0.6308	1.0000		
Breed		*JAK2*/e6/seq			Total	allele	
		AA	*AG*	*GG*		*A*	*G*
Pomeranian	*n*	55	59	24	138	0.6123	0.3877
	Frequency	0.3986	0.4275	0.1739	1.0000		
Suffolk	*n*	49	110	36	195	0.5333	0.4667
	Frequency	0.2513	0.5641	0.1846	1.0000		
Breed		*JAK2*/e24/*Hpy*188III			Total	allele	
		AA	*AG*	*GG*		*A*	*G*
Pomeranian	*n*	43	67	28	138	0.5544	0.4456
	Frequency	0.3116	0.4855	0.2029	1.0000		
Suffolk	*n*	49	128	18	195	0.5795	0.4205
	Frequency	0.2513	0.6564	0.0923	1.0000		

n, number of individuals.

In the case of the first polymorphic site (*JAK2/e6/Ear*I), for the Pomeranian breed, numerous occurrences of individuals of each genotype were noted, with the largest number of individuals being heterozygous (42%), followed by sheep with the *GG* genotype (34.8%), and slightly less often, individuals with *AA* genotype were identified (23.2%). Allele frequencies were similar (A, 44.2%; G, 55.8%). The genotype distribution was consistent with HWE (p-value = 0.08). However, in the flock of Suffolk sheep, the *GG* genotype (63%) and the G allele (74%) clearly dominated, and the distribution of genotypes was not consistent with HWE (p-value < 0.01).

In the case of the second polymorphic site (*JAK2*/e6/seq), heterozygotes (POM 42.7% and SUF 56.4%) were the most numerous in both breeds, followed by individuals with the *AA* genotype (39.9% and 25.1%, respectively) and with the *GG* genotype (17.4% and 18.5%, respectively). The A allele occurred more frequently than the G allele (61.2% vs. 53.3%). The distribution of genotypes in both populations was consistent with HWE (p-value = 0.24 for the Pomeranian breed and p-value = 0.06 for the Suffolk breed).

The polymorphism in exon 24 (*JAK2/e24/Hpy*188III) was also characterized by a slight dominance of heterozygous individuals (POM 48.6% and SUF 65.6%), followed by sheep with the *AA* genotype (31.2% and 25.1%, respectively) and the least frequent being sheep with the *GG* genotype (20.3% and 9.2%, respectively). The frequencies of the A allele were similar (55.4% and 57.8%). The distribution of genotypes in the Pomeranian sheep population was consistent with HWE (p-value = 0.84) and inconsistent with HWE in the Suffolk sheep herd (p-value < 0.01).

3.6. Ovine JAK2 Associations—Growth Traits

At the early stages of animal life, highly significant differences depending on the genotype were observed for most of the analyzed growth performance traits and for each polymorphic site (Table 7). The exception was the subsequent body weight at mating, which was significant only for the Suffolk breed.

Sheep with the *AA* genotype for the *JAK2/e6/Ear*I polymorphic site were significantly heavier than the other sheep on day 2 (+0.3–0.7 kg for Pomeranian and +0.7–0.8 kg for Suffolk), on day 30 (+1.2–1.3 kg; only Suffolk), and on the 56th day of life (Pomeranian +1.2–2.2 kg and Suffolk 2.8–3.6 kg) and at mating (+4.1 kg; only Suffolk), especially from the worst-performing individuals with the *GG* genotype ($p \leq 0.01$) compared to heterozygous sheep and sheep with the *AA* genotype (Pomeranian +16–27 g and Suffolk +40–53 g) between 2 and 56 days of age. For Suffolk sheep, between 30 and 56 days of age, even higher differences in average gains were noted (+62–90 g).

Table 7. Means and standard errors (in parentheses) of weights and gains traits in sheep with different *JAK2* genotypes of three polymorphic sites.

	Breed	Genotype	Body Weight at 2 Days (kg)	Body Weight at 30 Days (kg)	Body Weight at 56 Days (kg)	Average Daily Gains between 2–56 Days (g)	Average Daily Gains between 30–56 Days (g)	Body Weight at Mating (kg)
*JAK2/e6/Ear*I	Pomeranian	*AA*	4.3 AB (0.04)	n/a	19.3 AB (0.19)	278 AB (3.52)	n/a	54.4 (0.43)
		AG	4.0 BC (0.03)	n/a	18.1 BC (0.15)	262 Ba (2.78)	n/a	53.8 (0.28)
		GG	3.6 AC (0.02)	n/a	17.1 AC (0.18)	251 Aa (3.20)	n/a	53.2 (0.43)
	Suffolk	*AA*	4.8 AB (0.05)	14.2 AB (0.13)	23.5 AB (0.21)	347 AB (3.44)	359 AB (7.24)	62.0 AB (0.55)
		AG	4.1 B (0.02)	13.0 B (0.19)	20.7 Ba (0.25)	307 Ba (4.39)	297 Ba (9.86)	58.9 B (0.39)
		GG	4.0 A (0.03)	12.9 A (0.15)	19.9 Aa (0.18)	294 Aa (2.96)	269 Aa (4.95)	57.9 A (0.26)

Table 7. *Cont.*

	Breed	Genotype	Body Weight at 2 Days (kg)	Body Weight at 30 Days (kg)	Body Weight at 56 Days (kg)	Average Daily Gains between 2–56 Days (g)	Average Daily Gains between 30–56 Days (g)	Body Weight at Mating (kg)
JAK2/e6/seq	Pomeranian	*AA*	3.8 ^{A} (0.04)	n/a	17.9 ^{A} (0.18)	261 ^{A} (3.16)	n/a	53.8 (0.31)
		AG	3.9 ^{B} (0.04)	n/a	17.7 ^{B} (0.16)	256 ^{B} (2.73)	n/a	54.2 (0.34)
		GG	4.3 ^{AB} (0.05)	n/a	19.4 ^{AB} (0.24)	280 ^{AB} (4.01)	n/a	54.9 (0.53)
	Suffolk	*AA*	4.1 ^{A} (0.06)	12.7 ^{A} (0.25)	19.8 ^{A} (0.34)	292 ^{A} (5.32)	275 ^{A} (8.56)	58.4 ^{A} (0.54)
		AG	4.1 ^{B} (0.03)	13.1 ^{a} (0.14)	20.3 ^{B} (0.17)	301 ^{B} (2.93)	278 ^{B} (5.79)	58.4 ^{B} (0.25)
		GG	4.6 ^{AB} (0.06)	13.7 ^{Aa} (0.20)	22.5 ^{AB} (0.32)	333 ^{AB} (5.29)	338 ^{AB} (8.67)	60.2 ^{AB} (0.59)
JAK2/e24/*Hpy*188III	Pomeranian	*AA*	4.2 ^{AB} (0.05)	n/a	19.2 ^{AB} (0.20)	279 ^{AB} (3.83)	n/a	54.4 (0.43)
		AG	3.9 ^{AC} (0.04)	n/a	17.8 ^{AC} (0.13)	258 ^{AC} (2.14)	n/a	53.8 (0.28)
		GG	3.6 ^{BC} (0.05)	n/a	16.8 ^{BC} (0.15)	244 ^{BC} (2.98)	n/a	53.2 (0.43)
	Suffolk	*AA*	4.5 ^{AB} (0.06)	14.1 ^{AB} (0.16)	22.7 ^{AB} (0.26)	338 ^{AB} (4.07)	332 ^{AB} (9.04)	60.1 ^{Aa} (0.54)
		AG	4.1 ^{AC} (0.02)	13.0 ^{AC} (0.12)	20.2 ^{AC} (0.15)	298 ^{AC} (2.46)	278 ^{AC} (4.99)	58.7 ^{Ba} (0.23)
		GG	3.7 ^{BC} (0.05)	11.4 ^{BC} (0.40)	17.7 ^{BC} (0.35)	260 ^{BC} (5.98)	243 ^{BC} (11.28)	54.9 ^{AB} (0.46)

The means in the columns marked with the same letters within the same breed differ significantly at: small letters, $p \leq 0.05$; capitals, $p \leq 0.01$. n/a: for the Pomeranian breed, such data were not recorded.

In the case of the second polymorphic site (*JAK2*/e6/seq) in both breeds, sheep with the *GG* genotype had the greatest body weights on day 2 (+0.4–0.5 kg), day 30 (+0.6–1 kg; only Suffolk), and 56 days of age (Pomeranian +1.5–1.7 kg and Suffolk +2.2–2.7 kg) and daily gains for the period from 2 to 56 days of age (Pomeranian +19–24 g and Suffolk +32–41 g) and for the period from 30 to 56 days of age (+60–63 g; Suffolk only) ($p \leq 0.01$) compared to heterozygotes and *AA* genotypes. For the Suffolk breed, the weight of sheep with the *GG* genotype was also higher by 1.8 kg during mating than in sheep with the other genotypes ($p \leq 0.01$).

Polymorphism within exon 24 (*JAK2*/e24/*Hpy*188III) was also characterized by highly significant associations ($p \leq 0.01$), where individuals with the *AA* genotype had the highest body weights compared to heterozygotes and individuals with the *GG* genotype on the 2nd day of life (Pomeranian +0.3–0.6 kg and Suffolk +0.4–0.8 kg) as well as on day 30 (+1.1–2.7 kg; Suffolk only) and on day 56 (Pomeranian +1.4–2.4 kg and Suffolk 2.5–5 kg). It was similar with daily gains both in the period from 2 to 56 days (Pomeranian +21–35 g and Suffolk +40–78 g) and in the period from 30 to 56 days (+54–89 g; Suffolk breed only). Also, during mating, sheep with the *AA* genotype were characterized by a higher weight than the other individuals (+1.4–5.2 kg; only in the Suffolk breed).

3.7. Ovine JAK2 Associations-Reproductive Traits

Table 8 presents the values of reproduction traits for Pomeranian and Suffolk sheep as well as lamb rearing in relation to three polymorphic sites located in the *JAK2* sheep gene. Genetic origin differentiated the average values of all the analyzed reproductive characteristics of mothers. In the case of the polymorphism in exon 6 (*JAK2*/*Ear*I), the highest fertility was shown by heterozygous Pomeranian sheep (95.8%), while the lowest (82.34%) was found in sheep with the *GG* genotype (82.34%). Statistically significant differences ($p \leq 0.01$) were noted between individuals with the *GG* genotype and individuals that were homozygous *AA* (+11.96%) and heterozygous *AG* (+13.54%). In the case of the Suffolk breed, the highest fertility rate was shown in heterozygous individuals (94.41%) and the lowest in *AA* individuals (91.03%). However, no statistically significant differences were found in Suffolks. The prolificacy was similar among the genotypes within the breeds; however, statistically significant differences were noted in the Suffolk breed, where individuals with the *GG* genotype were characterized by the lowest prolificacy in comparison to individuals with the *AA* genotype ($p \leq 0.01$) and heterozygotes ($p \leq 0.05$).

Regardless of breed, the best lamb survival rates were noted for heterozygous sheep (94.28% Pomeranian; 96.34% Suffolk). The lowest lamb survival rates in sheep of both breeds were found in *GG* individuals, which was confirmed statistically ($p \leq 0.01$).

For the second polymorphic site (*JAK2*/e6/seq), there were clear differences between breeds. In Pomeranian sheep, the *GG* genotype was associated with the highest fertility (+7.52% *AG*, $p \leq 0.05$; +9.53% *AA*, $p \leq 0.01$), while in Suffolk sheep, similar fertility was observed in individuals with *AA* and *AG* genotypes, which significantly ($p \leq 0.05$) differed from mothers with the *GG* genotype (+4.28–4.4%).

The prolificacy of sheep in both groups, regardless of the genotype, did not exceed 1.18 and was similar. No statistically significant differences were observed. The Pomeranian sheep with the *GG* genotype reared the largest percentage of lambs (97.68%) compared to other mothers, and statistical differences were shown only for mothers with the *AA* genotype (+11.5%; $p \leq 0.01$). Different relationships were confirmed statistically ($p \leq 0.01$) in Suffolk sheep; the lamb survival was higher (92.09%) when ewes had the AG genotype than when the ewes had the GG genotype (difference +11.27%).

The relationship of the *JAK2*/e24/*Hpy188*III genotypes with fertility and lamb survival showed that, regardless of breed, sheep with *AA* genotype were characterized by the highest values of the analyzed traits ($p \leq 0.01$; $p \leq 0.05$). The difference in fertility between mothers with the *AA* genotype and individuals with the *GG* genotype was nearly 14% in Pomeranian sheep and 6% in Suffolk sheep, while the differences in lamb survival were approximately 12% for Pomeranian sheep and 19.5% for Suffolk sheep. In the case of the prolificacy, statistical differences were demonstrated in the Pomeranian breed, where individuals with the *GG* genotype differed significantly from the other individuals (0.14–0.19; $p \leq 0.01$).

Table 8. Influence of genotypes on reproductive performance of Pomeranian and Suffolk sheep breeds.

	Breed	Genotype	Fertility (%)	Prolificacy (*n*/ewe)	Lamb Survival (%)
JAK2/e6/*Ear*I	Pomeranian	*AA*	94.30 ^{A} (2.23)	1.15 (0.05)	93.55 ^{A} (2.21)
		AG	95.88 ^{B} (1.45)	1.11 (0.04)	94.28 ^{B} (1.67)
		GG	82.34 ^{AB} (2.70)	1.18 (0.05)	81.38 ^{AB} (3.86)
	Suffolk	*AA*	91.03 (2.89)	1.19 ^{A} (0.06)	95.40 ^{A} (1.92)
		AG	94.41 (1.11)	1.18 ^{a} (0.05)	96.34 ^{B} (1.48)
		GG	94.40 (1.66)	1.09 ^{Aa} (0.02)	84.52 ^{AB} (2.00)

Table 8. *Cont.*

	Breed	Genotype	Fertility (%)	Prolificacy (*n*/ewe)	Lamb Survival (%)
JAK2/e6/seq	Pomeranian	*AA*	88.30 ^{A} (2.35)	1.11 (0.04)	86.20 ^{A} (2.88)
		AG	90.31 ^{a} (2.05)	1.18 (0.04)	89.69 (2.64)
		GG	97.83 ^{Aa} (1.59)	1.13 (0.05)	97.68 ^{A} (1.66)
	Suffolk	*AA*	94.62 ^{a} (1.77)	1.13 (0.04)	87.06 (2.84)
		AG	94.74 ^{b} (1.11)	1.14 (0.03)	92.09 ^{A} (1.53)
		GG	90.34 ^{ab} (2.47)	1.06 (0.03)	80.82 ^{A} (4.23)
JAK2/e24/*Hpy*188III	Pomeranian	*AA*	95.17 ^{A} (1.76)	1.13 ^{A} (0.04)	95.77 ^{A} (1.66)
		AG	93.48 ^{B} (1.79)	1.08 ^{B} (0.03)	90.66 (1.93)
		GG	81.33 ^{AB} (3.74)	1.27 ^{AB} (0.08)	84.00 ^{A} (4.46)
	Suffolk	*AA*	95.05 ^{a} (1.44)	1.15 (0.03)	92.64 ^{A} (2.43)
		AG	94.17 (1.13)	1.11 (0.03)	89.44 ^{B} (1.66)
		GG	88.89 ^{a} (3.81)	1.11 (0.05)	73.15 ^{AB} (5.43)

The means in the columns marked with the same letters within the same breed differ significantly at: small letters, $p \leq 0.05$; capitals, $p \leq 0.01$. Numbers in parentheses are the standard errors of the means.

4. Discussion

Selection based on genetic markers focuses primarily on improving the performance characteristics of animals, such as milk yield and growth performance. The rate of genetic improvement can be increased by genotype-based selection [20]. Fertility traits are also increasingly taken into consideration during animal selection, as these traits directly influence production efficiency. The intensity of selection depends on factors such as the number of calves or lambs born per year and the interval between generations. In a production system that involves both milk and meat production, irregular reproduction patterns, such as long generation interval, can lead to economic unprofitability [21]. The use of SNP markers offers the possibility of selecting the appropriate genotype/haplotype or a combination of genotypes with a related phenotype, resulting in significant breeding benefits. An example of this is found in dairy and beef cattle-breeding programs in North America and Europe, where the use of molecular genetics contributes to the profitability of farms and the continuous improvement of production values [22].

4.1. GWAS Studies

Most of the target performance traits in livestock are polygenic, which are primarily suitable for testing using the genome-wide association study (GWAS) method [23]. Keogh et al. [24] performed GWAS (Charolaise and Limousin breeds) for SNPs involved in the reproductive potential of these breeds (calving interval, calving difficulty, and calf mortality) and selected production-related traits. The highest signals for carcass weight were noted for BTA2 (dominant effect of the myostatin gene) and BTA6 (non-SMC condensin I complex, subunit G (*NCAPG*)/ligand-dependent nuclear receptor corepressor-like (*LCORL*) locus). The ovine *JAK2* gene, located on the *Ovis aries* chromosome 2 (similarly to the ovine myostatin gene), does not appear directly as a strong signal in GWAS studies in sheep. In contrast, one region on OAR6 (13 SNPs) was associated with body weight (*NCAPG/LCORL* locus, similar to cattle on BTA6) [25]. An important methodological limitation of GWAS studies is the inclusion in the analyses of only frequent gene variants, i.e., SNPs that occur in more than 5% of individuals in the population, and indicating only strong signals above the conventional threshold line. Weaker signals that may be missed by GWAS analysis can be identified and described using traditional quantitative trait locus (QTL) mapping, as long as these signals are associated with genes involved in complex biological pathways and processes [26].

4.2. Physiological Implications

Body weight is the most crucial indicator of growth and development in cow and sheep production. It directly and indirectly affects meat and wool production as well as the reproduction of animals [27]. Growth regulation is controlled by multiple signaling pathways that change throughout the life of the animal. Doubtlessly, the somatotropic axis (GH/IGF-I along with their receptors and many signaling proteins) is involved in these complex signaling pathways. This pathway is also essential for cell growth, differentiation, and development, in particular for the modulation and amplification of gonadotropin, follicle-stimulating hormone (FSH), and luteinizing hormone (LH) activity during follicle growth in the ovary [28]. After birth, skeletal muscle development is controlled by the classical Wnt (portmanteau word formed by combining "Wingless" and "Integration-1") signaling pathway, while the nonclassical Wnt signaling pathway mainly mediates the self-renewal and the growth of muscle fibers [29]. There is also evidence that the JAK2/STAT3 pathway mediates the regulation of muscle satellite cell differentiation [30] and regulates the expression of genes related to skeletal muscle development and energy metabolism, especially affecting the expression of the *MyoD* and *Myf5* genes [31]. The IL-6/JAK2/STAT3 pathway may be a principal mediator in denervated skeletal muscle atrophy [32]. The JAK2-STAT5 pathway is activated by the pulsatile release of GH in response to acute aerobic exercise in human skeletal muscle [33]. Even a single amino acid variation in the JAK2 molecule can carry severe consequences. In humans, most patients with polycythemia vera (PV), one-half of the cases of essential thrombocythemia (ET), and other myeloproliferative neoplasms (MPNs) cases possess an JAK2 p.V617F mutation (exon 12) [34] leading to constitutive activity of the JAK2 gene-encoded tyrosine kinase. Identical mutations of the *JAK2* gene and consequences occur in dogs [35]. Of the five SNPs analyzed in this manuscript, only one (rs210148032 in exon 16 of the bovine *JAK2* gene) may have analogous physiological implications. However, there is a lack of research in this area. A more in-depth analysis is necessary in relation to genetic implications.

4.3. Genetic Implications

A total of five polymorphic sites (two in cattle and three in sheep) located in different regions of the *JAK2* gene were analyzed. To our knowledge, this is the first attempt to assess the relationship between these SNPs and selected cattle and sheep production traits. They are an example of two different types of gene polymorphism, which may determine their different partial influence on the observed differences in shaping growth performance and reproductive characteristics in these animal species.

The polymorphic site rs210148032 in exon 16 of the bovine *JAK2* gene is a typical example of a missense mutation (the first letter of the codon). This determines the replacement of the isoleucine amino acid with valine at position 704 of the amino acid chain in the JH2 domain (i.e., pseudokinase) of active JAK2. Such a change may therefore determine the function of JAK2, as the JH2 domain directly regulates the activity of the tyrosine kinase JH1 domain. The rate at which amino acids are seen to mutate to other residues in homologous proteins has been extensively studied [36]. The isoleucine–valine substitution (as well as leucine–valine and leucine–isoleucine) is quite common in nature. The severity of such amino acid substitution is estimated at the levels 2–4 (http://www.insilicase.com/Web/SubstitutionScore.aspx) (accessed on 14 May 2023), where, for example, tryptophan is highly conserved and rarely changes to any other amino acid (level 17). Alanine, valine, isoleucine, and leucine are closely related in physicochemical properties because they are small, hydrophobic residues similar to methionine [37]. Being hydrophobic, isoleucine/valine prefers to be buried in protein hydrophobic cores. This may partly explain the fact that ambiguous results were obtained in the three cattle breeds. Hereford and Limousin cows with the p.Ile704 were always the heaviest and had the best weight gains. However, Angus cows with the rare p.Val704 had higher weights, although for most traits, this was not statistically confirmed. Rather, this SNP may be linked to other SNPs of greater importance, and the significance of the change of these particular A/G nucleotides can be explained in a similar way as for silent mutations.

The second polymorphic site in exon 23 of the bovine *JAK2* gene (rs211067160) and both SNPs in exon 6 (rs160146162 and rs160146160) as well as in exon 24 (rs160146116) of the sheep *JAK2* gene are examples of silent mutations. Generally, those SNPs do not result in alteration of gene expression, and there are no changes in the function and structure of the mature protein. However, even a simple A/G change may result in posttranscriptional changes such as alterations in mRNA splicing and stability, followed by nucleocytoplasmic export through the nuclear pore complex (NPC) by receptor-mediated and Ran-GTP gradient-dependent active transport [38]. In addition, synonymous codons may influence ribosome occupancy time and thus can increase or decrease elongation rates during translation [39]. The location of the *JAK2*/e23/*Hae*III polymorphism is extremely important (codon for p.Pro1057), as it is the central part encoding the JH1 tyrosine kinase domain with the conserved tyrosine sites in JAK2: p.Tyr1007 and p.Tyr1008. Similarly, in sheep, a p.Leu1082 residue also in the JH1 domain (encoded, among others, by exon 24 where *JAK2*/*Hpy*188III SNP is positioned) may also play a crucial role in the function of kinase. The stability of this fragment determines the ability to transmit a signal inside the target cell after activation of the receptor–ligand, which is usually a member of the gp^{130} receptor family and class II cytokine–receptor family (such as interleukin 3 receptor family, erythropoietin receptor (EPO), growth hormone (GH) receptor, prolactin receptor, and thrombopoietin (TPO) receptor) [40]. Another silent A/G mutation (rs110298451) located in exon 20 of the bovine *JAK2* gene was described by Padzik and Szewczuk [17] at the third nucleotide of the lysine codon (AA<u>A</u> → AA<u>G</u>) at position 912 (p.K912) of the mature amino acid chain (also JH1 domain with tyrosine kinase activity). The influence of the polymorphism on growth traits varied depending on the breed: for the Hereford breed, the *GG* genotype was favored; for the Angus breed, the heterozygous genotype; and for the Limousin breed, the *AA* genotype. Therefore, synonymous SNP within exons encoding the JH1 kinase domain may be associated with other causative mutations in the *JAK2* gene or a completely different gene involved in developing growth performance traits in cattle. In sheep, two SNPs located in exon 6 (codons for p.Glu177 and p.Thr196) were analyzed. This exon encodes a fairly conservative FERM domain, which primarily determines the binding of the JAK2 molecule to the target receptor (receptor association). However, apart from binding to the receptor, this domain plays a role in the overall organization of the tyrosine kinase structure and its subsequent activation/deactivation: a third level of regulation of kinase activity [41]. Sheep of both breeds with a combination of *AA*/*GG*/*AA* genotypes of SNPs located in exon 6 and exon 24 appear to be heavier and have better daily gains compared to other individuals. Since these are the first association studies for these SNPs, the results cannot be compared to studies by other authors. Therefore, this observation needs to be confirmed statistically in separate analyses considering haplotypes or genotype combinations.

4.4. *Involvement of JAK2 in the Formation of Reproductive Traits*

The optimal level of reproduction in beef cattle herds is closely tied to the potential economic benefits, which largely rely on beef production [42,43]. Two of the important indicators determining the fertility of cows are the age of the first calving and the length of the calving interval. Earlier age at first calving reduces the cost of rearing heifers through earlier conception, which is influenced by body weight and condition of the cows [44], while the extension of the rearing period of heifers increases costs incurred by breeders [45,46].

Regardless of the polymorphic location, Hereford cows of the *GG* genotype calved the earliest and had the shortest calving interval compared to the *AA* genotype. In Angus cows, statistically significant differences were observed only for the *JAK2*/e16/*Rsa*I polymorphism and the calving interval, where, once more, individuals with the *GG* genotype had a shorter calving interval. Cows with a short birth intervals are usually characterized by the best fertility and the highest reproductive efficiency [45]. The length of the period in question is more important in the case of dairy cattle, while in the case of beef production, attention is paid to the calving rhythm, which means that breeders keep cows with regular calving intervals for further breeding [45]. The results obtained in this paper are difficult to relate to the studies of other authors in the context of the analyzed polymorphisms and the discussed reproductive traits. The only work on a similar subject is the study by Padzik and Szewczuk [17], which identified a silent mutation in exon 20 (dbSNP ID: rs110298451) in the *JAK2* gene in the same breeds of cattle. The age of cows at first calving was later (average days: Limousin: 1118, Hereford 1038, and Angus 1085, respectively). Moreover, the authors did not find a significant relationship between *JAK2*/e20/*Rsa*I polymorphism and age at first calving.

Reproductive traits in sheep have low heritability. Therefore, traditional selection by phenotype results in small annual genetic progress [47]. Identification of candidate genes is one strategy for improving these traits. These genes directly or indirectly affect fertility [48,49], prolificacy [50,51], and lamb rearing [51,52], which are of great importance and may contribute to increasing the rate of genetic improvement of these traits. It should be remembered that polygenic traits are traits that are constantly dispersed, referring to the existence of many genes that help in the expression of various gene traits (it is selective). Environmental elements, including animal stress, also significantly affect gene expression [27,53]. Therefore, in our study, the observations were made on the same farm, with the same management and feeding system for both sheep breeds. The sheep spent most of the year on pasture, where the nutritional properties of forage change depending on the season, which may affect gene expression. Metzler-Zebeli et al. [54] stated that different feeding levels can alter gene expression. Sheep grazing, stimulated by hot summers and cold winters, can also alter gene expression [55].

Because no reports are available in the literature on the relationships between the three polymorphic sites analyzed in our study and performance and reproduction traits of sheep, it is not possible to contrast and compare our results with those of other authors. Therefore, we discussed our results in terms of their economic importance and their influence on the genetic improvement of sheep.

In the case of polymorphisms located in exons 6 (*JAK2*/*Ear*I) and 24 (*JAK2*/*Hpy*188III) in the group of sheep with the *GG* genotype, a certain tendency was observed associated with the deterioration of fertility and lower rearing of lambs, which is also associated with the economic aspect of production [56,57].

An inverse relationship was found in sheep of the autochthonous Pomeranian breed in the case of the *JAK2*/e6/seq polymorphism, where the fertility and lamb survival in ewes of the *GG* genotype exceeded 97%, which indicates excellent reproduction. Suffolk ewes with the *GG* genotype were also characterized by high fertility (90.34%). It is assumed that the value of the indicator can be considered satisfactory when it is 90%, good if it exceeds 95%, and very good when it is close to 100% [58]. Preliminary analyses showed that ewes with the AA and AG genotypes had better fertility and lamb survival, apart from Pomeranian sheep in the case of the *JAK2*/e6/seq polymorphic site. Irrespective of the analyzed polymorphic site, the *GG* ewes reared fewer lambs than other mothers (except for sheep of the Pomeranian breed *JAK2*/e6/seq). The values of the index oscillate between 73.15–84.52%, which means that the losses of lambs in the analyzed groups of mothers with *GG* genotypes exceeded the acceptable level of 5% [59–61]. In these groups of sheep, there were lambs from twin pregnancies, which, according to Clune et al. [59], may be less developed and quite often show reduced viability immediately after birth [60], which delays or even prevents proper colostrum intake [62]. As a result, deaths may occur in the first hours of life. Losses during the rearing period may also be due to the lack of colostrum or its poorer quality and insufficient milk yield of mothers. For this reason, lambs are often fed with preparations containing powdered colostrum and, in further rearing, with milk replacers [61]. Therefore, it should be considered that greater care provided by breeders may improve lamb-rearing results in this breed of sheep. The selection and elimination of individuals with the *GG JAK2*/e6/seq genotype should also be considered.

Lamb survival is an important element of sheep production closely related to the prolificacy and profitability of production. The prolificacy of sheep in both breed groups was similar, regardless of the genotype, and did not exceed 1.18. The prolificacy of Suffolk sheep was relatively low (1.06–1.19). The results for PZO indicate it has good potential in terms of prolificacy [63].

The assumed profitability of the production of lambs for slaughter occurs from at least 1.5 lambs raised from the mother [64], which in practice is difficult in the case of native breeds, such as Pomeranian sheep, except for prolific breeds. This trait in sheep depends to a large extent on, e.g., the breed, nutrition, and age of the ewes [57,64]. Taking into account the possibilities and predispositions of the analyzed sheep breeds in the context of higher prolificacy, attention should be paid to the nutrition of ewes before and during breeding [65], which, according to Łozicki [66], could increase the number of maturing ova and, consequently, improve fertility by 10–30%. In addition, the introduction of biotechnological methods in reproduction, namely the stimulation of the functioning of the reproductive system but also the induction of superovulation, according to Skliarov et al. [67], suggests the possibility of a significant improvement in this indicator.

5. Conclusions

The phenotypic characteristics of cattle and sheep are the result of a complex interaction of many genetic and environmental factors, which usually act simultaneously, and it is difficult to determine the degree of influence of each of them. Therefore, early identification of genetic features in young animals enables more effective selection management and effective breeding in the herd. Production indicators, such as, body weight, daily gains of lambs and calves at different stages of their lives, as well as reproductive indices, are important factors that are crucial for successful ruminant production and thus translate into profitability of production. Identification of genomic regions and biological pathways that contribute to the understanding of variability in body weight traits and reproductive traits is important for selection purposes. Therefore, the identification of a number of genes with a moderate and low impact on the traits discussed in the paper will allow the use of genetic-marker-assisted selection for in-breed selection in order to achieve higher body weight, daily gains, and better fertility and rearing of lambs and calves in various housing and feeding conditions. The results obtained in the study are ambiguous, and additional association studies are needed on other herds and within different breeds of cattle and sheep. For this reason, an analysis of haplotypes and/or combined genotypes of the *JAK2* gene and selected polymorphic sites located in the genes encoding the somatotropic axis should be performed.

Supplementary Materials: The following supporting information can be downloaded at: https://www.mdpi.com/article/10.3390/ani13152470/s1, Figure S1: Electrophoretic separation of restriction fragments resulting from digestion with the RsaI restriction enzyme (ACRS-PCR; bovine JAK2/e16/RsaI polymorphism); lane 1 (M), mass standard pUC19 DNA/MspI; lane 2, AG genotype; lane 3, AA genotype; lane 4, GG genotype; Figure S2: Electrophoretic separation of restriction fragments resulting from digestion with the HaeIII restriction enzyme (ACRS-PCR; bovine JAK2/e23/HaeIII polymorphism); lane 1 (M), mass standard pUC19 DNA/MspI; lane 2, AG genotype; lane 3, AA genotype; lane 4, GG genotype; Figure S3: Electrophoretic separation of restriction fragments resulting from digestion with the EarI restriction enzyme (PCR-RFLP; ovine JAK2/e6/EarI polymorphism) polymorphic site; lane 1 (M), mass standard pUC19 DNA/MspI; lane 2, AG genotype; lane 3, AA genotype; lane 4, GG genotype; Figure S4: Electrophoretic separation of restriction fragments resulting from digestion with the Hpy188III restriction enzyme (PCR-RFLP; ovine JAK2/e24/Hpy188III polymorphism); lane 1 (M), mass standard pUC19 DNA/MspI; lane 2, AG genotype; lane 3, AA genotype; lane 4, GG genotype; Table S1: A summary of the JAK2 gene fragments used for DNA sequencing and subsequent identification of polymorphic sites.

Author Contributions: Conceptualization, N.O. and M.A.S.; data curation, M.A.S. and M.W.-D.; investigation, N.O.; methodology, N.O. and M.A.S.; resources, T.S. and B.B.; software, M.A.S.; supervision, M.A.S.; visualization, S.Z.; writing—original draft, N.O. and M.A.S.; writing—review and editing, N.O., M.A.S., S.Z., T.S. and B.B. All authors have read and agreed to the published version of the manuscript.

Funding: This research received no external funding.

Institutional Review Board Statement: The collection of the material (blood samples) has been approved by the Local Ethics Commission, permission No. 25/09 (cattle), and the Local Commission for Animal Experiments in Poznań: Resolution No. 20/2018 (sheep).

Informed Consent Statement: Not applicable.

Data Availability Statement: All data generated or analyzed during the study are included in this published article. The datasets used and/or analyzed in the current study are available from the corresponding author on reasonable request.

Conflicts of Interest: The authors declare no conflict of interest.

References

1. Crossa, J.; Pérez-Rodríguez, P.; Cuevas, J.; Montesinos-López, O.; Jarquín, D.; de Los Campos, G.; Burgueño, J.; González-Camacho, J.M.; Pérez-Elizalde, S.; Beyene, Y.; et al. Genomic Selection in Plant Breeding: Methods, Models, and Perspectives. *Trends Plant Sci.* **2017**, *22*, 961–975. [CrossRef]
2. Tong, H.; Küken, A.; Nikoloski, Z. Integrating molecular markers into metabolic models improves genomic selection for Arabidopsis growth. *Nat. Commun.* **2020**, *11*, 2410. [CrossRef]
3. Katoh, K.; Takahashi, T.; Kobauashi, Y.; Obara, Y. Somatropic Axis and Nutrition in Young Ruminants around Weaning Time. *Asian-Australas. J. Anim. Sci.* **2007**, *20*, 1156–1168. [CrossRef]
4. Chmielewska, P.; Barwiak, K.; Dańko, P.; Ciesielka, M.; Beń-Skowronek, I. Insulin-like growth factors and theirs role in pathogenesis od diabetic nephropathy and retinopathy. *Pediatr. Endocrinol.* **2018**, *17*, 191–200.
5. Rosenfeld, R.G.; Hwa, V. Biology of the somatotroph axis (after the pituitary). *Ann. Endocrinol.* **2017**, *78*, 80–82. [CrossRef]
6. Backeljauw, P.F.; Dattani, M.T.; Cohen, P.; Rosenfeld, R.G. *Disorders of Growth Hormone/Insulin-Like Growth Factor Secretion and Action*, 4th ed.; Elsevier Inc.: Amsterdam, The Netherlands, 2014; pp. 392–404.
7. He, K.; Loesch, K.; Cowan, J.W.; Li, X.; Deng, L.; Wang, X.; Jiang, J.; Frank, S.J. Janus kinase 2 enhances the stability of the mature growth hormone receptor. *Endocrinology* **2005**, *146*, 4755–4765. [CrossRef]
8. Babon, J.J.; Lucet, I.S.; Murphy, J.M.; Nicola, N.A.; Varghese, L.N. The molecular regulation of Janus kinase (JAK) activation. *Biochem. J.* **2014**, *462*, 1–13. [CrossRef]
9. Lacronique, V.; Boureux, A.; Valle, V.D.; Poirel, H.; Quang, C.T.; Mauchauffé, M.; Berthou, C.; Lessard, M.; Berger, R.; Ghysdael, J.; et al. A TEL-JAK2 fusion protein with constitutive kinase activity in human leukemia. *Science* **1997**, *278*, 1309–1312. [CrossRef]
10. Rane, S.G.; Reddy, E.P. Janus kinase: Components of multiple signaling pathways. *Oncogene* **2000**, *19*, 5662–5679. [CrossRef]
11. Khan, M.Z.; Khan, A.; Xiao, J.; Ma, Y.; Ma, J.; Gao, J.; Cao, Z. Role of the JAK-STAT Pathway in Bovine Mastitis and Milk Production. *Animals* **2020**, *10*, 2107. [CrossRef]
12. Brenaut, P.; Bangera, R.; Bevilacqua, C.; Rebours, E.; Cebo, C.; Martin, P. Validation of RNA isolated from milk fat globules to profile mammary epithelial cell expression during lactation and transcriptional response to a bacterial infection. *J. Dairy Sci.* **2012**, *95*, 6130–6144. [CrossRef]
13. Yamaji, D.; Kang, K.; Robinson, G.W.; Hennighausen, L. Sequential activation of genetic programs in mouse mammary epithelium during pregnancy depends on STAT5A/Bconcentration. *Nucleic Acids Res.* **2013**, *41*, 1622–1636. [CrossRef]
14. Groner, B.; Gouilleux, F. Prolactin-mediated gene activation in mammary epithelial cells. *Curr. Opin. Genet. Dev.* **1995**, *5*, 587–594. [CrossRef]
15. Bole-Feysot, C. Prolactin (PRL) and Its Receptor: Actions, Signal Transduction Pathways and Phenotypes Observed in PRL Receptor Knockout Mice. *Endocr. Rev.* **1998**, *19*, 225–268. [CrossRef] [PubMed]
16. Szewczuk, M. Association of a genetic marker at the bovine Janus kinase 2 locus (JAK2/RsaI) with milk production traits of four cattle breeds. *J. Dairy Res.* **2015**, *82*, 287–292. [CrossRef] [PubMed]

17. Padzik, N.; Szewczuk, M. Association between polymorphism in the JAK2 gene (JAK2/e20/RsaI) and selected performance parameters in beef cattle. *Anim. Sci. Pap. Rep.* **2020**, *38*, 333–344.
18. Padzik, N.; Szewczuk, M.; Ropka-Molik, K. Distribution of JAK2 genotypes across Suffolk and Pomeranian sheep. *Small Rumin Res.* **2021**, *194*, 106282. [CrossRef]
19. Strzetelski, J.A.; Brzóska, F.; Kowalski, Z.M.; Osięgłowski, S. *Nutritional Recommendations for Ruminants and Feed Value Tables*; Praca zbiorowa Instytut Zootechniki PIB: Kraków, Poland, 2014.
20. Polido, P.B.; Ferreira, G.F.; Alberton, O.; De Souza, S.G.H. Marcadores moleculares aplicados nomelhoramento genético de bovinos. *Arq. Ciênc. Vet. Zool.* **2012**, *15*, 161–169.
21. Grossi, D.A.; Venturini, G.C.; Paz, C.C.; Bezerra, L.A.; Lôbo, R.B.; Oliveira, J.A.; Munari, D.P. Genetic associations between age at first calving and heifer body weight and scrotal circumference in Nelore cattle. *J. Anim. Breed. Genet.* **2009**, *126*, 387–393 [CrossRef] [PubMed]
22. Menezes, I.P.P.; Gaiotto, F.; Suassuna, N.; Hoffmann, L.; Barroso, P. Susceptibility of Gossypium mustelinum Populations to the Main Cotton Diseases in Brazil. *J. Agric. Sci.* **2014**, *6*, 39–47. [CrossRef]
23. de Oliveira Silva, R.M.; Stafuzza, N.B.; de Oliveira Fragomeni, B.; de Camargo, G.M.F.; Ceacero, T.M.; Cyrillo, J.N.D.S.G.; Baldi, F. Boligon, A.A.; Mercadante, M.E.Z.; Lourenco, D.L.; et al. Genome-Wide Association Study for Carcass Traits in an Experimental Nelore Cattle Population. *PLoS ONE* **2017**, *12*, e0169860. [CrossRef] [PubMed]
24. Keogh, K.; Carthy, T.R.; McClure, M.C.; Waters, S.M.; Kenny, D.A. Genome-wide association study of economically important traits in Charolais and Limousin beef cows. *Animal* **2021**, *15*, 100011. [CrossRef] [PubMed]
25. Al-Mamun, H.A.; Kwan, P.; Clark, S.A.; Ferdosi, M.H.; Tellam, R.; Gondro, C. Genome-wide association study of body weight in Australian Merino sheep reveals an orthologous region on OAR6 to human and bovine genomic regions affecting height and weight. *Genet. Sel. Evol.* **2015**, *47*, 66. [CrossRef] [PubMed]
26. Abiola, O.; Angel, J.M.; Avner, P.; Bachmanov, A.A.; Belknap, J.K.; Bennett, B.; Blankenhorn, E.P.; Blizard, D.A.; Bolivar, V. Brockmann, G.A.; et al. Complex Trait Consortium. The nature and identification of quantitative trait loci: A community's view *Nat. Rev. Genet.* **2003**, *4*, 911–916. [CrossRef] [PubMed]
27. Lu, Z.; Yue, Y.; Yuan, C.; Liu, J.; Chen, Z.; Niu, C.; Sun, X.; Zhu, S.; Zhao, H.; Guo, T.; et al. Genome-Wide Association Study of Body Weight Traits in Chinese Fine-Wool Sheep. *Animals* **2020**, *10*, 170. [CrossRef] [PubMed]
28. Armstrong, D.G.; Webb, R. Ovarian follicular dominance: The role of intraovarian growth factors and novel proteins. *Rev. Reprod.* **1997**, *2*, 139–146. [CrossRef]
29. Rudnicki, M.A.; Williams, B.O. Wnt signaling in bone and muscle. *Bone* **2015**, *80*, 60–66. [CrossRef]
30. Trenerry, M.K.; Della Gatta, P.A.; Cameron-Smith, D. JAK/STAT signaling and human in vitro myogenesis. *BMC Physiol.* **2011**, *9*, 6. [CrossRef]
31. Liu, J.; Jing, X.; Gan, L.; Sun, C. The JAK2/STAT3 signal pathway regulates the expression of genes related to skeletal muscle development and energy metabolism in mice and mouse skeletal muscle cells. *Biosci. Biotechnol. Biochem.* **2012**, *76*, 1866–1870. [CrossRef]
32. Huang, Z.; Zhong, L.; Zhu, J.; Xu, H.; Ma, W.; Zhang, L.; Shen, Y.; Law, B.Y.; Ding, F.; Gu, X.; et al. Inhibition of IL-6/JAK/STAT3 pathway rescues denervation-induced skeletal muscle atrophy. *Ann. Transl. Med.* **2020**, *8*, 1681. [CrossRef]
33. Consitt, L.A.; Wideman, L.; Hickey, M.S.; Morrison, R.F. Phosphorylation of the JAK2-STAT5 pathway in response to acute aerobic exercise. *Med. Sci. Sports Exerc.* **2008**, *40*, 1031–1038. [CrossRef] [PubMed]
34. Zhao, S.; Zhang, X.; Xu, Y.; Feng, Y.; Sheng, W.; Cen, J.; Wu, D.; Han, Y. Impact of JAK2V617F Mutation Burden on Disease Phenotype in Chinese Patients with JAK2V617F-positive Polycythemia Vera (PV) and Essential thrombocythemia (ET). *Int. J. Med. Sci.* **2016**, *13*, 85–91. [CrossRef]
35. Beurlet, S.; Krief, P.; Sansonetti, A.; Briend-Marchal, A.; Kiladjian, J.J.; Padua, R.A.; Chomienne, C.; Cassinat, B. Identification of JAK2 mutations in canine primary polycythemia. *Exp. Hematol.* **2011**, *39*, 542–545. [CrossRef] [PubMed]
36. Yampolsky, L.Y.; Stoltzfus, A. The exchangeability of amino acids in proteins. *Genetics* **2005**, *170*, 1459–1472. [CrossRef] [PubMed]
37. Creixell, P.; Schoof, E.M.; Tan, C.S.; Linding, R. Mutational properties of amino acid residues: Implications for evolvability of phosphorylatable residues. *Philos. Trans. R. Soc. B Biol. Sci.* **2012**, *19*, 2584–2593. [CrossRef] [PubMed]
38. Robert, F.; Pelletier, J. Exploring the Impact of Single-Nucleotide Polymorphisms on Translation. *Front Genet.* **2018**, *9*, 507. [CrossRef]
39. Brule, C.E.; Grayhack, E.J. Synonymous Codons: Choose Wisely for Expression. *Trends Genet.* **2017**, *33*, 283–297. [CrossRef]
40. Hu, X.; Li, J.; Fu, M.; Zhao, X.; Wang, W. The JAK/STAT signaling pathway: From bench to clinic. *Signal Transduct. Target Ther.* **2021**, *6*, 402. [CrossRef]
41. Funakoshi-Tago, M.; Pelletier, S.; Moritake, H.; Parganas, E.; Ihle, J.N. Jak2 FERM domain interaction with the erythropoietin receptor regulates Jak2 kinase activity. *Mol. Cell Biol.* **2008**, *28*, 5792–5801. [CrossRef]
42. Boyer, C.N.; Griffith, A.P.; DeLong, K.L. Reproductive Failure and Long-Term Profitability of Spring- and Fall-Calving Beef Cows. *J. Agric. Resour. Econ.* **2020**, *45*, 78–91. [CrossRef]
43. Moorey, S.E.; Biase, F.H. Beef heifer fertility: Importance of management practices and technological advancements. *J. Anim. Sci.* **2020**, *11*, 2–12. [CrossRef] [PubMed]
44. Shin, S.; Lee, J.; Do, C. Genetic relationship of age at first calving with conformation traits and calving interval in Hanwoo cows. *J. Anim. Sci. Technol.* **2021**, *63*, 740–750. [CrossRef] [PubMed]

45. Brzáková, M.; Čítek, J.; Svitáková, A.; Veselá, Z.; Vostrý, L. Genetic Parameters for Age at First Calving and First Calving Interval of Beef Cattle. *Animals* **2020**, *10*, 2122. [CrossRef]
46. Krupová, Z.; Krupa, E.; Wolfová, M. Economic weights of current and new breeding objective traits in Aberdeen Angus. *J. Anim. Sci.* **2020**, *65*, 77–85. [CrossRef]
47. Ghiasi, H.; Abdollahi-Arpanahi, R. The candidate genes and pathways affecting litter size in sheep. *Small Rumin. Res.* **2021**, *205*, 106546. [CrossRef]
48. Luridiana, S.; Cosso, G.; Pulinas, L.; Di Stefano, M.V.; Curone, G.; Carcangiu, V.; Mura, M.C. New polymorphisms at MTNR1A gene and their association with reproductive resumption in sarda breed sheep. *Theriogenology* **2020**, *158*, 438–444. [CrossRef]
49. Sánchez-Ramos, R.; Trujano-Chavez, M.Z.; Gallegos-Sánchez, J.; Becerril-Pérez, C.M.; Cadena-Villegas, S.; Cortez-Romero, C. Detection of Candidate Genes Associated with Fecundity through Genome-Wide Selection Signatures of Katahdin Ewes. *Animals* **2023**, *13*, 272. [CrossRef]
50. Yang, Q.; Yan, H.; Li, J.; Xu, H.; Wang, K.; Zhu, H.; Lan, X. A novel 14-bp duplicated deletion within goat GHR gene is significantly associated with growth traits and litter size. *Anim. Genet.* **2017**, *48*, 499–500. [CrossRef]
51. Sebastiano, L.; Consuelo, M.M.; Veronica, D.S.M.; Luisa, P.; Giovanni, C.; Michella, N.; Vincenzo, C. Polymorphism of insulin-like growth factor 1 gene and its relationship with reproductive performances and milk yield in Sarda dairy sheep. *Vet. Anim. Sci.* **2019**, *15*, 100084. [CrossRef]
52. Abate, Z. Review of the reproductive performances of indigenous sheep in Ethiopia. *Glob. J. Biol. Agric.* **2016**, *6*, 117–126.
53. Núñez-Torres, O.P.; Almeida-Secaira, R.I. Quantitative genetics: Principles of farming in livestock production Genética cuantitativa: Principios de la crianza en la producción pecuaria. *J. Selva Andin. Anim.* **2022**, *9*, 23–36.
54. Metzler-Zebeli, B.U.; Hollmann, M.; Sabitzer, S.; Podstatzky-Lichtenstein, L.; Klein, D.; Zebeli, Q. Epithelial response to high-grain diets involves alteration in nutrient transporters and Na^+/K^+-ATPase Mrna expression in rumen and colon of goats. *J. Anim. Sci.* **2013**, *91*, 4256–4266. [CrossRef] [PubMed]
55. Lu, Z.; Chu, M.; Li, Q.; Jin, M.; Fei, X.; Ma, L.; Zhang, L.; Wei, C. Transcriptomic Analysis Provides Novel Insights into Heat Stress Responses in Sheep. *Animals* **2019**, *9*, 387. [CrossRef] [PubMed]
56. Gootwine, E. Invited review: Opportunities for genetic improvement toward higher prolificacy in sheep. *Small Rumin. Res.* **2020**, *186*, 106090. [CrossRef]
57. Ivanova, T.; Stoikova-Grigorova, R.; Bozhilova-Sakova, M.; Ignatova, M.; Dimitrova, I.; Koutev, V. Phenotypic and genetic characteristics of fecundity in sheep. A review. *Bulg. J. Agric. Sci.* **2021**, *27*, 1002–1008.
58. Lipecka, C.Z. Weaning and Care of Offspring. In *Breeding, Rearing and Use of Sheep*, 3rd ed.; Niżnikowski, R., Ed.; Wieś Jutra: Warszawa, Poland, 2011; pp. 89–103.
59. Clune, T.; Lockwood, A.; Hancock, S.; Thompson, A.N.; Beetson, S.; Campbell, A.J.D.; Glanville, E.; Brookes, D.; Trengove, C.; O'Handley, R.; et al. Abortion and Lamb Mortality between Pregnancy Scanning and Lamb Marking for Maiden Ewes in Southern Australia. *Animals* **2022**, *12*, 10. [CrossRef]
60. Hutchison, D.; Clarke, B.E.; Hancock, S.; Thompson, A.N.; Bowen, E.; Jacobson, C. Lower Reproductive Rate and Lamb Survival Contribute to Lower Lamb Marking Rate in Maiden Ewes Compared to Multiparous Ewes. *Animals* **2022**, *12*, 513. [CrossRef]
61. Fesseha, H.; Gebremichael, G.; Asefa, I.; Edaso, T. Study on incidence of lamb morbidity and mortality and associated risk factors in the mixed crop-livestock production system of Gewata District, Kaffa zone, southwestern Ethiopia. *Anim. Dis.* **2023**, *3*, 11. [CrossRef]
62. Martins, L.; Oliveira, L.C.C. Colostrum as the speed up key for ruminant newborn: What do we know and should further characterize. *J. Dairy Vet. Anim. Res.* **2020**, *9*, 95–98.
63. PZO. Polish Union of Sheep-Farmers. 2022. Available online: http://Pzow.Pl/ (accessed on 27 May 2023).
64. Paraponiak, P.; Wieczorek-Dąbrowska, M. Impact of organic farming on improving reproductive parameters in sheep. *Wiadomości Zootech.* **2019**, *4*, 30–40.
65. Lourencon, R.V.; Patra, A.K.; Puchala, R.; Dawson, L.J.; Ribeiro, L.P.D.S.; Encinas, F.; Goetsch, A.L. Effects of Nutritional Plane at Breeding on Feed Intake, Body Weight, Condition Score, Mass Indexes, and Chemical Composition, and Reproductive Performance of Hair Sheep. *Animals* **2023**, *13*, 735. [CrossRef] [PubMed]
66. Łozicki, A. Rules for Feeding Sheep. In *Breeding, Rearing and Use of Sheep*, 3rd ed.; Niżnikowski, R., Ed.; Wieś Jutra: Warszawa, Poland, 2011; pp. 120–129.
67. Skliarov, P.; Perez, C.; Petruscha, V.; Fedorenko, S.; Bilyi, D. Induction and synchronization of oestrus in sheep and goats. *J. Cent. Eur. Agric.* **2021**, *22*, 39–53. [CrossRef]

animals

Article

A Whole Genome Sequencing-Based Genome-Wide Association Study Reveals the Potential Associations of Teat Number in Qingping Pigs

Zezhang Liu [1], Hong Li [2], Zhuxia Zhong [1] and Siwen Jiang [1,*]

[1] Agricultural Ministry Key Laboratory of Swine Breeding and Genetics & Key Laboratory of Agricultural Animal Genetics, Breeding and Reproduction of Ministry of Education, Huazhong Agricultural University, Wuhan 430070, China; zzl19920507@hotmail.com (Z.L.); zzx05031319@163.com (Z.Z.)

[2] Novogene Bioinformatics Institute, Beijing 100083, China; lihong@novogene.com

* Correspondence: jiangsiwen@mail.hzau.edu.cn

Simple Summary: Teat number is important for the mothering ability of a sow, but this trait has seldom been investigated by high-depth genomic data-based genome-wide association study (GWAS). Here, we performed GWAS for the teat number-related traits in 100 Chinese native Qingping pigs by recording their left and right teat numbers and analyzing their genetic variations through 10-fold whole-genome sequencing. *T-Box Transcription Factor 3* (*TBX3*) on *Sus scrofa* chromosome (SSC) 14 and Wnt signaling pathway are revealed to be associated with teat number-related traits, with important roles in mammary gland morphogenesis and development.

Abstract: Teat number plays an important role in the reproductive performance of sows and the growth of piglets. However, the quantitative trait loci (QTLs) and candidate genes for the teat number-related traits in Qingping pigs remain unknown. In this study, we performed GWAS based on whole-genome single-nucleotide polymorphisms (SNPs) and insertions/deletions (Indels) for the total number of teats and five other related traits in 100 Qingping pigs. SNPs and Indels of all 100 pigs were genotyped using 10× whole genome resequencing. GWAS using General Linear Models (GLM) detected a total of 28 SNPs and 45 Indels as peak markers for these six traits. We also performed GWAS for the absolute difference between left and right teat number (ADIFF) using Fixed and random model Circulating Probability Unification (FarmCPU). The most strongly associated SNP and Indel with a distance of 562,788 bp were significantly associated with ADIFF in both GLM and FarmCPU models. In the 1-Mb regions of the most strongly associated SNP and Indel, there were five annotated genes, including *TRIML1*, *TRIML2*, *ZFP42*, *FAT1* and *MTNR1A*. We also highlighted *TBX3* as an interesting candidate gene for SSC14. Enrichment analysis of candidate genes suggested the Wnt signaling pathway may contribute to teat number-related traits. This study expanded significant marker-trait associations for teat number and provided useful molecular markers and candidate genes for teat number improvement in the breeding of sows.

Keywords: teat number; Qingping pig; whole genome resequencing; *TBX3*; Wnt signaling pathway

1. Introduction

Porcine teats are located from the anterior to posterior limb bud and symmetrical to the abdominal midline, and teat number is an important trait for the mothering ability of a sow, which can affect the piglets' weight gain and mortality. It has been shown that teat number is a quantitative trait with a medium level of heritability (0.32) [1]. Using genetic markers can speed up the genetic improvement of teat number. Thus far, 655 teat number quantitative trait loci (QTLs) in pigs have been reported and included in the PigQTL database [2]. Previous genome-wide association studies (GWAS) found microsatellite markers or single nucleotide polymorphisms (SNPs) in popular breeds, including

Durocs [1,3], Large Whites [4], and their crosses with Meishan pigs [5,6]. GWAS were also performed for teat number-related traits in several Chinese native pig breeds, including Erhualian [7], Sushan [8] and Beijing black [9]. However, the genetic architecture for teat number in Qingping pig, a Chinese native pig breed, is still not clear.

In 2000, Wada et al. revealed two QTLs for teat number on SSC1 and SSC7 by QTL analysis of 265 F2 offspring of the Meishan and Göttingen miniature pig [10]. Since then, an increasing number of studies have used microsatellite markers to confirm the QTL on SSC7 in other pig populations, including Meishan × Duroc F2 resource population [5], F2 populations of Yorkshire boars and Meishan sows [11], and Meishan × Large White F2 pigs [12]. Additionally, a number of studies detected associated SNPs for teat number near or within this QTL on SSC7 in Duroc pig [1,3,13,14], White Duroc × Erhualian F2 resource population [15], a commercial swine population [16], and Large White pig [17]. Notably, *VRTN* (located at 103.4 Mb on SSC7 on *Sscrofa*10.2) was a credible candidate gene in this major QTL for teat number on SSC7. These studies also indicated that QTLs for teat number-related traits are distributed across the genome on every chromosome. Several other genes have also been annotated as teat number-associated genes, such as Lysine Demethylase 6B (*KDM6B*) [15], TOX High Mobility Group Box Family Member 3 (*TOX3*) [17], Estrogen Receptor 1 (*ESR1*) and Nuclear Receptor Subfamily 5 Group A Member 1 (*NR5A1*) [8]. However, most previous studies have not used the high-density single-nucleotide polymorphisms (SNPs) detected by high-throughput sequencing data to investigate teat number-related traits.

The purpose of the current study was to detect genome-wide associations for teat number-related traits in Qingping pigs using whole-genome SNPs and insertions/deletions (Indels) based on whole-genome resequencing.

2. Materials and Methods

2.1. Sample and Sequencing

In this study, all 100 pigs were raised indoors in the Qingping pig Conservation Farm in Yichang, Hubei, China. Ear tissues were collected and stored in liquid nitrogen until further analysis. Genomic DNA samples were extracted from ear tissues, using a standard phenol–chloroform method. Sequencing libraries were constructed and sequenced by the Novogene Bioinformatics Institute (Novogene, Beijing, China). High-throughput sequencing was performed as paired-end 150 sequencing using a HiSeq 4000 sequencing system (Illumina, San Diego, CA, USA).

2.2. Phenotypic Data

In this study, three independent persons observed all 100 pigs and counted the teat number of the left and right lines. A total of six teat number-related traits were obtained: (i) the number of teats on the left side (LTN); (ii) the number of teats on the right side (RTN); (iii) the total number of teats (TNUM = LTN + RTN); (iv) the maximum number of teats in LTN and RTN (MAXAP); (v) the difference between the two sides (L-R = LTN − RTN) and (vi) the absolute difference between left and right teat number (ADIFF = | LTN − RTN |). The mean, standard deviation, minimum value, maximum value, and coefficient of variance for each trait were calculated using R (version 3.6.0) (R Core Team, Vienna, Austria).

2.3. Genotyping and Quality Control

Clean reads from all 100 pigs were aligned to the *Sscrofa11.1* reference genome by the BWA software (version: 0.7.8) (Wellcome Trust Sanger Institute, Hinxton, UK) [18]. To reduce mismatches generated by PCR amplification before sequencing, duplicated reads were removed using SAMtools (Wellcome Trust Sanger Institute, Hinxton, UK) [19]. SNPs and Indels calling were initially performed to generate a gvcf file using UnifiedGenotyper in GATK (version 3.6) (Broad Institute, Cambridge, MA, USA) [20]. SNPs and Indels were divided using SelectVariants in GATK. Hard filtering of SNPs was applied to the raw variant set using "QUAL < 30.0 || QD < 2.0 || FS > 60.0 || MQ < 40.0 || SOR > 3.0 ||

ReadPosRankSum < −8.0". Hard filtering of Indels was applied to the raw variant set using "QUAL < 30.0 | | QD < 2.0 | | FS > 200.0 | | SOR > 10.0 | | ReadPosRankSum < −20.0 | | MQ < 40.0 | | MQRankSum < −12.5". SNPs and Indels in the VCF files were quality-filtered using VCFtools (v0.1.16) (Wellcome Trust Sanger Institute, Hinxton, UK) to remove variants with sequencing depth less than 3 [21]. SNPs and Indels were further filtered using PLINK 2.0 (Complete Genomics, Mountain View, CA, USA) [22]. Missing genotypes were imputed using beagle.03Jul19.b33.jar [23], followed by filtering SNPs and Indels again using PLINK 2.0 to obtain the high-quality common SNPs and Indels of 100 pigs for further analysis (individual call rate > 0.90; minor allele frequency > 0.05; call rate > 0.90, SNPs and Indels in Hardy–Weinberg equilibrium ($p > 1 \times 10^{-6}$) and excluding SNPs and Indels located on the sex chromosomes).

2.4. SNP-Based Heritability

The phenotypic variance explained by genome-wide SNPs (SNP-based heritability) was estimated using GREML in Genome-wide Complex Trait Analysis (GCTA) [24]. Briefly, the genetic relationships between pairwise individuals from all the autosomal SNPs were estimated using the genetic relationship matrix (GRM) based on high-quality common SNPs, followed by GRM and phenotype for restricted maximum likelihood (REML) analysis to estimate the variance explained by the SNPs.

2.5. Principal Component Analysis

Principal component analysis (PCA) was used to explore the population structure of Qingping pigs and determine whether principal components (PCs) should be added to the GWAS. PCA was performed using MVP.Data.PC in rMVP R (version 3.6.0) (Huazhong Agricultural University, Wuhan, China) [25] package based on high quality common SNPs. PC1 and PC2 were visualized using MVP.PCAplot in rMVP R (version 3.6.0) package.

2.6. GWAS Using General Linear Model (GLM)

In the present study, GLM in rMVP R (version 3.6.0) package was used to perform the GWAS for teat number-related traits based on high-quality common SNPs and Indels [26]. Principal component analysis showed no discernible clustering. Therefore, no principal component was adjusted in the subsequent association analysis. Each SNP or Indel for teat number-related traits was tested by GLM as follows:

$$y = Xb + e$$

where y is the vector of each teat number-related trait in Qingping pigs; X, a matrix of test SNP or Indel; b, an incidence vector for X; e, a vector of residuals following a normal distribution with a mean of zero and $I\sigma_e^2$ covariance, where I is the identity matrix and σ_e^2 is the residual variance.

2.7. GWAS Using FarmCPU

The Fixed Effect Model (FEM) and the Random Effect Model (REM) are used iteratively in FarmCPU [27], and FarmCPU in rMVP R (version 3.6.0) package was also used to perform GWAS for ADIFF. In the GLM model, some genetic markers were significantly associated with ADIFF with a whole-genome significant p-value (2.16×10^{-9} for SNPs and 2.44×10^{-8} for Indels). These markers can be used to define kinship in the REM step of FarmCPU to avoid the model over-fitting problem in FEM. The FEM is used to test each genetic marker, one at a time. Pseudo QTNs are included as covariates to control false positives. Specifically, the FEM could be expressed by the following equation:

$$y_i = M_{i1}b_1 + M_{i2}b_2 + \cdots + M_{it}b_t + S_{ij}d_j + e_i$$

where y_i is the phenotype of the ith individual; M_{i1}, M_{i2}, ... , M_{it}, the genotypes of t. pseudo QTNs, initiated with no QTN; b_1, b_2, ... , b_t, the corresponding effects of the pseudo

QTNs; S_{ij}, the genotype of the ith individual and jth genetic marker; d_j, the corresponding effect of the jth genetic marker; e_i, the residual having a distribution with zero mean and variance of σ_e^2.

The REM is used to optimize the selection of pseudo QTNs from markers with whole genome significant p-values (2.16×10^{-9} for SNPs and 2.44×10^{-8} for Indels) and positions by using the SUPER algorithm [28]. The REM could be expressed by the following equation:

$$y_i = u_i + e_i$$

where y_i and e_i. are the same as in FEM, and u_i indicates the total additive genetic effect of the ith individual. The expectations of the individuals' total genetic effects are zeros. The variance and covariance matrix of the individuals' total genetic effects can be expressed by $G = 2K\sigma_a^2$, where σ_a^2 is an unknown genetic variance and K is the kinship matrix calculated by pseudo QTNs. The FEM and REM are iterated until no new pseudo QTNs are added, or the specified maximum number of iterations is reached.

2.8. Comparison with Known QTLs and Haplotype Analysis

Pig QTLs based on *Sscrofa11.1* were downloaded from the PigQTL database. The information of QTLs for teat number-related traits was obtained using R (version 3.6.0), followed by comparing the significant SNPs and Indels with these QTLs using R (version 3.6.0). For the strongest significant SNP (rs322863105) for ADIFF on SSC17, SNPs with suggestive significant p-values (1/557,540, 1.79×10^{-6}) around rs322863105 were used for haplotype block analysis to evaluate the linkage disequilibrium (LD) patterns of selected SNPs within this region using Haploview version 4.2 [29]. The effects of this SNP on ADIFF were plotted using ggplot2 R (version 3.6.0) package.

2.9. Annotation of Candidate Genes and Functional Enrichment Analysis

Candidate genes, including or close to the significant SNPs and Indels, were annotated using the biomaRt [30] R (version 3.6.0) package based on *Sscrofa11.1*. Candidate genes located in 1-Mb regions of significant SNPs and Indels were also annotated. Gene ontology (GO) and Kyoto encyclopedia of genes and genomes (KEGG) pathway enrichment analyses and visualizations were performed using the clusterProfiler R (version 3.6.0) package [31].

3. Results

3.1. Genotyping and Phenotypic Statistics

After whole-genome sequencing, 2.76 TB of sequences were generated, with a mean coverage of 98.47% at an average of 9.66-fold depth for 100 Qingping pigs in this study (Table S1). Clean data were mapped to the pig reference genome (*Sscrofa*11.1), with 36,482,281 SNPs and 4,859,001 Indels being called with GATK. After filtering, 23,193,931 SNPs and 2,053,221 Indels, with a distribution roughly proportional to autosomal chromosomes of pigs, were retained for subsequent analyses (Figure 1).

Figure 1. The distribution of SNPs and Indels on autosomal chromosomes of pigs.

Table 1 shows the descriptive statistics of teat number-related traits of Qingping pigs. The average teat number [standard deviation (SD)] was seen to be 14.78 (0.97), ranging from 14 to 17, which was higher than the values of Beijing Black Pig (13.6) [9], Japanese Duroc (13.73) [3], American (10.90) and Canadian Duroc (10.92) [14], but lower than the value of Erhualian pigs (19.13) [7]. The meansvalues (SD) for the other teat number-related traits, were 7.36 (0.48), 7.42 (0.61), 7.50 (0.59), −0.06 (0.51) and 0.22 (0.46) for LTN, RTN, MAXAP, L-R and ADIFF, with coefficient of variation (CV) values of 6.55, 8.17, 6.56 and 7.93% for LTN, RTN, TNUM and MAXAP, respectively.

In Table 1, it was shown that the values of SNP-based heritability (h^2_{SNP}) for teat number-related traits were 0.29, 0.19, 0.36, 0.38, 0.00 and 0.24 for LTN, RTN, TNUM, MAXAP, L-R and ADIFF in Qingping pigs, respectively. Unfortunately, the standard errors (SE) were large, which reflects the low accuracy of the heritability estimates. Therefore, we compared the results of the teat number-related heritability estimates in Qingping pigs with those of other pig breeds. In Qingping pigs, the narrow-sense heritability for TNUM (0.36) (only considering the contribution of additive genetic effects) was consistent with the values of purebred Korean Yorkshire pigs (0.37) [4] and Duroc pigs (0.34 ± 0.05) [3]. In previous studies, the h^2_{SNP} values of LTN, RTN, MAXAP and L-R were 0.16, 0.26, 0.261 and 0.00, which were consistent with our results [17]. Unexpectedly, ADIFF had a medium h^2_{SNP} value (0.24), which was virtually null in previous studies [16,17]. These results suggest that estimates of the heritability of teat number-related traits in Qingping pigs were reliable, but caution is required due to the large standard errors.

Table 1. Summary of phenotypic data in terms of mean, standard deviation, minimum, maximum, coefficient of variance, and SNP-based heritability and standard error (SE) for each trait.

Trait	N	Mean	SD	Min	Max	C.V.	h^2_{SNP}	SE
LTN	100	7.36	0.48	7.00	8.00	6.55	0.29	0.21
RTN	100	7.42	0.61	6.00	9.00	8.17	0.19	0.21
TNUM	100	14.78	0.97	14.00	17.00	6.56	0.36	0.23
MAXAP	100	7.50	0.59	7.00	9.00	7.93	0.38	0.22
L-R	100	−0.06	0.51	−2.00	2.00	NA	0.00	0.20
ADIFF	100	0.22	0.46	0.00	2.00	NA	0.24	0.20

LTN: the number of teats on the left side; RTN: the number of teats on the right side; TNUM: the total number of teats (TNUM = LTN + RTN); MAXAP: the maximum number of teats in LTN and RTN (MAXAP); L-R: the difference between the two sides (L-R = LTN − RTN); ADIFF: the absolute difference between left and right teat number (ADIFF = | LTN − RTN |).

3.2. GLM GWAS for Teat Number-Related Traits

PCA results indicated that Qingping pigs could not be clustered into groups (Figure S1), so GWAS was performed using GLM with no principal component. The Bonferroni correction assumes that each of the tests is independent, an is thereby inherently conservative when considering SNPs in LD. We calculated the effectively independent tests based on the estimated number of independent markers [32]. A total of 557,540 SNP and 28,629 Indel independent tests were suggested, with the threshold p-value of 8.97×10^{-8} (0.05/557,540) for SNPs and 1.75×10^{-6} (0.05/28,629) for Indels.

Figures 2 and 3 show the Manhattan plots for LTN, RTN, TNUM, MAXAP, L-R and ADIFF, and Figure S2 presents the Q-Q plots for these traits. A total of 28 significant SNPs and 45 significant Indels were detected as peak associated variants for LTN, RTN, TNUM, MAXAP, L-R and ADIFF (Table 2). Among the 28 SNPs identified, 10 SNPs were located within 11 genes and the others (18 SNPs) were located at 3798 to 46,070,668 bp from the nearest genes (Table 2). Among the 45 Indels identified, 21 Indels were located within 22 genes and the others (24 Indels) were located 2660 to 137,303 bp from the nearest genes (Table 2).

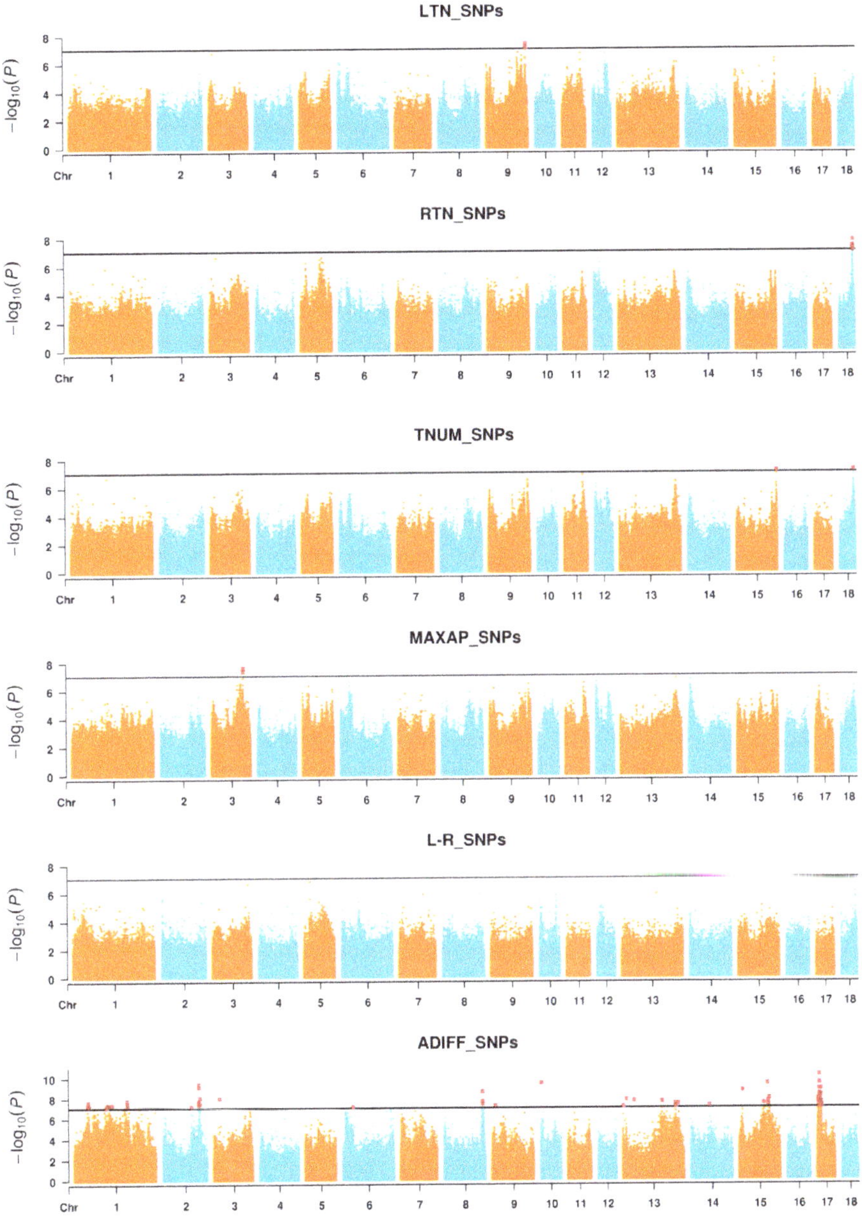

Figure 2. Manhattan plots of GLM GWAS for teat number-related traits in Qingping pigs, including LTN, RTN, TNUM, MAXAP, L-R and ADIFF based on SNPs. LTN: the number of teats on the left side; RTN: the number of teats on the right side; TNUM: the total number of teats (TNUM = LTN + RTN); MAXAP: the maximum number of teats in LTN and RTN (MAXAP); L-R: the difference between the two sides (L-R = LTN − RTN); ADIFF: the absolute difference between left and right teat number (ADIFF = | LTN − RTN |).

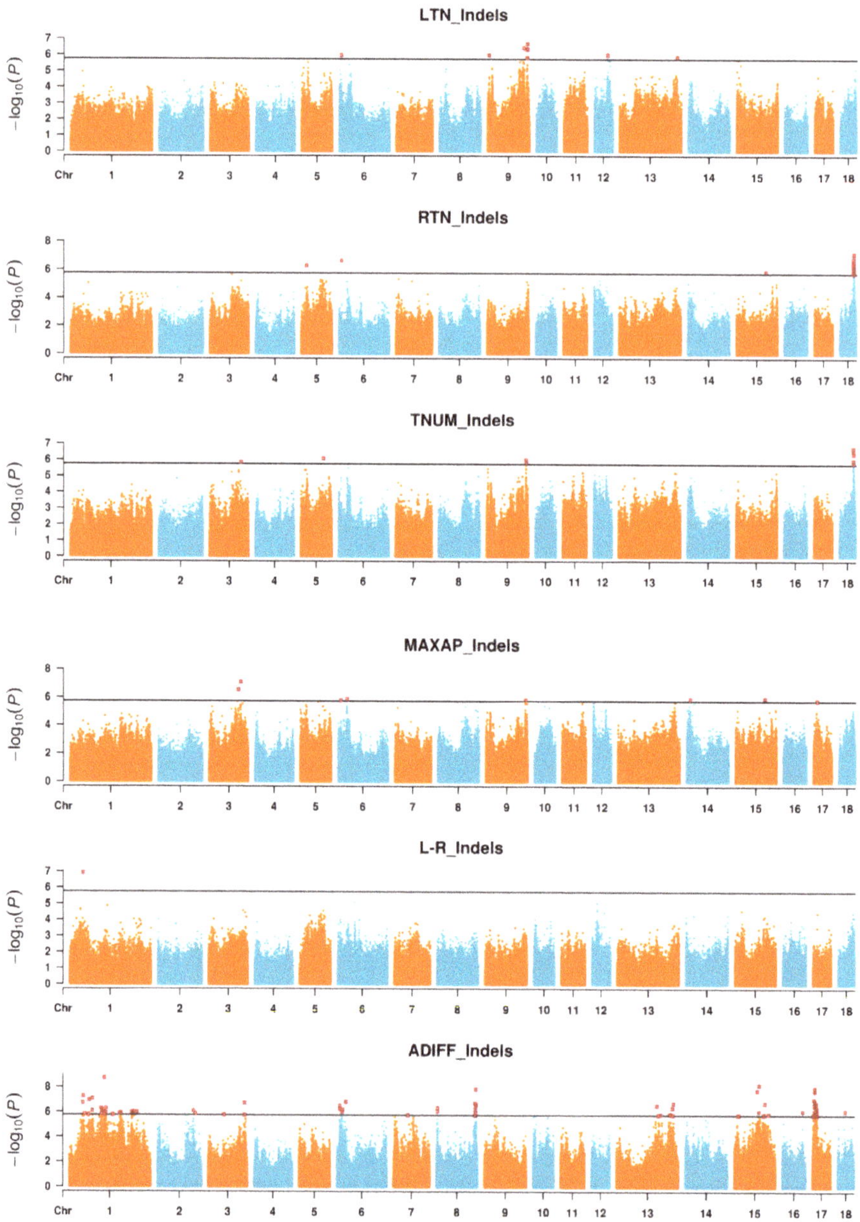

Figure 3. Manhattan plots of GLM GWAS for teat number-related traits in Qingping pigs, including LTN, RTN, TNUM, MAXAP, L-R and ADIFF based on Indels. LTN: the number of teats on the left side; RTN: the number of teats on the right side; TNUM: the total number of teats (TNUM = LTN + RTN); MAXAP: the maximum number of teats in LTN and RTN (MAXAP); L-R: the difference between the two sides (L-R = LTN − RTN); ADIFF: the absolute difference between left and right teat number (ADIFF = | LTN − RTN |).

Table 2. Significant SNPs and Indels of GLM GWAS for teat number-related traits.

SNP	Trait	SSC	Position	Effect	*p*-Value	QTLs *	Annotation	Gene (Distance from the Gene in bp)
chr9:130956708	LTN	9	130956708	0.64	3.85×10^{-8}		intronic	*PACC1(within), NENF(within)*
rs345573243	RTN	18	47399908	0.62	1.60×10^{-8}	24,290, 7470	intronic	*OSBPL3(within)*
rs703282466	MAXAP	3	106788730	0.45	2.29×10^{-8}	5224, 8797, 8798, 4250, 4256	intergenic	*LTBP1(63796), ENSSSCG00000050704(9889)*
chr15:137183045	TNUM	15	137183045	−0.79	7.02×10^{-8}	223,293	intronic	*MLPH(within)*
rs345573243	TNUM	18	47399908	0.96	6.25×10^{-8}	24,290, 7470	intronic	*OSBPL3(within)*
rs1108940033	ADIFF	1	45796983	0.61	2.19×10^{-8}		intergenic	*PHF3(500583),ENSSSCG00000049526(273685)*
rs321204530	ADIFF	1	177056841	0.59	1.44×10^{-8}	5223, 5255, 822, 845, 1250	intronic	*MDGA2(within)*
rs318957512	ADIFF	2	97849348	0.50	5.79×10^{-8}		intronic	*ADGRV1(within)*
rs326371568	ADIFF	2	123586226	0.62	3.87×10^{-10}	4255	intergenic	*FAM170A(119517),PRR16(613282)*
rs342451777	ADIFF	2	127224122	0.57	8.31×10^{-9}	4255	intergenic	*ENSSSCG00000042143(13463),ENSSSCG00000040936(55836)*
rs325963999	ADIFF	3	19301837	0.66	9.15×10^{-9}	5224, 7455, 7472	intronic	*KATNIP(within)*
rs326276043	ADIFF	6	34051848	0.55	6.98×10^{-8}	24,289	intergenic	*ENSSSCG00000050973(273191),CYLD(8093)*
rs338649298	ADIFF	8	129552162	0.59	2.22×10^{-9}		intergenic	*SNCA(163855),ENSSSCG00000043431(95940)*
rs321470648	ADIFF	9	10769337	0.45	5.38×10^{-8}		intergenic	*ENSSSCG00000046278(36887),ENSSSCG00000045225(26362)*
rs1109963100	ADIFF	10	2911179	0.80	3.37×10^{-10}		intergenic	*ENSSSCG00000042899(210778),BRINP3(52348)*
rs339887165	ADIFF	13	12135388	0.68	1.46×10^{-8}	7479	intergenic	*ENSSSCG00000044771(16544),ENSSSCG00000051554(31818)*
chr13:39266305	ADIFF	13	39266305	0.72	1.88×10^{-8}	7479	intronic	*DNAH12(within)*
rs701874475	ADIFF	13	134665423	−0.51	2.19×10^{-8}	7479	intergenic	*LMLN(8675),ENSSSCG00000050583(21280)*
rs338558804	ADIFF	13	179683904	−0.45	3.49×10^{-8}		intergenic	*ENSSSCG00000038062(120975),NRIP1(140958)*
rs343864506	ADIFF	13	187708682	−0.54	3.72×10^{-8}		intergenic	*ENSSSCG00000047308(444803),ENSSSCG00000050420(500018)*
rs334271954	ADIFF	14	62959726	0.52	5.70×10^{-8}		intergenic	*FAM13C(33877),SLC16A9(66937)*
rs1109225784	ADIFF	15	12581324	0.64	1.96×10^{-9}		intergenic	*U6(133156),U6(186459)*
rs326978910	ADIFF	15	84015934	0.38	3.38×10^{-8}	7468	intronic	*OSBPL6(within)*
rs334746473	ADIFF	15	97200323	0.75	4.23×10^{-10}	7468	intergenic	*ENSSSCG00000046205(186079),U2(688281)*
rs322863105	ADIFF	17	7610979	−0.88	6.01×10^{-11}		intergenic	*ENSSSCG00000045345(132678),ENSSSCG00000047202(438984)*
chr17:8221026	ADIFF	17	8221026	0.79	1.47×10^{-9}		intergenic	*U6(19092),FAT1(227734)*
rs330045817	ADIFF	17	8536301	−0.71	3.76×10^{-10}		ncRNA_intronic	*FAT1(within)*
rs324534432	ADIFF	17	13364668	0.79	1.47×10^{-9}		intergenic	*PSD3(100469),ENSSSCG00000046441(3798)*
Indel								
chr6:7472906	LTN	6	7472906	−0.41	1.06×10^{-6}	24,289	intronic	*CDYL2(within)*
rs695882779	LTN	9	3690507	0.39	9.84×10^{-7}		ncRNA_intronic	*ENSSSCG00000049604(within)*
chr9:119650540	LTN	9	119650540	−0.30	3.62×10^{-7}		intergenic	*ENSSSCG00000044083(337170),ENSSSCG00000050832(4698)*
rs792699200	LTN	9	130899869	0.51	4.41×10^{-7}		intronic	*PACC1(within)*
chr12:44292044	LTN	12	44292044	−0.37	9.55×10^{-7}	5227, 5261, 6472, 6479, 595, 2929	intronic	*NOS2(within)*
chr13:194617904	LTN	13	194617904	−0.31	1.30×10^{-6}		intergenic	*KRTAP11-1(111382),ENSSSCG00000047315(5813)*
rs709659410	RTN	5	17969771	0.42	5.28×10^{-7}	2927	intergenic	*KRT73(2682),KRT2(14544)*
chr6:9186279	RTN	6	9186279	0.89	2.28×10^{-7}	24,289	intronic	*WWOX(within)*

Table 2. *Cont.*

SNP	Trait	SSC	Position	Effect	*p*-Value	QTLs *	Annotation	Gene (Distance from the Gene in bp)
rs790747253	RTN	15	100810568	−0.61	1.28×10^{-6}	7468	intronic	*PGAP1(within)*
chr18:48316684	RTN	18	48316684	0.41	7.00×10^{-8}	24,290, 7470	intronic	*STK31(within)*
chr3:98429885	MAXAP	3	98429885	0.50	2.93×10^{-7}	5224, 8797, 8798	intergenic	*ENSSSCG00000045166(59604),ENSSSCG00000046007(449247)*
rs793312568	MAXAP	3	106792080	0.43	8.28×10^{-8}	5224, 8797, 8798, 4250, 4256	intergenic	*LTBP1(67153),ENSSSCG00000050704(6530)*
chr6:9186279	MAXAP	6	9186279	0.82	1.59×10^{-6}	24,289	intronic	*WWOX(within)*
chr6:30642639	MAXAP	6	30642639	0.43	1.31×10^{-6}	24,289	intergenic	*ENSSSCG00000047270(70049),ENSSSCG00000041426(189382)*
chr9:131996847	MAXAP	9	131996847	−0.49	1.46×10^{-6}		intronic	*ENSSSCG00000040650(within)*
chr14:12750600	MAXAP	14	12750600	0.43	1.31×10^{-6}		intronic	*HMBOX1(within)*
rs790747253	MAXAP	15	100810568	−0.60	1.17×10^{-6}	7468	intronic	*PGAP1(within)*
rs711984029	MAXAP	17	13041965	−0.47	1.68×10^{-6}		intronic	*PSD3(within)*
rs793312568	TNUM	3	106792080	0.63	1.41×10^{-6}	5224, 8797, 8798, 4250, 4256	intergenic	*LTBP1(67153),ENSSSCG00000050704(6530)*
chr5:75592729	TNUM	5	75592729	0.98	8.38×10^{-7}		intronic	*NELL2(within)*
rs792699200	TNUM	9	130899869	1.00	9.67×10^{-7}		intronic	*PACC1(within)*
chr18:48316684	TNUM	18	48316684	0.62	4.01×10^{-7}	24,290, 7470	intronic	*STK31(within)*
chr1:44096236	LR	1	44096236	0.69	1.25×10^{-7}		intergenic	*ENSSSCG00000042072(159101),ENSSSCG00000045405(51729)*
chr1:44973455	ADIFF	1	44973455	0.70	1.96×10^{-7}		intronic	*ZUP1(within),RSPH4A(within)*
chr1:46840905	ADIFF	1	46840905	0.71	5.46×10^{-8}		intergenic	*ENSSSCG00000050391(123491),U6(392319)*
chr1:65957275	ADIFF	1	65957275	0.43	1.14×10^{-7}		intergenic	*FBXL4(21578),FAXC(274880)*
chr1:77875820	ADIFF	1	77875820	0.42	8.57×10^{-8}		intergenic	*FYN(65243),U6(107381)*
chr1:118273285	ADIFF	1	118273285	−0.64	1.96×10^{-9}	5223, 6481, 5255	intergenic	*ENSSSCG00000049391(9019),ENSSSCG00000045826(5556)*
rs1113667849	ADIFF	2	123584099	0.55	8.50×10^{-7}	4255	intergenic	*FAM170A(117391),PRR16(615405)*
chr3:54080763	ADIFF	3	54080763	0.44	1.70×10^{-6}	5224, 7455, 7472, 6465	intergenic	*LONRF2(99093),REV1(54761)*
chr3:122749868	ADIFF	3	122749868	0.76	1.91×10^{-7}		intergenic	*LRATD1(17941),ENSSSCG00000045589(179131)*
chr6:9702570	ADIFF	6	9702570	0.61	3.15×10^{-7}	24,289	intronic	*WWOX(within)*
chr6:29725029	ADIFF	6	29725029	−0.41	1.54×10^{-7}	24,289	intergenic	*ENSSSCG00000034192(132519),CES5A(117783)*
chr7:48226837	ADIFF	7	48226837	0.67	1.74×10^{-6}	5257	intronic	*RASGRF1(within)*
chr8:3878010	ADIFF	8	3878010	−0.48	4.61×10^{-7}	7477	intronic	*ENSSSCG00000027349(within)*
chr8:131927486	ADIFF	8	131927486	0.68	1.46×10^{-8}		intergenic	*AFF1(2666),ENSSSCG00000032190(41799)*
chr13:134712100	ADIFF	13	134712100	−0.50	2.83×10^{-7}	7479	intergenic	*ENSSSCG00000050583(20384),OSBPL11(6327)*
chr13:188309252	ADIFF	13	188309252	0.35	4.51×10^{-7}		intergenic	*ENSSSCG00000050420(95234),ENSSSCG00000043493(500979)*
chr13:191714830	ADIFF	13	191714830	0.76	1.91×10^{-7}		intergenic	*ENSSSCG00000051384(248028),ENSSSCG00000048685(53357)*
chr15:12585471	ADIFF	15	12585471	0.67	1.74×10^{-6}		intergenic	*U6(137304),U6(182308)*
chr15:84277014	ADIFF	15	84277014	0.53	6.57×10^{-9}	7468	intergenic	*ENSSSCG00000036052(46519),ENSSSCG00000038561(122000)*
rs792656057	ADIFF	16	69550278	−0.60	8.69×10^{-7}	5228	intronic	*GRIA1(within)*
rs793561441	ADIFF	17	5622328	−0.33	9.14×10^{-8}		intronic	*PCM1(within)*
rs700363122	ADIFF	17	8173767	0.60	1.13×10^{-8}		intergenic	*ENSSSCG00000047202(116838),U6(28063)*
rs789477433	ADIFF	18	25541841	0.43	8.35×10^{-7}	24,290	intergenic	*ENSSSCG00000048651(270433),FAM3C(27695)*

* QTL number in PigQTL database. Effect means additive effect. LTN: the number of teats on the left side; RTN: the number of teats on the right side; TNUM: the total number of teats (TNUM = LTN + RTN); MAXAP: the maximum number of teats in LTN and RTN (MAXAP); L-R: the difference between the two sides (L-R = LTN − RTN); ADIFF: the absolute difference between left and right teat number (ADIFF = |LTN − RTN|).

Compared with known QTLs for teat number-related traits in the PigQTL database, 14 significant SNPs and 24 significant Indels overlapped with known QTLs (Table 2). Interestingly, Manhattan plots showed similar trends between SNPs and Indels. In Table 2, there were several significant SNPs and Indels were located at the same teat number-related traits QTLs in the PigQTL database. For RTN, significant SNP (rs345573243) and Indel (chr18:48316684) on SSC18 were located at QTL 7470 and QTL 24290, and coincidentally, rs345573243 and chr18:48316684 also showed significant associations with TNUM. For MAXAP, significant SNP (rs703282466) and Indel (rs793312568) on SSC3 were located at QTL 4250 and 4256, and coincidentally, rs793312568 also exhibited a significant association with TNUM. For ADIFF, significant SNPs (rs326371568 and rs342451777) and Indel (rs1113667849) on SSC2 were located at QTL 4255; significant SNP (rs701874475) and Indel (chr13:134712100) on SSC13 at QTL 7479; significant SNP (rs326978910) and Indel (chr15:84277014) on SSC15 at QTL 7468. Additionally, five new significant SNPs were closed to significant Indels, including rs1108940033 was closed to chr1:44973455 and chr1:46840905 on SSC1, rs338649298 was closed to chr8:131927486 on SSC8, rs343864506 was closed to chr13:188309252 on SSC13, rs1109225784 was closed to chr15:12585471 on SSC15, chr17:8221026 was closed to rs700363122 on SSC17.

3.3. *FarmCPU GWAS for ADIFF*

The GLM GWAS results of ADIFF showed significant associations even using whole genome SNPs or Indels to decide the thresholds (SNP: 0.05/2,319,3931; Indels: 0.05/2,053,221). However, Q-Q plots and genomic inflation factors (λ_{SNP} = 1.37, λ_{Indel} = 1.36) indicated possible false positives (Figure S2). FarmCPU, a powerful and efficient GWAS model, was used to control false positives and retain true positives. Q-Q plots and genomic inflation factors (λ_{SNP} = 0.98, λ_{Indel} = 0.86) were improved by FarmCPU (Figure S3). In Figure 4a and Table 3, 9 SNPs and 9 Indels were shown to be significantly associated variants for ADIFF on SSC1, 2, 3, 6, 8, 10, 11, 12, 13, 14, 15 and 17, with six SNPs and five Indels included in known QTLs (Table 3).

Three of the 9 significant SNPs were located within four genes and six SNPs were located 15,848 to 133,156 bp from the nearest genes (Table 3). Three of the 9 significant Indels were located within three genes and six Indels were located 3267 to 99,873 bp from the nearest genes (Table 3). Compared with the GLM GWAS results for ADIFF, 4 SNPs (rs325963999, rs1109963100, rs1109225784, rs322863105) and four Indels (chr1:77875820, chr1:118273285, chr15:84277014, rs700363122) were duplicated in the FarmCPU GWAS results, suggesting the reliability of the results (Table 3).

The strongest significant SNP in GLM for ADIFF on SSC17 (rs322863105, p-value = 6.01×10^{-11}) also showed significant association with ADIFF in FarmCPU (p-value = 3.34×10^{-13}). The effect of rs322863105 on the ADIFF was estimated by genotyping Qingping pigs for this SNP. Individuals with the TT genotype had a lower ADIFF, suggesting LTN and RTN were more symmetrical (Figure 4b). Linkage analysis of the suggestive significant SNPs around this SNP identified one haplotype block of 8 kb between rs338532551 and rs322792299, including rs322863105 (Figure 4c). Three annotated genes were contained in the 1-Mb region around rs322863105, including *tripartite motif family like 1* (*TRIML1*), *tripartite motif family like 2* (*TRIML2*), and *ZFP42 zinc finger protein* (*ZFP42*). Moreover, a peak Indel (rs700363122) close to this SNP showed significant associations with ADIFF in the results of both GLM (1.13×10^{-8}) and FarmCPU (9.10×10^{-16}), with two annotated genes in the 1-Mb region around this Indel, including *FAT Atypical Cadherin 1* (*FAT1*), and *Melatonin Receptor 1A* (*MTNR1A*).

Table 3. Significant SNPs and Indels of FarmCPU GWAS for ADIFF.

SNP	SSC	Position	Effect	*p*-Value	QTLs *	Annotation	Gene (Distance from the Gene in bp)
rs325963999 #	3	19301837	0.27	7.82×10^{-15}	5224, 7455, 7472	intronic	*KATNIP(within)*
rs693622708	6	39540583	0.17	8.58×10^{-9}	24,289	intergenic	*UQCRFS1(166710), ENSSSCG00000050718(41760)*
rs326134805	8	88543901	0.13	3.81×10^{-15}	7477, 1100	intergenic	*ENSSSCG00000044017(86061),SLC7A11(63983)*
rs1109963100 #	10	2911179	0.33	3.74×10^{-13}		intergenic	*ENSSSCG00000042899(210778),BRINP3(52348)*
rs1113875395	12	11463144	0.16	7.51×10^{-9}	5227, 1128	intergenic	*ABCA8(15848),ENSSSCG00000045738(20026)*
rs343773900	13	110154062	0.24	5.42×10^{-15}	7479	intronic	*PLD1(within)*
rs333970515	13	132158230	−0.18	3.99×10^{-13}	7479	ncRNA_exonic	*ENSSSCG00000047632(within)*
rs1109225784 #	15	12581324	0.32	1.46×10^{-16}		intergenic	*U6(133156),U6(186459)*
rs322863105 #	17	7610979	−0.34	3.34×10^{-13}		intergenic	*ENSSSCG00000045345(132678),ENSSSCG00000047202(438984)*
Indel							
chr1:77875820 #	1	77875820	0.17	1.02×10^{-7}		intergenic	*FYN(65243),U6(107381)*
chr1:118273285 #	1	118273285	−0.35	1.74×10^{-10}	5223, 6481, 5255	intergenic	*ENSSSCG00000049391(9019),ENSSSCG00000045826(5556)*
rs788352632	2	62598411	−0.12	3.81×10^{-10}	909	ncRNA_intronic	*ENSSSCG00000048292(within)*
chr11:6896071	11	6896071	−0.13	1.22×10^{-8}	5260	ncRNA_intronic	*ENSSSCG00000036846(within)*
rs787621311	13	110898097	0.23	2.27×10^{-7}	7479	intronic	*FNDC3B(within)*
rs701717756	14	38222283	0.26	9.42×10^{-10}		intergenic	*RBM19(17536),ENSSSCG00000042669(3247)*
chr14:91416556	14	91416556	−0.30	1.34×10^{-9}		intergenic	*ENSSSCG00000047278(204580),CXCL12(99865)*
chr15:84277014 #	15	84277014	0.22	2.92×10^{-9}	7468	intergenic	*ENSSSCG00000036052(46519),ENSSSCG00000038561(122000)*
rs700363122 #	17	8173767	0.40	9.10×10^{-16}		intergenic	*ENSSSCG00000047202(116838),U6(28063)*

* QTL number in in PigQTL database. # Duplicate signals between GLM and FarmCPU. Effect means additive effect. ADIFF: the absolute difference between left and right teat number (ADIFF = |LTN − RTN|).

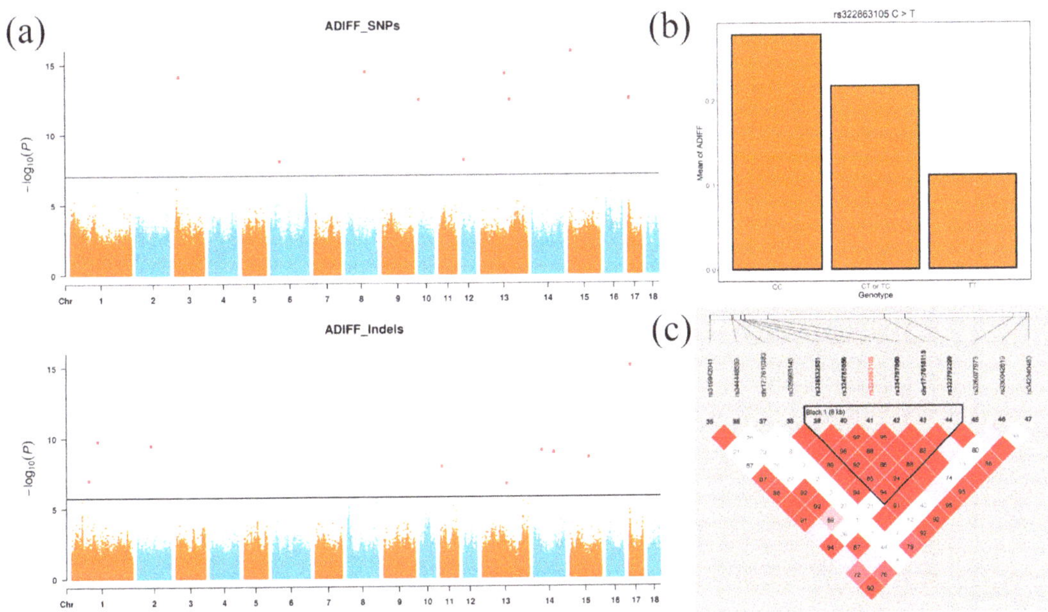

Figure 4. FarmCPU GWAS for ADIFF. (**a**) Manhattan plots of the GWAS based on SNPs and Indels. (**b**) Difference analysis of the strongest significant SNP (rs322863105) on SSC17 in GLM, which was retested in FarmCPU. (**c**) Haplotype block analysis of selected suggestive significant SNPs associated with ADIFF on SSC17 in GLM, including rs322863105. ADIFF: the absolute difference between left and right teat number (ADIFF = | LTN − RTN |).

3.4. Functional Enrichment of Candidate Genes

Annotated genes within 1-Mb regions of significant SNPs and Indels were defined as candidate genes. A total of 397 annotated genes were found in these regions (Table S2). In Figure 5a, GO enrichment analysis showed the enrichment of these candidate genes in epidermis development ($p = 2.31 \times 10^{-7}$), epidermal cell differentiation ($p = 4.61 \times 10^{-9}$), and skin development ($p = 1.12 \times 10^{-7}$). In Figure 5b KEGG pathway analysis revealed the enrichment of candidate genes in the pathways, such as the Sphingolipid signaling pathway ($p = 1.27 \times 10^{-3}$), ECM-receptor interaction ($p = 4.79 \times 10^{-3}$), and Glycine, serine and threonine metabolism ($p = 5.43 \times 10^{-3}$). Furthermore, we also paid attention to the Wnt signaling pathway (Figure 5c), due to its important role in initiating mammary morphogenesis and all subsequent stages of mammary formation as previously reported [33].

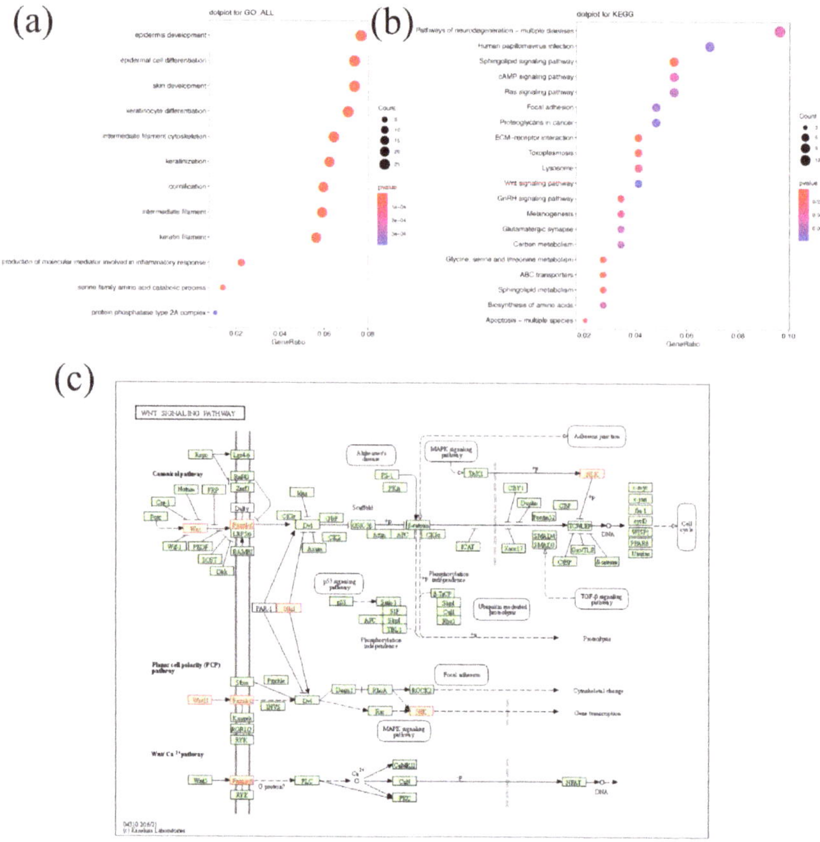

Figure 5. Enrichment results of candidate genes within 1-Mb regions of significant SNPs and Indels. (**a**) Dotplot of GO term enrichment. (**b**) Dotplot of KEGG pathway enrichment. (**c**) Wnt signaling pathway.

4. Discussion

During the first month after birth, piglets only have sow milk as a source of nutrients, which contributes to the regulation of their basal metabolism and temperature. Therefore, the sow's ability to produce milk can influence the health and growth of piglets, probably with a long-term effect post-weaning. The sows' lactation performance can be improved by enhancing the growth of the mammary gland and sows with a low prolactin/progesterone ratio before farrowing were reported to have a higher colostrum yield [34]. Additionally, milk production could be increased by adding L-arginine to the diets of lactating primiparous sows [35]. Moreover, for primiparous sows, teat suckling only for the first 2 days postpartum ensures the optimal mammary development and milk yield in the next lactation [36]. Furthermore, the lactating ability of sows can also be improved by increasing the teat number. As shown in the present study, teat number is heritable, with a genetic correlation to numerous microsatellite sites and SNPs (especially a QTL on SSC7).

The h^2_{SNP} of most teat number-related traits in Qingping pigs was moderate, except for L-R, suggesting it is feasible to improve teat number in pigs through genetic selection. SNPs and Indels associated with teat number-related traits might play an essential role in teat number improvement. Several candidate genes were reported to be related to mammary gland development and breast cancer. *CDYL2*, including an Indel significantly associated with LTN on SSC6, positively regulates breast cancer cell migration, invasion and epithelial-to-mesenchymal transition through p65/NF-κB and STAT3 [37]. *FAM3C*, which is in the 1-Mb region of an Indel on SSC18 and associated with ADIFF, encodes Interleukin-Like Epithelial-Mesenchymal Transition Inducer for the proliferation and migration of breast cancer cells [38]. *WWOX*, including an Indel (chr6:9186279) significantly associated with RTN, is known to play a role in breast cancer [39]. *TRIML2* and *MTNR1A* were close to SNP (rs330045817) and Indel (rs700363122), respectively, on SSC17. As mentioned above, these two different types of variants were close to each other and significantly associated with ADIFF in both GLM and FarmCPU models. *TRIML2* was significantly associated with prognosis, with a higher expression in triple-negative breast cancer cell lines than in normal mammary cell lines [40]. Common variants in *MTNR1A* may contribute to breast cancer susceptibility [41]. *TBX5*, encoding *T-Box Transcription Factor 5*, was close to the significant Indel (rs701717756) for ADIFF on SSC14. In a large German family, *TBX3* and *TBX5* duplication was reported to be associated with Ulnar-Mammary syndrome [42]. Importantly, *TBX3* was the placode marker required for the formation of mammary placodes and the development of fetal mammary glands in all mammals [43]. Although not located in the 1-Mb region of rs701717756, *TBX3* was close to this Indel with a distance of 625,516 bp.

Unlike earlier studies, the present study failed to detect significant association signals in genome regions around *VRTN* on SSC7, which was reported as a credible candidate gene for the teat number and the vertebra number [17,44]. Zhuang et al. suggested that the genetic heterogeneity of variants in *VRTN* may exist in different populations and *VRTN* may not be a strong or the only causal gene for teat number based on their finding that *VRTN* mutation was significantly associated with the teat number in Canadian Duroc pigs, but not in American Duroc pigs [14]. Moreover, *VRTN* mutation on SSC7 was also not significantly associated with the teat number in Chinese pig breeds, including Beijing Black pig and Sushan pig [8,9], but significantly associated with the vertebra number in Beijing Black pig. Furthermore, *VRTN* was reported to modulate somite segmentation [44]. These reports suggested that *VRTN* plays a more important role in vertebra number and varies in its role in teat number among populations with different genetic backgrounds. Therefore, the teat number in Qingping pigs is speculated to involve other genes or pathways.

The GO enrichment results showed significant enrichment in skin development, epidermis development, and epidermal cell differentiation. The mammary gland is an epithelial organ, and epithelial–stromal crosstalk is a key aspect of mammary morphogenesis [45]. In KEGG enrichment analysis, the Wnt signaling pathway did not reach a significance level of 0.05 (0.078). Interestingly, candidate genes, including *WNT11*, *WNT16* and *FZD3*, were located at key positions in this pathway (Figure 5c). *WNT16* and *FZD3* were also involved in GO terms, skin development and epidermis development. The Wnt signaling cascade is implicated in almost all stages of mammary development and is pivotal for the specification and morphogenesis of the mammary gland [46]. *WNT11* was expressed in stromal cells and basal cells in the adult mammary gland [46]. Chu et al. reported that *WNT11* and *FZD3* were expressed in mammary buds at E12.5 and E15.5 [47]. Importantly, Wnt signaling interacted with *TBX3* in mammary placode development [47]. These reports suggested that candidate genes might affect teat number during mammary gland morphogenesis. Unlike other breeds, the teat number of Qingping pigs showed a medium heritability (0.24) for ADIFF, implying the potential involvement of candidate genes in mammary gland morphogenesis. Therefore, our candidate genes related to mammary gland morphogenesis and development can be assumed to contribute to teat number improvement and even may influence milk production.

5. Conclusions

In this study, GLM and FarmCUP GWAS were carried out to detect associated SNPs and Indels for 6 teat number-related traits. We found a total of 33 SNPs and 50 Indels for teat number. The most significant SNP and Indel were located on SSC17. Six candidate genes were enriched in the Wnt signaling pathway with an important role in mammary gland morphogenesis and development. A novel candidate gene on SSC14, *TBX3*, was detected as a mammary placode marker. These findings contribute to our understanding of the genetic architecture of teat number and provide genetic markers for genetic improvement of teat number in Qingping pigs.

Supplementary Materials: The following supporting information can be downloaded at: https://www.mdpi.com/article/10.3390/ani12091057/s1. Figure S1: Principal component analysis of Qingping pigs; Figure S2: The Q-Q plots of the GLM GWAS for teat number-related traits in Qingping pigs, including LTN, RTN, TNUM, MAXAP, L-R and ADIFF. (a–f) The Q-Q plots of GLM GWAS based on SNPs. (g–l) The Q-Q plots of GLM GWAS based on Indels; Figure S3: The Q-Q plots of the FarmCPU GWAS for ADIFF in Qingping pigs. (a) The Q-Q plot of FarmCPU GWAS based on SNPs. (b) The Q-Q plot of FarmCPU GWAS based on Indels; Table S1: Sequencing reads, alignment statistics, and mean coverage for each sample in this study; Table S2: Annotated genes within 1-Mb regions of significant SNPs and Indels.

Author Contributions: Conceptualization, Z.L. and S.J.; methodology, Z.L. and H.L.; software, Z.L.; formal analysis, Z.L.; investigation, Z.L.; resources, Z.L. and Z.Z.; data curation, Z.L.; writing—original draft preparation, S.J.; writing—review and editing, Z.L.; visualization, S.J.; supervision, S.J.; project administration, S.J.; funding acquisition, S.J. All authors have read and agreed to the published version of the manuscript.

Funding: S.J. was funded by the National Key R & D Program of China (2021YFD1301200, 2017YFD0502000), China Agriculture Research System (CARS-36), The Fundamental Research Funds for the Central Universities (2662020DKPY012), Research Funds for Hubei Key Laboratory of Animal Embryo Engineering and Molecular Breeding (KLAEMB-2019-01) and Key R & D pro-jects of Hubei Province (2021BBA082, 2020BBB069, 2020ABA016).

Institutional Review Board Statement: The experimental protocol used in this study was reviewed and approved by the Animal Experimental Ethical Inspection of Laboratory Animal Centre, Huazhong Agriculture University. All sample collection was conducted under a permit (No. HZAUSW-2017-004) approved by the Attitude of the Animal Management and Ethics Committee.

Informed Consent Statement: Not applicable.

Data Availability Statement: All raw sequences for the 100 QP pigs have been deposited into the NCBI Sequence Read Archive under PRJNA489520 and will be available on 30 December 2023. Phenotypes and genotypes are deposited in OSF of Center for Open Science (https://osf.io/szu86/) (will be accessible on 30 December 2023).

Acknowledgments: We thank the Qingping pig Conservation Farm for providing samples and information. The computations in this paper were run on the bioinformatics computing platform of the National Key Laboratory of Crop Genetic Improvement, Huazhong Agricultural University.

Conflicts of Interest: The authors declare no conflict of interest.

References

1. Li, Y.; Pu, L.; Shi, L.; Gao, H.; Zhang, P.; Wang, L.; Zhao, F. Revealing New Candidate Genes for Teat Number Relevant Traits in Duroc Pigs Using Genome-Wide Association Studies. *Animals* **2021**, *11*, 806. [CrossRef] [PubMed]
2. Pig Quantitative Trait Locus (Qtl) Database (Pig Qtldb). Available online: https://www.animalgenome.org/cgi-bin/QTLdb/SS/index (accessed on 27 December 2021).
3. Arakawa, A.; Okumura, N.; Taniguchi, M.; Hayashi, T.; Hirose, K.; Fukawa, K.; Ito, T.; Matsumoto, T.; Uenishi, H.; Mikawa, S. Genome-Wide Association Qtl Mapping for Teat Number in a Purebred Population of Duroc Pigs. *Anim. Genet.* **2015**, *46*, 571–575. [CrossRef] [PubMed]
4. Lee, J.; Lee, S.; Park, J.E.; Moon, S.H.; Choi, S.W.; Go, G.W.; Lim, D.; Kim, J.M. Genome-Wide Association Study and Genomic Predictions for Exterior Traits in Yorkshire Pigs1. *J. Anim. Sci.* **2019**, *97*, 2793–2802. [CrossRef] [PubMed]
5. Sato, S.; Atsuji, K.; Saito, N.; Okitsu, M.; Sato, S.; Komatsuda, A.; Mitsuhashi, T.; Nirasawa, K.; Hayashi, T.; Sugimoto, Y.; et al. Identification of Quantitative Trait Loci Affecting Corpora Lutea and Number of Teats in a Meishan X Duroc F2 Resource Population. *J. Anim. Sci.* **2006**, *84*, 2895–2901. [CrossRef]
6. Hernandez, S.C.; Finlayson, H.A.; Ashworth, C.J.; Haley, C.S.; Archibald, A.L. A Genome-Wide Linkage Analysis for Reproductive Traits in F2 Large White X Meishan Cross Gilts. *Anim. Genet.* **2014**, *45*, 191–197. [CrossRef]
7. Wang, L.; Zhang, Y.; Zhang, T.; Zhang, L.; Yan, H.; Liu, X.; Wang, L. Genotyping by Sequencing Reveals a New Locus for Pig Teat Number. *Anim. Genet.* **2017**, *48*, 470–472. [CrossRef]

8. Zhou, L.; Zhao, W.; Fu, Y.; Fang, X.; Ren, S.; Ren, J. Genome-Wide Detection of Genetic Loci and Candidate Genes for Teat Number and Body Conformation Traits at Birth in Chinese Sushan Pigs. *Anim. Genet.* **2019**, *50*, 753–756. [CrossRef]
9. Niu, N.; Wang, H.; Shi, G.; Liu, X.; Liu, H.; Liu, Q.; Yang, M.; Wang, L.; Zhang, L. Genome Scanning Reveals Novel Candidate Genes for Vertebral and Teat Number in the Beijing Black Pig. *Anim. Genet.* **2021**, *52*, 734–738. [CrossRef]
10. Wada, Y.; Akita, T.; Awata, T.; Furukawa, T.; Sugai, N.; Inage, Y.; Ishii, K.; Ito, Y.; Kobayashi, E.; Kusumoto, H.; et al. Quantitative Trait Loci (Qtl) Analysis in a Meishan X Göttingen Cross Population. *Anim. Genet.* **2000**, *31*, 376–384. [CrossRef]
11. Jinghu, Z.; Xiong, Y.; Zuo, B.; Lei, M.; Jiang, S.; Li, F.; Zheng, R.; Li, J.; Xu, D. Detection of Quantitative Trait Loci Associated with Several Internal Organ Traits and Teat Number Trait in a Pig Population. *J. Genet. Genom.* **2007**, *34*, 307–314.
12. Bidanel, J.P.; Rosendo, A.; Iannuccelli, N.; Riquet, J.; Gilbert, H.; Caritez, J.C.; Billon, Y.; Amigues, Y.; Prunier, A.; Milan, D. Detection of Quantitative Trait Loci for Teat Number and Female Reproductive Traits in Meishan X Large White F2 Pigs. *Animal* **2008**, *2*, 813–820. [CrossRef] [PubMed]
13. Tan, C.; Wu, Z.; Ren, J.; Huang, Z.; Liu, D.; He, X.; Prakapenka, D.; Zhang, R.; Li, N.; Da, Y.; et al. Genome-Wide Association Study and Accuracy of Genomic Prediction for Teat Number in Duroc Pigs Using Genotyping-by-Sequencing. *Genet. Sel. Evol.* **2017**, *49*, 35. [CrossRef] [PubMed]
14. Zhuang, Z.; Ding, R.; Peng, L.; Wu, J.; Ye, Y.; Zhou, S.; Wang, X.; Quan, J.; Zheng, E.; Cai, G.; et al. Genome-Wide Association Analyses Identify Known and Novel Loci for Teat Number in Duroc Pigs Using Single-Locus and Multi-Locus Models. *BMC Genom.* **2020**, *21*, 344. [CrossRef] [PubMed]
15. Tang, J.; Zhang, Z.; Yang, B.; Guo, Y.; Ai, H.; Long, Y.; Su, Y.; Cui, L.; Zhou, L.; Wang, X.; et al. Identification of Loci Affecting Teat Number by Genome-Wide Association Studies on Three Pig Populations. *Asian-Australas. J. Anim. Sci.* **2017**, *30*, 1–7. [CrossRef]
16. Rohrer, G.A.; Nonneman, D.J. Genetic Analysis of Teat Number in Pigs Reveals Some Developmental Pathways Independent of Vertebra Number and Several Loci Which Only Affect a Specific Side. *Genet. Sel. Evol.* **2017**, *49*, 4. [CrossRef]
17. Moscatelli, G.; Dall'Olio, S.; Bovo, S.; Schiavo, G.; Kazemi, H.; Ribani, A.; Zambonelli, P.; Tinarelli, S.; Gallo, M.; Bertolini, F.; et al. Genome-Wide Association Studies for the Number of Teats and Teat Asymmetry Patterns in Large White Pigs. *Anim. Genet.* **2020**, *51*, 595–600. [CrossRef]
18. Li, H.; Durbin, R. Fast and Accurate Short Read Alignment with Burrows-Wheeler Transform. *Bioinformatics* **2009**, *25*, 1754–1760. [CrossRef]
19. Li, H.; Handsaker, B.; Wysoker, A.; Fennell, T.; Ruan, J.; Homer, N.; Marth, G.; Abecasis, G.; Durbin, R.; Subgroup Genome Project Data Processing. The Sequence Alignment/Map Format and Samtools. *Bioinformatics* **2009**, *25*, 2078–2079. [CrossRef]
20. DePristo, M.A.; Banks, E.; Poplin, R.; Garimella, K.V.; Maguire, J.R.; Hartl, C.; Philippakis, A.A.; del Angel, G.; Rivas, M.A.; Hanna, M.; et al. A Framework for Variation Discovery and Genotyping Using Next-Generation DNA Sequencing Data. *Nat. Genet.* **2011**, *43*, 491–498. [CrossRef]
21. Danecek, P.; Auton, A.; Abecasis, G.; Albers, C.A.; Banks, E.; DePristo, M.A.; Handsaker, R.E.; Lunter, G.; Marth, G.T.; Sherry, S.T.; et al. The Variant Call Format and Vcftools. *Bioinformatics* **2011**, *27*, 2156–2158. [CrossRef]
22. Chang, C.C.; Chow, C.C.; Tellier, L.C.; Vattikuti, S.; Purcell, S.M.; Lee, J.J. Second-Generation Plink: Rising to the Challenge of Larger and Richer Datasets. *Gigascience* **2015**, *4*, 7. [CrossRef] [PubMed]
23. Browning, B.L.; Zhou, Y.; Browning, S.R. A One-Penny Imputed Genome from Next-Generation Reference Panels. *Am. J. Hum. Genet.* **2018**, *103*, 338–348. [CrossRef] [PubMed]
24. Yang, J.; Benyamin, B.; McEvoy, B.P.; Gordon, S.; Henders, A.K.; Nyholt, D.R.; Madden, P.A.; Heath, A.C.; Martin, N.G.; Montgomery, G.W.; et al. Common Snps Explain a Large Proportion of the Heritability for Human Height. *Nat. Genet.* **2010**, *42*, 565–569. [CrossRef] [PubMed]
25. Team, R Core. *R: A Language and Environment for Statistical Computing*; R Foundation for Statistical Computing: Vienna, Austria, 2020.
26. Yin, L.; Zhang, H.; Tang, Z.; Xu, J.; Yin, D.; Zhang, Z.; Yuan, X.; Zhu, M.; Zhao, S.; Li, X.; et al. Rmvp: A Memory-Efficient, Visualization-Enhanced, and Parallel-Accelerated Tool for Genome-Wide Association Study. *Genom. Proteom. Bioinform.* **2021**, *19*, 619–628. [CrossRef] [PubMed]
27. Liu, X.; Huang, M.; Fan, B.; Buckler, E.S.; Zhang, Z. Iterative Usage of Fixed and Random Effect Models for Powerful and Efficient Genome-Wide Association Studies. *PLoS Genet.* **2016**, *12*, e1005767. [CrossRef] [PubMed]
28. Wang, Q.; Tian, F.; Pan, Y.; Buckler, E.S.; Zhang, Z. A Super Powerful Method for Genome Wide Association Study. *PLoS ONE* **2014**, *9*, e107684. [CrossRef]
29. Barrett, J.C.; Fry, B.; Maller, J.; Daly, M.J. Haploview: Analysis and Visualization of Ld and Haplotype Maps. *Bioinformatics* **2005**, *21*, 263–265. [CrossRef]
30. Durinck, S.; Spellman, P.T.; Birney, E.; Huber, W. Mapping Identifiers for the Integration of Genomic Datasets with the R/Bioconductor Package Biomart. *Nat. Protoc.* **2009**, *4*, 1184–1191. [CrossRef]
31. Yu, G.; Wang, L.G.; Han, Y.; He, Q.Y. Clusterprofiler: An R Package for Comparing Biological Themes among Gene Clusters. *OMICS J. Integr. Biol.* **2012**, *16*, 284–287. [CrossRef]
32. Nicodemus, K.K.; Liu, W.; Chase, G.A.; Tsai, Y.Y.; Fallin, M.D. Comparison of Type I Error for Multiple Test Corrections in Large Single-Nucleotide Polymorphism Studies Using Principal Components Versus Haplotype Blocking Algorithms. *BMC Genet.* **2005**, *6* (Suppl. 1), S78. [CrossRef]
33. Spina, E.; Cowin, P. Embryonic Mammary Gland Development. *Semin. Cell Dev. Biol.* **2021**, *114*, 83–92. [CrossRef] [PubMed]

34. Loisel, F.; Farmer, C.; van Hees, H.; Quesnel, H. Relative Prolactin-to-Progesterone Concentrations around Farrowing Influence Colostrum Yield in Primiparous Sows. *Domest. Anim. Endocrinol.* **2015**, *53*, 35–41. [CrossRef] [PubMed]
35. Kim, S.W.; Wu, G. Regulatory Role for Amino Acids in Mammary Gland Growth and Milk Synthesis. *Amino Acids* **2009**, *37*, 89–95. [CrossRef] [PubMed]
36. Farmer, C. Review: Mammary Development in Lactating Sows: The Importance of Suckling. *Animal* **2019**, *13*, s20–s25. [CrossRef] [PubMed]
37. Siouda, M.; Dujardin, A.D.; Barbollat-Boutrand, L.; Mendoza-Parra, M.A.; Gibert, B.; Ouzounova, M.; Bouaoud, J.; Tonon, L.; Robert, M.; Foy, J.P.; et al. Cdyl2 Epigenetically Regulates Mir124 to Control Nf-Kappab/Stat3-Dependent Breast Cancer Cell Plasticity. *iScience* **2020**, *23*, 101141. [CrossRef]
38. Yang, W.; Feng, B.; Meng, Y.; Wang, J.; Geng, B.; Cui, Q.; Zhang, H.; Yang, Y.; Yang, J. Fam3c-Yy1 Axis Is Essential for Tgfbeta-Promoted Proliferation and Migration of Human Breast Cancer Mda-Mb-231 Cells Via the Activation of Hsf1. *J. Cell. Mol. Med.* **2019**, *23*, 3464–3475. [CrossRef]
39. Pospiech, K.; Pluciennik, E.; Bednarek, A.K. Wwox Tumor Suppressor Gene in Breast Cancer, a Historical Perspective and Future Directions. *Front. Oncol.* **2018**, *8*, 345. [CrossRef]
40. Song, X.; Zhang, C.; Liu, Z.; Liu, Q.; He, K.; Yu, Z. Characterization of Cerna Network to Reveal Potential Prognostic Biomarkers in Triple-Negative Breast Cancer. *PeerJ* **2019**, *7*, e7522. [CrossRef]
41. Deming, S.L.; Lu, W.; Beeghly-Fadiel, A.; Zheng, Y.; Cai, Q.; Long, J.; Shu, X.O.; Gao, Y.T.; Zheng, W. Melatonin Pathway Genes and Breast Cancer Risk among Chinese Women. *Breast Cancer Res. Treat.* **2012**, *132*, 693–699. [CrossRef]
42. Cenni, C.; Andres, S.; Hempel, M.; Strom, T.M.; Thomas, E.; Davies, A.; Timoney, N.; Frigiola, A.; Logan, M.; Holder-Espinasse, M. Tbx3 and Tbx5 Duplication: A Family with an Atypical Overlapping Holt-Oram/Ulnar-Mammary Syndrome Phenotype. *Eur. J. Med. Genet.* **2021**, *64*, 104213. [CrossRef]
43. Slepicka, P.F.; Somasundara, A.V.H.; Santos, C.O.D. The Molecular Basis of Mammary Gland Development and Epithelial Differentiation. *Semin. Cell Dev. Biol.* **2021**, *114*, 93–112. [CrossRef] [PubMed]
44. Duan, Y.; Zhang, H.; Zhang, Z.; Gao, J.; Yang, J.; Wu, Z.; Fan, Y.; Xing, Y.; Li, L.; Xiao, S.; et al. Vrtn Is Required for the Development of Thoracic Vertebrae in Mammals. *Int. J. Biol. Sci.* **2018**, *14*, 667–681. [CrossRef] [PubMed]
45. Gjorevski, N.; Nelson, C.M. Integrated Morphodynamic Signalling of the Mammary Gland. *Nat. Rev. Mol. Cell Biol.* **2011**, *12*, 581–593. [CrossRef] [PubMed]
46. Yu, Q.C.; Verheyen, E.M.; Zeng, Y.A. Mammary Development and Breast Cancer: A Wnt Perspective. *Cancers* **2016**, *8*, 65. [CrossRef] [PubMed]
47. Chu, E.Y.; Hens, J.; Andl, T.; Kairo, A.; Yamaguchi, T.P.; Brisken, C.; Glick, A.; Wysolmerski, J.J.; Millar, S.E. Canonical Wnt Signaling Promotes Mammary Placode Development and Is Essential for Initiation of Mammary Gland Morphogenesis. *Development* **2004**, *131*, 4819–4829. [CrossRef]

animals MDPI

Article

Genome-Wide Association Study and Identification of Candidate Genes for Intramuscular Fat Fatty Acid Composition in Ningxiang Pigs

Qinghua Zeng [1,†], Hu Gao [1,2,†], Shishu Yin [1], Yinglin Peng [3], Fang Yang [1], Yawei Fu [1,2], Xiaoxiao Deng [2], Yue Chen [2], Xiaohong Hou [2], Qian Wang [1,2], Zhao Jin [1,2], Gang Song [1,2], Jun He [1], Yulong Yin [1,2,4,*] and Kang Xu [2,4,*]

[1] Animal Nutrition Genome and Germplasm Innovation Research Center, College of Animal Science and Technology, Hunan Agricultural University, Changsha 410128, China; chuweixiang168168@163.com (Q.Z.); gaohu_20190008@163.com (H.G.); yinshishu2019@126.com (S.Y.); y621829@163.com (F.Y.); fuyw2020@163.com (Y.F.); wangq0130@163.com (Q.W.); jz2725552543@163.com (Z.J.); sg19971109@163.com (G.S.); hejun@hunau.edu.cn (J.H.)

[2] Laboratory of Animal Nutrition Physiology and Metabolism, The Institute of Subtropical Agriculture, Chinese Academy of Sciences, Changsha 410125, China; 2017102040003@whu.edu.cn (X.D.); 18837025618@126.com (Y.C.); 17865669807@163.com (X.H.)

[3] Hunan Institute of Animal & Veterinary Science, Changsha 410131, China; 13907487646@126.com

[4] Guangdong Laboratory for Lingnan Modern Agriculture, Guangzhou 510642, China

* Correspondence: yinyulong@isa.ac.cn (Y.Y.); xukang2020@163.com (K.X.)

† These authors contributed equally to this work.

Simple Summary: In this study, we conducted a comprehensive investigation of the fatty acid composition in Ningxiang pigs using a genome-wide association study. Our findings revealed a combination of previously reported and novel candidate genes associated with saturated fatty acids (SFAs), monounsaturated fatty acids (MUFAs), and polyunsaturated fatty acids (PUFAs). Notably, we identified significant single-nucleotide polymorphisms (SNPs) that are closely linked to specific fatty acids, and some of these genes explained substantial phenotypic variance. These noteworthy discoveries have the potential to significantly improve meat quality and fat deposition in Ningxiang pigs through targeted breeding approaches. Our research provides valuable insights into the intricate composition of fatty acids, thus offering practical implications for elevating meat quality and ultimately benefiting both the pig industry and consumers. The significance of this study is underscored by its potential to drive positive changes in society by promoting healthier and superior-quality pork products.

Abstract: Ningxiang pigs exhibit a diverse array of fatty acids, making them an intriguing model for exploring the genetic underpinnings of fatty acid metabolism. We conducted a genome-wide association study using a dataset comprising 50,697 single-nucleotide polymorphisms (SNPs) and samples from over 600 Ningxiang pigs. Our investigation yielded novel candidate genes linked to five saturated fatty acids (SFAs), four monounsaturated fatty acids (MUFAs), and five polyunsaturated fatty acids (PUFAs). Significant associations with SFAs, MUFAs, and PUFAs were found for 37, 21, and 16 SNPs, respectively. Notably, some SNPs have significant PVE, such as ALGA0047587, which can explain 89.85% variation in Arachidic acid (C20:0); H3GA0046208 and DRGA0016063 can explain a total of 76.76% variation in Elaidic Acid (C18:1n-9(t)), and the significant SNP ALGA0031262 of Arachidonic acid (C20:4n-6) can explain 31.76% of the variation. Several significant SNPs were positioned proximally to previously reported genes. In total, we identified 11 candidate genes (*hnRNPU, CEPT1, ATP1B1, DPT, DKK1, PRKG1, EXT2, MEF2C, IL17RA, ITGA1* and *ALOX5*), six candidate genes (*ALOX5AP, MEDAG, ISL1, RXRB, CRY1*, and *CDKAL1*), and five candidate genes (*NDUFA4L2, SLC16A7, OTUB1, EIF4E* and *ROBO2*) associated with SFAs, MUFAs, and PUFAs, respectively. These findings hold great promise for advancing breeding strategies aimed at optimizing meat quality and enhancing lipid metabolism within the intramuscular fat (IMF) of Ningxiang pigs.

Keywords: *longissimus dorsi*; saturated fatty acids; monounsaturated fatty acids; polyunsaturated fatty acids; GWAS

1. Introduction

Ningxiang is a famous breed of pig in China, with a history spanning more than 1000 years. As an obese breed, Ningxiang is superior to lean meat breeds in terms of intramuscular fat (IMF). Its market weight is approximately 74 kg, and the carcass slaughter rate is approximately 74 percent [1]. Ningxiang pork products are popular due to their unique flavor and nutritional value. Research has shown that the flavor and nutritional value of meat are closely related to the IMF content and fatty acid composition [2–4].

Fatty acids play a crucial role in the flavor of pork as fat-soluble flavor precursors [5,6]. Oleic acid (C18:1) is a major monounsaturated fatty acid (MUFA) in the fatty acid composition and is the most abundant, accounting for approximately 40% of the total fatty acid content. C18:1 in beef fat provides the meat with good tenderness, flavor, and antioxidant capacity [7,8]. Moreover, the type and content of fatty acids in meat diets can impact human health. For example, linoleic acid aids in slowing the progression of atherosclerosis [9]. Furthermore, the ratios of n-3, n-6, and other polyunsaturated fatty acids (PUFAs) are closely related to human diseases, such as cardiovascular disease and depression, as well as growth and development [10–13]. In summary, the fatty acid profile of pork is a crucial evaluation criterion for meat quality.

Genome-wide association studies (GWAS) can efficiently and accurately identify candidate genes related to target traits using single nucleotide polymorphisms (SNPs) as genetic molecular markers. In recent years, researchers have discovered numerous candidate genes associated with porcine IMF deposition and fatty acid composition through GWAS. A GWAS of the IMF in the Italian White breed by Davoli et al. revealed seven new SNPs, and three new candidate genes were annotated. These genes were not related to the IMF content [14]. Van et al. [15] used data from the Axiom pig 660K array to conduct a GWAS on the IMF of 454 Duroc and 659 Landrace boars and identified two quantitative trait locus (QTL) regions for newly synthesized fatty acid traits on SSC4 and SSC14 in Duroc pigs. Viterbo et al. [16] performed a GWAS on the fatty acid composition of 480 purebred Duroc pigs, identifying 25, 29, and 16 SNP loci significantly associated with stearic acid, oleic acid, and saturated fatty acids (SFAs), respectively. The genetic factors contributing to the fatty acid composition of Ningxiang pigs remain unknown. This study aimed to examine significant candidate genes for specific fatty acid compositions and explore potential biological pathways by conducting a GWAS on the fatty acid composition of Ningxiang pigs.

2. Materials and Methods

2.1. Animal Harvest and Sample Collection

The feeding and dietary conditions of all Ningxiang pigs were the same, and samples were collected in two batches (July 2019 and August 2020). *Longissimus dorsi* samples (taken from the 12th ribs) were collected from 691 Ningxiang pigs that were slaughtered at a predetermined age (180 ± 5 days age) at the Ningxiang Chu Weixiang Slaughterhouse and Meat Processing, LLC (Ningxiang City, Hunan Province, China). A sample of approximately 2 × 1 × 1 cubic centimeters was quickly taken and stored in a self-sealing bag with dry ice for IMF measurement. Additionally, the samples used for DNA extraction were simultaneously stored in a liquid nitrogen tank.

2.2. Determining Intramuscular Fat and Fatty Acids

In this study, we used Soxhlet extraction to measure the IMF contents of the 691 *longissimus dorsi* samples according to the standard "Meat and Meat Products-Determination of Free Fat Content" (GB/T 9695.1-2008) [17]. The composition of fatty acids was determined via gas chromatography (GC), with the specific procedure detailed below. First, 0.5 g of sample powder was accurately weighed, dried to a constant weight and placed in a 10 mL centrifuge tube. Next, 2 mL of a benzene and petroleum ether mixture (1:1 by volume) was added, and the tube was wrapped in tin foil and left in a dark place for 24 h. Then, 2 mL of a 0.4 mol/L KOH methanol solution was added for methylation. The sample was shaken well and left to stand for 15 min. Double-distilled water was added to a volume of 10 mL, and the sample was centrifuged at 10,000 rpm for 10 min to obtain 100 μL of supernatant. The supernatant was finally diluted with hexane. A gas chromatograph (Agilent 7890A, Santa Clara, CA, USA) was used to determine the content of medium- to long-chain fatty acids. The GC analysis conditions were as follows: the chromatographic column was an SP-2560 (100 m × 0.25 mm, 0.20 μm) capillary column, and high-purity nitrogen was used as the carrier gas. The heating program was as follows: initial temperature of 140 °C maintained for 5 min, then increased to 240 °C at 4 °C/min; sample inlet temperature of 260 °C; flame ionization detector (FID) temperature of 260 °C; split ratio of 100:1; and injection volume of 1 μL. A total of 25 fatty acids were detected (Table S1), and only those present in more than 80% of individuals (N $\geq$ 533) were retained (Table 1).

Table 1. Detailed information on 14 fatty acids present in the IMF of the *longissimus dorsi* of Ningxiang pigs.

Fatty Acid	N	Relative Content (±SE) (%)	h^2 (±SE)
IMF (Intramuscular fat)	691	3.65 (±0.04)	0.72 (±0.02)
SFA (Saturated fatty acid)	691	39.35 (±0.38)	0.61 (±0.04)
C14:0 (Myristic acid)	649	1.42 (±0.01)	0.80 (±0.03)
C16:0 (Palmitic acid)	651	26.38 (±0.06)	0.76 (±0.03)
C17:0 (Margaric acid)	645	0.15 (±0.001)	0.45 (±0.08)
C18:0 (Stearic acid)	651	13.42 (±0.06)	0.89 (±0.02)
C20:0 (Arachidic acid)	649	0.21 (±0.002)	0.87 (±0.04)
MUFA (Monounsaturated fatty acid)	691	41.88 (±0.42)	0.84 (±0.02)
C16:1 (Palmitoleic acid)	650	3.28 (±0.40)	0.87 (±0.03)
C18:1n-9(t) (Elaidic acid)	627	0.11 (±0.001)	0.37 (±0.09)
C18:1n-9(c) (Oleic acid)	651	40.28 (±0.12)	0.74 (±0.03)
C20:1 (Eicosenoic acid)	650	0.80 (±0.01)	0.78 (±0.03)
PUFA (Polyunsaturated fatty acid)	691	12.78 (±0.18)	0.60 (±0.03)
C18:2n-6(c) (Linoleic acid)	651	10.12 (±0.10)	0.61 (±0.03)
C18:3n-3 (α-Linolenic acid: ALA)	572	0.36 (±0.01)	0.63 (±0.08)
C20:2 (Eicosa-11,14-dienoic acid)	649	0.33 (±0.003)	0.67 (±0.05)
C20:3n-6 (Dihomo-γ-linolenic acid)	647	0.38 (±0.01)	0.88 (±0.03)

Table 1. *Cont.*

Fatty Acid	N	Relative Content (±SE) (%)	h^2 (±SE)
C20:4n-6 (Arachidonic acid)	649	2.42 (±0.04)	0.55 (±0.05)

N is the number of individuals with this fatty acid in the population. Relative content is the mean value of the content, and SE is the standard error. h^2 is heritability.

2.3. DNA Extraction, Genotyping and Quality Control

2.3.1. DNA Extraction and Genotyping

Genomic DNA was extracted from muscle tissue using the standard phenol-chloroform method, and the DNA was dissolved in TE buffer. The concentration and purity of the DNA samples were measured using a Nanodrop One spectrophotometer (Thermo Fisher Scientific, Waltham, MA, USA). Samples with an $A_{260/280}$ ratio of 1.7~2.0 were genotyped using the GeneSeek Genomic Profiling (GGP) version 2 Porcine 50K SNP chip (Neogen Corporation, Lincoln, NE, USA), which comprises 50,697 SNP loci.

2.3.2. Quality Control and Genotype Imputation

First, SNPs located on the X and Y chromosomes and unknown or duplicate locations were removed. To decrease the missing genotype rate, Beagle 5.1 software [18] was employed to impute the remaining 38,817 SNPs. Then, quality control was conducted using PLINK v1.9 [19] with the following criteria: (1) SNP call rate $\geq$ 90%; (2) minor allele frequency (MAF) $\geq$ 1%; and (3) Hardy–Weinberg equilibrium (HWE) testing p value $\geq 10^{-6}$. After quality control, 12,201 SNPs were removed due to missing genotype rates, MAFs, and HWE. Finally, 691 individuals and 25,809 SNPs remained for subsequent analysis (Figure 1).

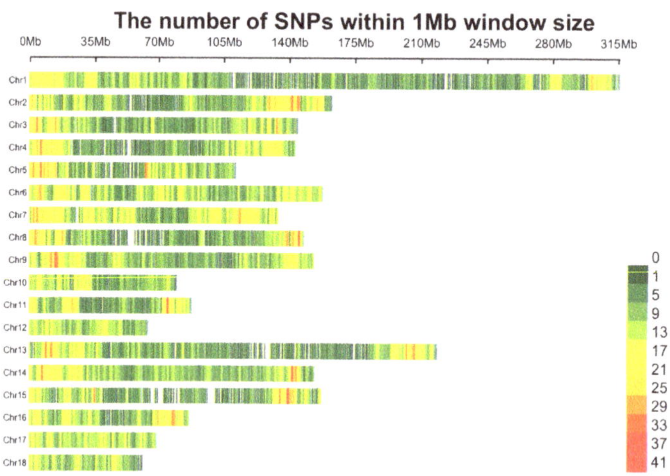

Figure 1. Distribution plot of SNPs after quality control.

2.4. Population Stratification

To mitigate the risk of concealed population stratification leading to spurious results in the GWAS, we conducted principal component analysis (PCA) using imputed genotypes (25,809 SNPs) with PLINK v1.9 (command: --pca). As depicted in Figure 2a, the population exhibited significant population stratification, necessitating the incorporation of the first principal components (PCs) for correction.

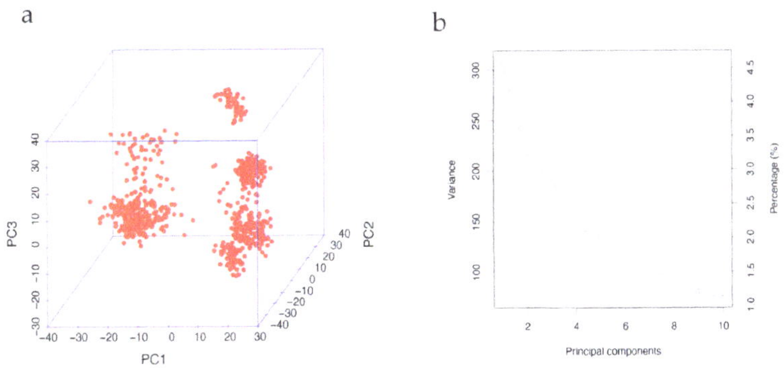

Figure 2. Principal component analysis. (**a**) Visualization of the first three PC values showing the existence of population stratification. (**b**) Screen plot of the first 10 PC values. Decreasing trends indicate that the first five PCs can be appropriately used to correct the population stratification.

2.5. Genome-Wide Association Study

The association between SNPs and fatty acids was examined using the BLINK model (Bayesian-information and Linkage-disequilibrium Iteratively Nested Keyway) using the GAPIT 3.0 package [20]. BLINK, an enhanced version of FarmCPU, enhances statistical power by relaxing the assumption of even distribution of trait-related genes across the genome and incorporates the Bayesian information criterion (BIC) in fixed effects models to improve computational efficiency [21]. The model integrates Equations (1)–(3) according to the BIC strategy, iteratively calculating and excluding all pseudo-quantitative trait nucleotides (QTNs) to identify significant loci.

$$y = Xa + Zb + e \tag{1}$$

$$y = Xa + Zb + Qp + e \tag{2}$$

$$y = Xa + Qp + e \tag{3}$$

where y is a vector of phenotypic data; a is the vector of fixed effects or covariates, including IMF content and the first five PCs; b is a vector of marker effects; p is the effect of pseudo-QTNs; X, Z, and Q are the incidence matrices corresponding to a, b, and p, respectively; and e is the vector of residual errors. The BLINK model uses Equation (1) to define pseudo-QTNs as a covariate for Equation (2). The SNP obtained from Equation (2) determines the information of QTNs according to linkage disequilibrium (LD) and then employs Equation (3) to perform accuracy detection of QTNs using the BIC strategy. The false discovery rate (FDR) method of multiple testing, as described by Benjamini-Hochberg, was utilized to measure the statistical significance of association studies at a genome-wide level. The cut-off for considering SNPs as significant was set at FDR ≤ 0.1. The phenotypic variance explained (PVE) by genetic effects was calculated as follows [22]:

$$PVE = \frac{2\alpha^2 MAF(1 - MAF)}{2\alpha^2 MAF(1 - MAF) + (se(\alpha^2)2N \times MAF(1 - MAF)} \tag{4}$$

where MAF is the minor allele frequency for the SNP, α^2 is the effect of the SNP marker, N is the sample size, and $se(\alpha^2)$ is the standard error of α^2.

2.6. Estimation of Heritability and Genetic Correlation

Heritability (h^2) was estimated using the following Formula (5) in HIBLUP [23]:

$$h^2 = \frac{\sigma_a^2}{\sigma_a^2 + \sigma_e^2} \quad (5)$$

where σ_a^2 and σ_e^2 are the additive genetic variance and the residual variance, respectively. The phenotypic and genetic correlation (r_p and r_g) between IMF and FAs were estimated using HIBLUP software [23].

$$r_P = \frac{COV_{P_{xy}}}{\sqrt{\sigma_{P_x}^2 \times \sigma_{P_y}^2}}; \; r_G = \frac{COV_{G_{xy}}}{\sqrt{\sigma_{G_x}^2 \times \sigma_{G_y}^2}} \quad (6)$$

where $COV_{P_{xy}}$ and $COV_{G_{xy}}$ represent the phenotypic and genetic covariance, respectively. σ_P^2 and σ_G^2 are the phenotypic and genetic variance, respectively.

2.7. Identification of Candidate Genes

Candidate genes were identified based on their physical positions and functions according to the Sus scrofa 10.2 reference genome assembly. The SNP-containing or nearest annotated genes for each potential SNP were obtained from the Sus scrofa (10.2) gtf file (http://ftp.ensembl.org/pub/release-80/gtf/sus_scrofa/Sus_scrofa.Sscrofa10.2.80.gtf.gz accessed on 1 May 2015) and taken as candidate genes.

2.8. Functional Enrichment Analysis

The g:Profiler website [24] was used for Gene Ontology (GO) and Kyoto Encyclopedia of Genes and Genomes (KEGG) enrichment analysis. A tailor-made algorithm was chosen for multiple testing correction (adjusted p value < 0.05).

3. Results

3.1. Descriptive Statistics for IMF and Fatty Acid Composition

The statistical analysis results of IMF and 25 fatty acids are presented in Table S1. The IMF content in the *longissimus dorsi* of Ningxiang pigs was determined to be 3.65%. In the overall fatty acid distribution, MUFAs had the highest proportion at 41.88%, followed by SFAs at 39.35%, and PUFAs at 12.78%. Eleven fatty acids, including C22:6n-3, C17:1, and C15:0, were removed due to a sample size below 553 individuals (<80%). The remaining 14 fatty acids and IMF were used for subsequent analysis (Table 1). Among these fatty acids, oleic acid (C18:1n-9(c)) exhibited the highest content (40.28%), while elaidic acid (C18:1n-9(t)) had the lowest content (0.11%).

3.2. Estimation of Genetic Parameters

The results of the correlation analysis between fatty acids and IMF are presented in Figure 3a,b. In the phenotypic correlation analysis, the majority of fatty acids, except for SFA, C20:0, and C18:3n-3, exhibited a significant correlation with IMF ($p < 0.001$). SFA demonstrated significant positive phenotypic correlations with MUFAs and PUFAs ($p < 0.001$) but showed negative genetic correlations with four MUFAs and five PUFAs ($p < 0.05$). Furthermore, SFA, MUFA, and PUFA were significantly positively correlated with five SFAs, four MUFAs, and five PUFAs, respectively ($p < 0.05$). In the genetic correlation analysis, C16:1 had the highest correlation with IMF (r_g = 0.64), and SFA was negatively correlated with most MUFAs and PUFAs, except C16:1 and C18:1n-9(t). All fatty acids had moderate to high heritabilities (0.37~0.89). Fatty acids C17:0 and C18:1n-9(t) had the lowest heritability estimates (0.45 and 0.37, respectively). Most fatty acids exhibited heritability estimates above 0.6. Notably, C18:0 exhibited the highest heritability estimate of 0.89 (Table 1).

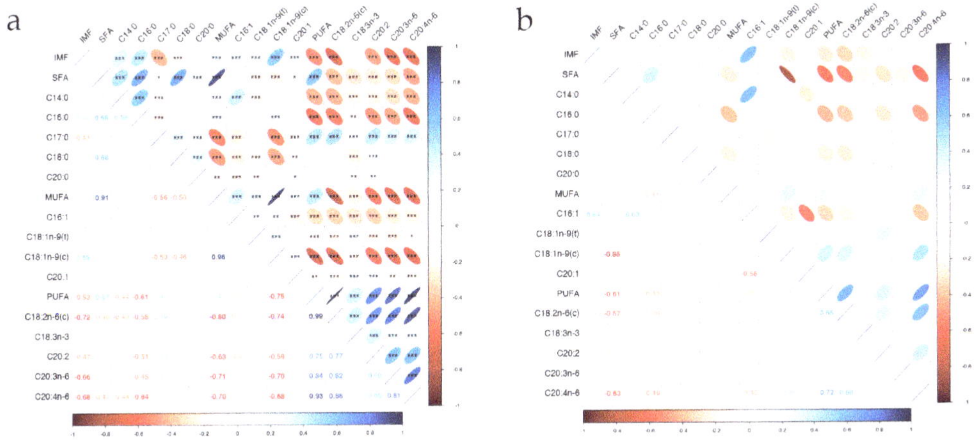

Figure 3. Correlations among IMF and 14 fatty acids in the *longissimus dorsi* of Ningxiang pigs. "***" represents p value < 0.001, "**" represents p value < 0.01, "*" represents p value < 0.05. (**a**) Phenotypic correlation; (**b**) genetic correlation.

3.3. Genome-Wide Association Study Results for Fatty Acids

After quality control, 25,809 SNPs for 691 Ningxiang pigs were retained for the GWAS. In total, 74 genome-wide level SNPs were identified for 14 FAs in this study.

3.3.1. SFA

Thirty-seven genome-wide significant SNPs were identified for five SFAs (C14:0, C16:0, C17:0, C18:0, and C20:0). C20:0 had the most loci, which were located on SSC2, SSC3, SSC4, SSC5, SSC7, SSC8, SSC13, SSC14, and SSC16 (Figure S1). Moreover, some loci, such as ALGA0047587 (89.85%), ASGA0059505 (11.39%), and DRGA0011206 (6.85%) (Table S2), explained a large portion of the phenotypic variance.

3.3.2. MUFA

The Manhattan plots showed that 21 genome-wide significant loci were identified on 12 chromosomes (SSC1, SSC2, SSC3, SSC4, SSC5, SSC7, SSC9, SSC11, SSC13, SSC14, SSC16, and SSC17) for four MUFAs (Figure S2). H3GA0046208 explained the largest portion of the phenotypic variance (45.24%) for C18:1n-9(t) (Table S2).

3.3.3. PUFA

The Manhattan plots showed that 16 genome-wide significant loci on seven chromosomes (SSC1, SSC2, SSC5, SSC8, SSC9, SSC13, and SSC16) were identified for five PUFAs (Figure S3).

3.4. Identification of Candidate Genes

Four hundred and fifty-three genes were identified within a 500 kb region upstream and downstream of the significant SNPs (Table S2). For SFAs, 354 genes were found in a 1 Mb genomic region; 40 genes were close to 37 loci, of which 11 SNPs (WU_10.2_3_116903421, ASGA0074106, WU_10.2_4_119395133, M1GA0024654, WU_10.2_3_142168876, ALGA0010606, WU_10.2_11_3591593, ALGA0080940, H3GA0041501, MARC0054269, and WU_10.2_16_59778879) were located within 11 genes (*ALK*, *MFAP3*, *CEPT1*, *NPEPPS*, *ENSSSCG00000008655*, *SWI5*, *CDK8*, *AVPI1*, *ALOX5*, *ITGA1*, and *SLIT3*). For MUFAs, six SNPs (ALGA0015731, ASGA0064960, H3GA0025990, H3GA0046208, WU_10.2_5_13180559, and ASGA0031521) were intragenic variants. For PUFAs, 16 SNP loci were identified in 16 genes.

3.5. Functional Enrichment of Candidate Genes

The GO and KEGG enrichment analyses were performed using the g:Profiler website for all fatty acid traits. The functional genes were significantly enriched in 262 GO terms (p_adj < 0.05) (see File S1). The top 10 molecular functions were involved in binding. The top 10 cellular components were cellular anatomical entities (GO:0110165) and membrane-bound organelles (GO:0043227). There were some GO terms associated with lipid and fatty acid metabolism in the biological process category (Figure 4a), such as fatty acid biosynthetic process (GO:0006633), unsaturated fatty acid metabolic process (GO:0033559), lipid biosynthetic process (GO:0008610), and lipid metabolic process (GO:0006629). In this study, annotated genes were significantly enriched in 12 KEGG pathways, such as metabolic pathways (KEGG:01100), inflammatory bowel disease (KEGG:05321), and Th17 cell differentiation (KEGG:04659) (Figure 4b). These KEGG pathways can be categorized into three groups on the KEGG website: Metabolism, Organismal Systems, and Human Diseases (https://www.kegg.jp/kegg/pathway.html (accessed on 1 October 2023)).

a

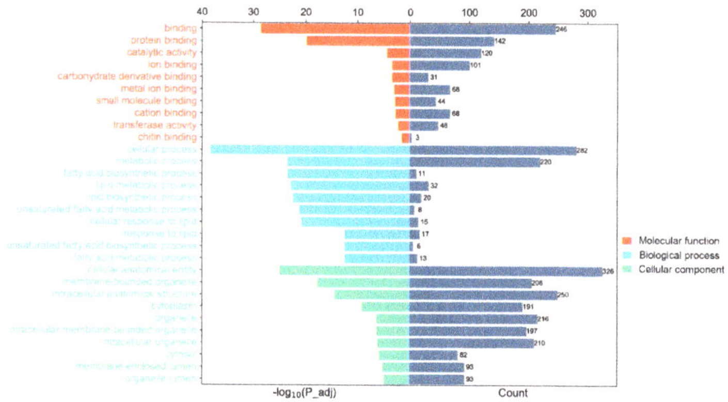

b

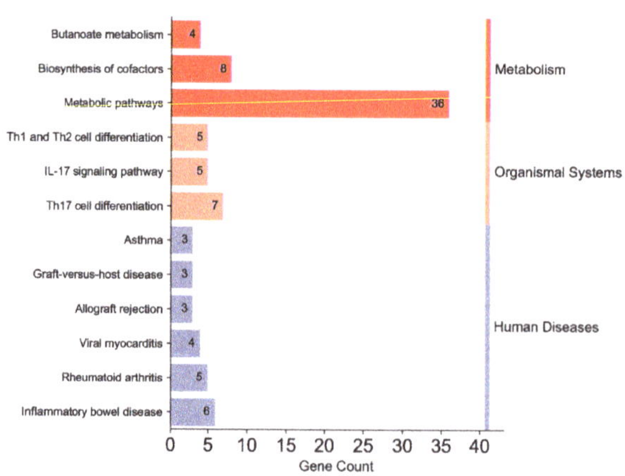

Figure 4. Gene functional enrichment analysis. (**a**) The top 10 enriched molecular functions and cellular components and 10 biological processes associated with fatty acid and lipid metabolism. (**b**) The 12 enriched KEGG pathways, which can be divided into three categories on the KEGG website (Metabolism, Organismal Systems, and Human Diseases).

4. Discussion

4.1. Phenotypic and Genetic Correlations

In this study, a total of 25 fatty acid species were detected in the *longissimus dorsi* of Ningxiang pigs, with 14 species commonly found in the population. SFA was the most abundant among them, while the MUFA relative content was the highest. According to research, the distribution pattern of fatty acids in different varieties of pork is similar [25]. However, there are significant differences in the composition of fatty acids in different tissues [26], which is also influenced by sex [27]. Interestingly, docosahexaenoic acid (DHA) has only been found in the tissues of Ningxiang pigs [26,28], but not in any other pig breeds or tissues [29,30]. DHA belongs to the omega-3 family alongside alpha-linolenic acid (ALA) and eicosapentaenoic acid (EPA), which are essential nutrients integral to human life. Both linoleic acid (h^2 = 0.88) and eicosa-11,14-dienoic acid (h^2 = 0.63) from the omega-3 family exhibit high heritability in the *longissimus dorsi* of Ningxiang pigs, which is an interesting phenomenon [31]. To ensure the safety and health of pork products, pig industry breeders and researchers have long sought to enhance the PUFA content and distribution in pork [32,33].

As an important indicator of pork quality, IMF has consistently garnered significant attention. Numerous studies have provided evidence for the impact of IMF on meat quality [34]. In recent years, the hypothesis that IMF deposition in muscle is impacted by fatty acid structure was verified [35]. By conducting a correlation analysis between IMF and various fatty acid components in the *longissimus dorsi* of Ningxiang pigs (Figure 3), it was observed that IMF exhibits a notably low correlation with SFAs. Furthermore, IMF exhibited a significant positive correlation with MUFAs and a significant negative correlation with PUFAs. Realini et al. [36] investigated the relationship between IMF deposition and fatty acid composition in New Zealand sheep and found results consistent with those of our study, revealing a negative correlation between MUFAs and IMF deposition, a positive correlation between PUFAs and IMF deposition, and a correlation coefficient of −0.72 between linoleic acid and IMF deposition. This phenomenon might be attributed to the endogenous synthesis rate and desaturation sequence of SFAs.

4.2. Candidate Genes for Fatty Acid Composition

Currently, the molecular mechanisms underlying fat deposition are a topic of interest. Fat deposition not only directly impacts the growth, development, and meat production traits of animals but also holds valuable implications for addressing human diseases. Ningxiang pigs are an excellent model for obesity research, and an increasing number of genes, regulatory factors, and metabolic pathways related to fat deposition in Ningxiang pigs have been identified [37]. In this study, we performed a comprehensive GWAS focusing on 14 fatty acids in the *longissimus dorsi* of Ningxiang pigs, and most of them exhibited significant correlations with IMF content. Seventy-four genome-wide significant SNPs were identified in this study, most of which were intronic and intergenic variations (Table S2). Only WU_10.2_2_9630034 is a missense mutation, and there is no research on the gene (*SDHAF2*) related to fatty acids.

A total of 40 genes were identified from 37 significant loci associated with SFAs which included *hnRNPU* [38], *CEPT1* [39], *ATP1B1*, *DPT*, *DKK1* [40], *PRKG1*, *EXT2* [41], *MEF2C*, *IL17RA* [42], *ITGA1* [43] and *ALOX5*. Among these, heteronuclear heterogeneous ribonucleoprotein particles (hnRNPs) represent a group of proteins with diverse functions, playing pivotal roles in RNA biogenesis, cellular localization, and transport [44]. Specifically, hnRNP U is involved in the *Blnc1/hnRNPU/EBF2* heterogeneous ribonucleoprotein particle complex, thus promoting the expression of brown adipocyte genes [38]. Additionally, Dickkopf WNT signaling pathway inhibitor 1 (*DKK1*) can regulate placental lipid metabolism through the WNT signaling pathway [40]. Surprisingly, muscle cell enhancer factor 2C (*MEF2C*) not only plays a crucial regulatory role in skeletal muscle cells but also exerts significant regulatory effects on adipose tissue deposition [45]. Elias et al. found that both *ALOX5* and *ALOX5AP* are involved in the browning of white adipose tissue through lipotoxin A4 [46]. In our study, arachidonic acid 5-nenenebb lipoxygenase (*ALOX5*) and its partner, arachidonic acid 5-lipoxygenase activator protein (*ALOX5AP*) were genes near the C20:0 and C16:1 significant SNP loci, respectively. In Ossabaw pig epicardial adipose tissue, *ALOX5* is positively correlated with n-3 PUFAs [47].

Twenty-five genes were annotated in the upstream and downstream regions of twenty-one loci related to MUFAs. Among them, genes such as *ALOX5AP* [46], *MEDAG* [48], *ISL1* [49], *RXRB* [50], *CRY1* [51], and *CDKAL1* [52] were found to be related to lipid synthesis and metabolism. Interestingly, *ALOX5AP* and *MEDAG* are two genes upstream and downstream of the MARC0099145 locus. Additionally, *MEDAG* is involved in fat metabolism, playing a crucial role in backfat deposition in most Western pig breeds [48]. Retinol-X receptor β (*RXRB*) is a member of the nuclear receptor superfamily of retinoic acid X receptors (RXRs) and is expressed in almost all tissues. In the liver, the activation of *RXRB* leads to increased expression of stearyl CoA desaturation (*SCD*) and *CD36* fatty acid transferase [50], *RXRB* is consistent with the results of this study and a study on the black Iberian pig [53]. Cryptochrome gene 1 (*CRY1*) is a member of the circadian clock gene family and plays a vital role in adipocyte biology. *CRY1* is regulated by the classic Wnt/β-catenin signaling pathway, which influences fat differentiation [54]. Moreover, *CRY1* contains two interaction regions that regulate its degradation to achieve diurnal blood glucose control [55], further affecting conditions such as obesity [51,56].

Nineteen genes were annotated in the upstream and downstream regions of 16 loci related to PUFAs. Genes such as *NDUFA4L2*, *SLC16A7*, *OTUB1*, *EIF4E* [57], and *ROBO2* [58] were found to be associated with lipid synthesis and metabolism. Mitochondrial dysfunction can cause an increase in *NDUFA4L2* expression, leading to lipid accumulation in renal cells [59]. In addition, *ssc-mir-708* is associated with fatty acids, but *miR-708* has only been reported in the regulation of cardiomyocyte proliferation to date [60]. Finally, eukaryotic translation initiation factor 4E (*EIF4E*) enhances the translation of various messenger RNAs involved in lipid metabolism processing and storage pathways, leading to weight gain after a high-fat diet [57]. These genes may influence the fatty acid composition. The majority of the fatty acids in Ningxiang pigs exhibit medium to high heritability, indicating that candidate genes are likely to impact the fatty acid composition of the *longissimus dorsi*. Therefore, the candidate genes related to fatty acids identified in this study can be considered for use in enhancing the fatty acid composition of imported or commercial pig meat, thereby improving the meat quality of commercial pigs.

5. Conclusions

This GWAS identified 74 genome-wide SNPs associated with 14 fatty acids in the *longissimus dorsi*. Some SNPs were located within or near reported genes, but some were novel for fatty acid composition. In total, twenty-two genes, such as *hnRNPU*, *ALOX5AP*, and *NDUFA4L2*, can be used as candidate genes for fatty acid composition in Ningxiang pigs. Our findings will be helpful for understanding the genetic basis of fatty acid composition and providing new targets for further breeding of pigs.

Supplementary Materials: The following supporting information can be downloaded at https://www.mdpi.com/article/10.3390/ani13203192/s1, Table S1: Detailed information on 25 fatty acids present in IMF of the *longissimus dorsi* of Ningxiang pigs; Table S2. Genome-wide significant SNPs loci for SFAs, MUFAs and PUFAs in Ningxiang pigs; File S1: GO and KEGG enrichment analysis; Figure S1. Manhattan plot for five SFAs (C14:0, C16:0, C17:0, C18:0, and C20:0). The x-axis represents the chromosomes, and y-axis represents the −log10 (*p*_value). The green line is genome-wide level threshold, the green dashed line is chromosome-level threshold; Figure S2. Manhattan plot for four MUFAs (C16:1, C18:1n-9(c), C18:1n-9(t), and C20:1). The x-axis represents the chromosomes, and y-axis represents the −log10 (*p*_value). The green line is genome-wide level threshold, the green dashed line is chromosome-level threshold; Figure S3. Manhattan plot for five PUFAs (C18:2n-6(c), C18:3n-3, C20:2, C20:3n-6, and C20:4n-6). The x-axis represents the chromosomes, and y-axis represents the −log10 (*p*_value). The green line is genome-wide level threshold, the green dashed line is chromosome-level threshold.

Author Contributions: Conceptualization, Q.Z., H.G. and Y.Y.; methodology, H.G. and K.X.; software, Q.Z., G.S. and X.D.; validation, H.G. and S.Y.; formal analysis, Q.Z. and F.Y.; investigation, Y.P., Y.C. and X.H.; resources, J.H., Y.Y. and K.X.; data curation, F.Y., Y.F., Q.W., S.Y. and Z.J.; writing—original draft preparation, Q.Z., H.G., S.Y. and F.Y.; writing—review and editing, J.H., Y.P., Y.Y. and K.X. visualization, H.G. and S.Y.; supervision, Y.Y. and K.X.; project administration, Y.Y. and K.X.; funding acquisition, Y.Y. and K.X. All authors have read and agreed to the published version of the manuscript.

Funding: This research was funded by the Laboratory of Lingnan Modern Agriculture Project (NT2021005), the Strategic Priority Research Program of Chinese Academy of Sciences (Grant No. XDA24030204), the Special Funds for the Construction of Innovative Provinces in Hunan (Grant numbers 2021NK1009 and 2021NK1012); the Natural Science Foundation of Hunan Province Project (2023JJ20043 and 2020JJ5635).

Institutional Review Board Statement: The experimental protocol used in this study was reviewed and approved by the Animal Experimental Ethical Inspection of Laboratory Animal Centre, Hunan Agricultural University. All sample collection was conducted under a permit (No. 2020047) approved by the Attitude of the Animal Management and Ethics Committee.

Informed Consent Statement: Not applicable.

Data Availability Statement: Not applicable.

Conflicts of Interest: The authors declare no conflict of interest.

References

1. Chen, J.; Chen, F.; Lin, X.; Wang, Y.; He, J.; Zhao, Y. Effect of Excessive or Restrictive Energy on Growth Performance, Meat Quality, and Intramuscular Fat Deposition in Finishing Ningxiang Pigs. *Animals* **2020**, *11*, 27. [CrossRef] [PubMed]
2. Uemoto, Y.; Abe, T.; Tameoka, N.; Hasebe, H.; Inoue, K.; Nakajima, H.; Shoji, N.; Kobayashi, M.; Kobayashi, E. Whole-genome association study for fatty acid composition of oleic acid in Japanese Black cattle. *Anim. Genet.* **2011**, *42*, 141–148. [CrossRef] [PubMed]
3. Fortin, A.; Robertson, W.M.; Tong, A.K. The eating quality of Canadian pork and its relationship with intramuscular fat. *Meat Sci.* **2005**, *69*, 297–305. [CrossRef] [PubMed]
4. Michas, G.; Micha, R.; Zampelas, A. Dietary fats and cardiovascular disease: Putting together the pieces of a complicated puzzle. *Atherosclerosis* **2014**, *234*, 320–328. [CrossRef] [PubMed]
5. Khan, M.I.; Jo, C.; Tariq, M.R. Meat flavor precursors and factors influencing flavor precursors—A systematic review. *Meat Sci.* **2015**, *110*, 278–284. [CrossRef]
6. Grassi, S.; Benedetti, S.; Opizzio, M.; Nardo, E.D.; Buratti, S. Meat and Fish Freshness Assessment by a Portable and Simplified Electronic Nose System (Mastersense). *Sensors* **2019**, *19*, 3225. [CrossRef]
7. Melton, S.L.; Amiri, M.; Davis, G.W.; Backus, W.R. Flavor and chemical characteristics of ground beef from grass-, forage-grain- and grain finished steers. *J. Anim. Sci.* **1982**, *55*, 77–87. [CrossRef]
8. Westerling, D.B.; Hedrick, H.B. Fatty acid composition of bovine lipids as influenced by diet, sex and anatomical location and relationship to sensory characteristics. *J. Anim. Sci.* **1979**, *48*, 1343–1348. [CrossRef]
9. de Oliveira Otto, M.C.; Wu, J.H.; Baylin, A.; Vaidya, D.; Rich, S.S.; Tsai, M.Y.; Jacobs, D.R., Jr.; Mozaffarian, D. Circulating and dietary omega-3 and omega-6 polyunsaturated fatty acids and incidence of CVD in the Multi-Ethnic Study of Atherosclerosis. *J. Am. Heart Assoc.* **2013**, *2*, e000506. [CrossRef]
10. Merino, J.; Guasch-Ferré, M.; Ellervik, C.; Dashti, H.; Sharp, S.; Wu, P.; Overvad, K.; Sarnowski, C.; Kuokkanen, M.; Lemaitre, R.; et al. Quality of dietary fat and genetic risk of type 2 diabetes: Individual participant data meta-analysis. *BMJ (Clin. Res. Ed.)* **2019**, *366*, l4292. [CrossRef]
11. Lands, B. Historical perspectives on the impact of n-3 and n-6 nutrients on health. *Prog. Lipid Res.* **2014**, *55*, 17–29. [CrossRef] [PubMed]
12. Lands, B.; Bibus, D.; Stark, K.D. Dynamic interactions of n-3 and n-6 fatty acid nutrients. *Prostaglandins Leukot. Essent. Fat. Acids* **2018**, *136*, 15–21. [CrossRef] [PubMed]
13. Bibus, D.; Lands, B. Balancing proportions of competing omega-3 and omega-6 highly unsaturated fatty acids (HUFA) in tissue lipids. *Prostaglandins Leukot. Essent. Fat. Acids* **2015**, *99*, 19–23. [CrossRef] [PubMed]
14. Davoli, R.; Luise, D.; Mingazzini, V.; Zambonelli, P.; Braglia, S.; Serra, A.; Russo, V. Genome-wide study on intramuscular fat in Italian Large White pig breed using the PorcineSNP60 BeadChip. *J. Anim. Breed. Genet.* **2016**, *133*, 277–282. [CrossRef]
15. van Son, M.; Enger, E.G.; Grove, H.; Ros-Freixedes, R.; Kent, M.P.; Lien, S.; Grindflek, E. Genome-wide association study confirm major QTL for backfat fatty acid composition on SSC14 in Duroc pigs. *BMC Genom.* **2017**, *18*, 369. [CrossRef]
16. Viterbo, V.S.; Lopez, B.I.M.; Kang, H.; Kim, H.; Song, C.W.; Seo, K.S. Genome wide association study of fatty acid composition in Duroc swine. *Asian-Australas J. Anim. Sci.* **2018**, *31*, 1127–1133. [CrossRef]
17. *GB/T 9695.1-2008*; Meat and Meat Products—Determination of Free Fat Content. AQSIQ, SAC: Beijing, China, 2008.

18. Purcell, S.; Neale, B.; Todd-Brown, K.; Thomas, L.; Ferreira, M.A.; Bender, D.; Maller, J.; Sklar, P.; de Bakker, P.I.; Daly, M.J.; et al. PLINK: A tool set for whole-genome association and population-based linkage analyses. *Am. J. Hum. Genet.* **2007**, *81*, 559–575. [CrossRef]
19. Chang, C.C.; Chow, C.C.; Tellier, L.C.; Vattikuti, S.; Purcell, S.M.; Lee, J.J. Second-generation PLINK: Rising to the challenge of larger and richer datasets. *Gigascience* **2015**, *4*, 7. [CrossRef]
20. Wang, J.; Zhang, Z. GAPIT Version 3: Boosting Power and Accuracy for Genomic Association and Prediction. *Genom. Proteom. Bioinform.* **2021**, *19*, 629–640. [CrossRef]
21. Pook, T.; Mayer, M.; Geibel, J.; Weigend, S.; Cavero, D.; Schoen, C.C.; Simianer, H. Improving Imputation Quality in BEAGLE for Crop and Livestock Data. *G3 (Bethesda)* **2020**, *10*, 177–188. [CrossRef]
22. Teslovich, T.; Musunuru, K.; Smith, A.; Edmondson, A.; Stylianou, I.; Koseki, M.; Pirruccello, J.; Ripatti, S.; Chasman, D.; Willer, C.; et al. Biological, Clinical, and Population Relevance of 95 Loci for Blood Lipids. *Nature* **2010**, *466*, 707–713. [CrossRef] [PubMed]
23. Yin, L.; Zhang, H.; Tang, Z.; Yin, D.; Fu, Y.; Yuan, X.; Li, X.; Liu, X.-L.; Zhao, S. HIBLUP: An integration of statistical models on the BLUP framework for efficient genetic evaluation using big genomic data. *Nucleic Acids Res.* **2023**, *51*, 3501–3512. [CrossRef] [PubMed]
24. Raudvere, U.; Kolberg, L.; Kuzmin, I.; Arak, T.; Adler, P.; Peterson, H.; Vilo, J. g:Profiler: A web server for functional enrichment analysis and conversions of gene lists (2019 update). *Nucleic Acids Res.* **2019**, *47*, W191–W198. [CrossRef] [PubMed]
25. Crespo-Piazuelo, D.; Criado-Mesas, L.; Revilla, M.; Castello, A.; Noguera, J.L.; Fernandez, A.I.; Ballester, M.; Folch, J.M. Identification of strong candidate genes for backfat and intramuscular fatty acid composition in three crosses based on the Iberian pig. *Sci. Rep.* **2020**, *10*, 13962. [CrossRef] [PubMed]
26. Jiang, Q.; Li, C.; Yu, Y.; Xing, Y.; Xiao, D.; Zhang, B. Comparison of fatty acid profile of three adipose tissues in Ningxiang pigs. *Anim. Nutr.* **2018**, *4*, 256–259. [CrossRef] [PubMed]
27. Cui, L.; Zhang, J.; Ma, J.; Guo, Y.; Li, L.; Xiao, S.; Ren, J.; Yang, B.; Huang, L. Sexually dimorphic genetic architecture of complex traits in a large-scale F2 cross in pigs. *Genet. Sel. Evol.* **2014**, *46*, 76. [CrossRef]
28. Lee, J.B.; Kang, Y.J.; Kim, S.G.; Woo, J.H.; Shin, M.C.; Park, N.G.; Yang, B.C.; Han, S.H.; Han, K.M.; Lim, H.T.; et al. GWAS and Post-GWAS High-Resolution Mapping Analyses Identify Strong Novel Candidate Genes Influencing the Fatty Acid Composition of the Longissimus dorsi Muscle in Pigs. *Genes* **2021**, *12*, 1323. [CrossRef]
29. Popova, T.; Givko, N. Fatty acid profile of the backfat layers in four pig breeds. *Food Sci. Appl. Biotechnol.* **2019**, *2*, 24. [CrossRef]
30. Zhang, W.; Zhang, J.; Cui, L.; Ma, J.; Chen, C.; Ai, H.; Xie, X.; Li, L.; Xiao, S.; Huang, L.; et al. Genetic architecture of fatty acid composition in the longissimus dorsi muscle revealed by genome-wide association studies on diverse pig populations. *Genet. Sel. Evol.* **2016**, *48*, 5. [CrossRef]
31. Demets, R.; Gheysen, L.; Van Loey, A.; Foubert, I. Antioxidative capacity of microalgal carotenoids for stabilizing n-3LC-PUFA rich oil: Initial quantity is key. *Food Chem.* **2023**, *406*, 135044. [CrossRef]
32. Mayer, C.; Côme, M.; Ulmann, L.; Martin, I.; Zittelli, G.C.; Faraloni, C.; Ouguerram, K.; Chénais, B.; Mimouni, V. The Potential of the Marine Microalga Diacronema lutheri in the Prevention of Obesity and Metabolic Syndrome in High-Fat-Fed Wistar Rats. *Molecules* **2022**, *27*, 4246. [CrossRef] [PubMed]
33. Calder, P.C. Very long-chain n-3 fatty acids and human health: Fact, fiction and the future. *Proc. Nutr. Soc.* **2018**, *77*, 52–72. [CrossRef] [PubMed]
34. Waszkiewicz-Robak, B.; Szterk, A.; Rogalski, M.; Rambuszek, M.; Kruk, M.; Rokowska, E. Nutritional value of raw pork depending on the fat type contents in pigs feed. *Acta Sci. Pol. Technol. Aliment.* **2015**, *14*, 153–163. [CrossRef] [PubMed]
35. Zhang, Z.; Zhang, Z.; Oyelami, F.O.; Sun, H.; Xu, Z.; Ma, P.; Wang, Q.; Pan, Y. Identification of genes related to intramuscular fat independent of backfat thickness in Duroc pigs using single-step genome-wide association. *Anim. Genet.* **2021**, *52*, 108–113. [CrossRef]
36. Realini, C.E.; Pavan, E.; Purchas, R.W.; Agnew, M.; Johnson, P.L.; Bermingham, E.N.; Moon, C.D. Relationships between intramuscular fat percentage and fatty acid composition in M. longissimus lumborum of pasture-finished lambs in New Zealand. *Meat Sci.* **2021**, *181*, 108618. [CrossRef]
37. Criado-Mesas, L.; Ballester, M.; Crespo-Piazuelo, D.; Castello, A.; Benitez, R.; Fernandez, A.I.; Folch, J.M. Analysis of porcine IGF2 gene expression in adipose tissue and its effect on fatty acid composition. *PLoS ONE* **2019**, *14*, e0220708. [CrossRef]
38. Mi, L.; Zhao, X.Y.; Li, S.; Yang, G.; Lin, J.D. Conserved function of the long noncoding RNA Blnc1 in brown adipocyte differentiation. *Mol. Metab.* **2017**, *6*, 101–110. [CrossRef]
39. Zayed, M.A.; Jin, X.; Yang, C.; Belaygorod, L.; Engel, C.; Desai, K.; Harroun, N.; Saffaf, O.; Patterson, B.W.; Hsu, F.F.; et al. CEPT1-Mediated Phospholipogenesis Regulates Endothelial Cell Function and Ischemia-Induced Angiogenesis Through PPARα. *Diabetes* **2021**, *70*, 549–561. [CrossRef]
40. Strakovsky, R.S.; Pan, Y.X. A decrease in DKK1, a WNT inhibitor, contributes to placental lipid accumulation in an obesity-prone rat model. *Biol. Reprod.* **2012**, *86*, 81. [CrossRef]
41. Pedrosa, V.B.; Schenkel, F.S.; Chen, S.Y.; Oliveira, H.R.; Casey, T.M.; Melka, M.G.; Brito, L.F. Genomewide Association Analyses of Lactation Persistency and Milk Production Traits in Holstein Cattle Based on Imputed Whole-Genome Sequence Data. *Genes* **2021**, *12*, 1830. [CrossRef]

42. Shinjo, T.; Iwashita, M.; Yamashita, A.; Sano, T.; Tsuruta, M.; Matsunaga, H.; Sanui, T.; Asano, T.; Nishimura, F. IL-17A synergistically enhances TNFα-induced IL-6 and CCL20 production in 3T3-L1 adipocytes. *Biochem. Biophys. Res. Commun.* **2016**, *477*, 241–246. [CrossRef] [PubMed]
43. Williams, A.S.; Kang, L.; Zheng, J.; Grueter, C.; Bracy, D.P.; James, F.D.; Pozzi, A.; Wasserman, D.H. Integrin α1-null mice exhibit improved fatty liver when fed a high fat diet despite severe hepatic insulin resistance. *J. Biol. Chem.* **2015**, *290*, 6546–6557. [CrossRef] [PubMed]
44. Hitachi, K.; Kiyofuji, Y.; Nakatani, M.; Tsuchida, K. Myoparr-Associated and -Independent Multiple Roles of Heterogeneous Nuclear Ribonucleoprotein K during Skeletal Muscle Cell Differentiation. *Int. J. Mol. Sci.* **2021**, *23*, 108. [CrossRef] [PubMed]
45. Xu, Z.; Wu, J.; Zhou, J.; Zhang, Y.; Qiao, M.; Sun, H.; Li, Z.; Li, L.; Chen, N.; Oyelami, F.O.; et al. Integration of ATAC-seq and RNA-seq analysis identifies key genes affecting intramuscular fat content in pigs. *Front. Nutr.* **2022**, *9*, 1016956. [CrossRef] [PubMed]
46. Elias, I.; Ferré, T.; Vilà, L.; Muñoz, S.; Casellas, A.; Garcia, M.; Molas, M.; Agudo, J.; Roca, C.; Ruberte, J.; et al. ALOX5AP Overexpression in Adipose Tissue Leads to LXA4 Production and Protection Against Diet-Induced Obesity and Insulin Resistance. *Diabetes* **2016**, *65*, 2139–2150. [CrossRef] [PubMed]
47. Walker, M.E.; Matthan, N.R.; Goldbaum, A.; Meng, H.; Lamon-Fava, S.; Lakshman, S.; Jang, S.; Molokin, A.; Solano-Aguilar, G.; Urban, J.F., Jr.; et al. Dietary patterns influence epicardial adipose tissue fatty acid composition and inflammatory gene expression in the Ossabaw pig. *J. Nutr. Biochem.* **2019**, *70*, 138–146. [CrossRef]
48. Gozalo-Marcilla, M.; Buntjer, J.; Johnsson, M.; Batista, L.; Diez, F.; Werner, C.R.; Chen, C.Y.; Gorjanc, G.; Mellanby, R.J.; Hickey, J.M.; et al. Genetic architecture and major genes for backfat thickness in pig lines of diverse genetic backgrounds. *Genet. Sel. Evol.* **2021**, *53*, 76. [CrossRef]
49. Li, H.; Heilbronn, L.K.; Hu, D.; Poynten, A.M.; Blackburn, M.A.; Shirkhedkar, D.P.; Kaplan, W.H.; Kriketos, A.D.; Ye, J.; Chisholm, D.J. Islet-1: A potentially important role for an islet cell gene in visceral fat. *Obesity* **2008**, *16*, 356–362. [CrossRef]
50. Singh Ahuja, H.; Liu, S.; Crombie, D.L.; Boehm, M.; Leibowitz, M.D.; Heyman, R.A.; Depre, C.; Nagy, L.; Tontonoz, P.; Davies, P.J. Differential effects of rexinoids and thiazolidinediones on metabolic gene expression in diabetic rodents. *Mol. Pharmacol.* **2001**, *59*, 765–773. [CrossRef]
51. Griebel, G.; Ravinet-Trillou, C.; Beeské, S.; Avenet, P.; Pichat, P. Mice deficient in cryptochrome 1 (cry1(−/−)) exhibit resistance to obesity induced by a high-fat diet. *Front. Endocrinol.* **2014**, *5*, 49. [CrossRef]
52. Choi, W.J.; Jin, H.S.; Kim, S.S.; Shin, D. Dietary Protein and Fat Intake Affects Diabetes Risk with CDKAL1 Genetic Variants in Korean Adults. *Int. J. Mol. Sci.* **2020**, *21*, 5607. [CrossRef] [PubMed]
53. Pena, R.N.; Noguera, J.L.; García-Santana, M.J.; González, E.; Tejeda, J.F.; Ros-Freixedes, R.; Ibáñez-Escriche, N. Five genomic regions have a major impact on fat composition in Iberian pigs. *Sci. Rep.* **2019**, *9*, 2031. [CrossRef] [PubMed]
54. Sun, S.; Zhou, L.; Yu, Y.; Zhang, T.; Wang, M. Knocking down clock control gene CRY1 decreases adipogenesis via canonical Wnt/β-catenin signaling pathway. *Biochem. Biophys. Res. Commun.* **2018**, *506*, 746–753. [CrossRef] [PubMed]
55. Toledo, M.; Batista-Gonzalez, A.; Merheb, E.; Aoun, M.L.; Tarabra, E.; Feng, D.; Sarparanta, J.; Merlo, P.; Botrè, F.; Schwartz, G.J.; et al. Autophagy Regulates the Liver Clock and Glucose Metabolism by Degrading CRY1. *Cell Metab.* **2018**, *28*, 268–281.e264. [CrossRef] [PubMed]
56. Sardon Puig, L.; Pillon, N.J.; Näslund, E.; Krook, A.; Zierath, J.R. Influence of obesity, weight loss, and free fatty acids on skeletal muscle clock gene expression. *Am. J. Physiol. Endocrinol. Metab.* **2020**, *318*, E1–E10. [CrossRef]
57. Conn, C.S.; Yang, H.; Tom, H.J.; Ikeda, K.; Oses-Prieto, J.A.; Vu, H.; Oguri, Y.; Nair, S.; Gill, R.M.; Kajimura, S.; et al. The major cap-binding protein eIF4E regulates lipid homeostasis and diet-induced obesity. *Nat. Metab.* **2021**, *3*, 244–257. [CrossRef]
58. Mohammadi, H.; Farahani, A.H.K.; Moradi, M.H.; Mastrangelo, S.; Di Gerlando, R.; Sardina, M.T.; Scatassa, M.L.; Portolano, B.; Tolone, M. Weighted Single-Step Genome-Wide Association Study Uncovers Known and Novel Candidate Genomic Regions for Milk Production Traits and Somatic Cell Score in Valle del Belice Dairy Sheep. *Animals* **2022**, *12*, 1155. [CrossRef]
59. Laursen, K.B.; Chen, Q.; Khani, F.; Attarwala, N.; Gross, S.S.; Dow, L.; Nanus, D.M.; Gudas, L.J. Mitochondrial Ndufa4l2 Enhances Deposition of Lipids and Expression of Ca9 in the TRACK Model of Early Clear Cell Renal Cell Carcinoma. *Front. Oncol.* **2021**, *11*, 783856. [CrossRef]
60. Deng, S.; Zhao, Q.; Zhen, L.; Zhang, C.; Liu, C.; Wang, G.; Zhang, L.; Bao, L.; Lu, Y.; Meng, L.; et al. Neonatal Heart-Enriched miR-708 Promotes Proliferation and Stress Resistance of Cardiomyocytes in Rodents. *Theranostics* **2017**, *7*, 1953–1965. [CrossRef]

animals

MDPI

Article

The Novel-miR-659/*SPP1* Interaction Regulates Fat Deposition in Castrated Male Pigs

Lianmei Xiao [1,†], Qiao Xu [1,2,†], Ximing Liu [1], Shuheng Chan [1], Yabiao Luo [1], Shuaihan He [1] and Meiying Fang [1,3,*]

1 Department of Animal Genetics and Breeding, National Engineering Laboratory for Animal Breeding, MOA Laboratory of Animal Genetics and Breeding, Beijing Key Laboratory for Animal Genetic Improvement, College of Animal Science and Technology, China Agricultural University, Beijing 100193, China; 17785245395@163.com (L.X.); xuqiao987@163.com (Q.X.); lximing2018@163.com (X.L.); 15380345581@163.com (S.C.); yabiaoluo2021@163.com (Y.L.); heshuaihan@cau.edu.cn (S.H.)
2 School of Life Sciences, University of Chinese Academy of Sciences, Beijing 100049, China
3 Sanya Institute of China Agricultural University, Sanya 572025, China
* Correspondence: meiying@cau.edu.cn; Tel./Fax: +86-10-62734943
† These authors contributed equally to this work.

Simple Summary: Castration is a standard method for eliminating boar taint in industrial hog production, but it also causes enormous fat accumulation in the carcass. Secreted phosphoprotein 1 (*SPP1*) was selected to investigate its functions on the regulation of adipose deposition based on our previous data. In the present study, *SPP1* overexpression and interference bidirectionally verified that *SPP1* inhibited adipogenic differentiation of porcine bone marrow mesenchymal stem cells (pBMSCs). Testosterone-treated cell models were used to simulate the androgen status of intact pigs, and testosterone addition influenced *SPP1* mRNA levels during the differentiation of pBMSCs. Moreover, we identified novel-miR-659 and targeted the 3′ untranslated region of *SPP1* based on bioinformatics analysis and dual-luciferase assays, and found that the novel-miR-659 upregulation promoted adipogenesis while novel-miR-659 downregulation suppressed adipogenesis in pBMSCs detected by Oil Red O staining and adipogenic markers. Collectively, the interaction between novel-miR-659 and *SPP1* can regulate adipose accumulation in castrated male pigs. Our data provide a theoretical basis for further study on the fat deposition mechanism caused by castration.

Abstract: Castration is usually used to remove boar taint in commercial pork production, but the adipose accumulation was increased excessively, which affected the meat quality of pigs. Based on our previous study, secreted phosphoprotein 1 (*SPP1*) was significantly differentially expressed between castrated and intact male pigs. However, the role of *SPP1* in regulating adipose growth and fat storage caused by castration is unknown. In this study, *SPP1* was identified to inhibit adipogenesis by the expression of adipogenic markers *PPARγ* and *FABP4* as well as Oil red staining assay during differentiation of porcine bone marrow mesenchymal stem cells (pBMSCs). Subsequently, testosterone was used to treat pBMSCs to simulate the androgen status of intact pigs. Compared with the control groups without testosterone, the *SPP1* expression in the testosterone group was markedly increased in the late stage of pBMSCs differentiation. Furthermore, novel-miR-659 was predicted by TargetScan and miRDB to target *SPP1* and verified through a dual-luciferase reporter assay. Oil Red O staining assay indicated that novel-miR-659 overexpression significantly promoted adipogenesis, whereas novel-miR-659 inhibition suppressed adipogenesis. The expressions of adipogenic markers *PPARγ* and *FABP4* showed the same tendency. Taken together, our study found that the targeted interaction between novel-miR-659 and *SPP1* is involved in regulation of fat deposition in castrated male pigs.

Keywords: *SPP1*; castration; fat deposition; novel-miR-659

1. Introduction

Adipose tissue is a loose connective tissue, which is primarily made up of adipocyte and plays a significant function in energy storage. Moreover, adipose tissue is an active endocrine organ that could secrete a variety of factors that regulate adipogenesis via paracrine signals [1]. Obesity is a global health issue [2], which is linked to an increased risk of death from diseases such as type 2 diabetes, hyperlipidemia, and steatohepatitis [3]. Current studies have shown that sex hormones also influence the metabolism and deposition of adipose tissue [4]. Multiple studies have shown that androgens such as testosterone and dihydrotestosterone could stimulate steatolysis and release energy, and that low testosterone causes obesity in animals, implying that androgen is a principal player in visceral adiposity [5,6]. Traditional surgical castration and immunocastration based on synthetic gonadoliberin were typical procedures for reducing boar taint in industrial pork production, both of which have been widely reported to promote the fat deposition in male pigs since they dramatically limit the secretion of androgens [7–12]. Castrated boars have rapid fat deposition, similar organ size, metabolic characteristics, cardiovascular system, and high homology with humans; therefore, it is the best animal model for studying human androgen-deficiency-related disorders [13–15]. However, the underlying molecular basis between androgen and fat deposition in organisms is unknown.

Secreted phosphoprotein 1 (*SPP1*), also known as Osteopontin (*OPN*), was originally discovered in osteoblasts and functions similarly to BMPs proteins [16]. *SPP1* is a secreted extracellular matrix (ECM) glycoprotein implicated in a wide range of pathogenesis. including insulin resistance, inflammation, tumor metastasis, and bone remodeling [17,18], and can be used as a biomarker for cancer and other intractable diseases [19,20]. *SPP1* has a critical role in bone mineralization and calcification inhibition. In addition, *SPP1* is highly expressed in normal, nonmineralized epithelial tissues such as kidneys, livers, lactating breast, immune cells, and salivary glands [21,22]. *SPP1* is divided into two types in mammals, based on its structure and function: secreted-*SPP1* (s*OPN*) and intracellular-*SPP1* (i*OPN*) [23,24]. Currently, a lot of research has displayed that *SPP1* is overexpressed in obese people, implying that *SPP1* may play a significant regulatory function in energy homeostasis [25–27]. Numerous studies have identified that *SPP1* is synthesized by adipocytes, and both adipocytes and adipose tissue are the main targets for *SPP1* function [26,28]. Additionally, high levels of *SPP1* are related to fat deposit, lipid accumulation, and inflammation in adipose tissue, and *SPP1* expression rose markedly in visceral adipose tissue in obese individuals [29,30]. Recently, *SPP1* can accelerate the development of white preadipocytes into brown adipocytes via CD44-dependent pathways [31–33], and the interaction of *SPP1* with integrin $\alpha v/\beta 1$ inhibits the adipogenesis of mesenchymal stem cells (MSCs) [34]. It suggests that *SPP1* has varied roles in the regulation of fat deposition, depending on the circumstances.

Other recent studies have also shown that the *SPP1* gene has an effect on fat deposition [35,36]. Castration is known to significantly reduce androgen levels in pigs [7,37]. *SPP1* was significantly differentially expressed between castrated and intact male pigs based on our previous data, and we hypothesized that the expression of *SPP1* may be regulated by androgen. Therefore, we aimed to study the relationship between the *SPP1* gene and obesity caused by androgen deficiency. Our research contributes to the development of a molecular basis for androgen regulation of fat metabolism.

2. Materials and Methods

2.1. Cell Culture and Differentiation

pBMSCs were cultivated in a medium (DMEM, 10% FBS and 1% Ps) at 37 °C in a 5% CO_2 humidified environment. When the cell density was 100% confluent, pBMSCs were incubated in an adipocyte differentiation medium containing 0.5 mM 3-isobutyl-1-methylxanthine (IBMX), 60 mM indomethacin, 100 nM dexamethasone, and 10 mg/mL insulin for 3 days. Next, they were changed to an adipocyte maintenance medium with 10 mg/mL insulin for another 1 day. This was performed three times.

2.2. Plasmid Constructs, RNA Interference, and Cell Transfection

SPP1 CDS was amplified by PCR using full length *SPP1* as a template, and subcloned into the PCDH vector. The sequences used here are as follows: sense (5′-TGACCTCCATAG AAGATT ATGAGAATTGCAGTGATAGCC-3′) and antisense (5′-TAAATTCGAATTCGCTA GTCAGTTGATCTCAGAAGACGC-3′). SiRNA was provided from GenePharma (Jiangsu, China); its sequence information is described in Table S1. Furthermore, cell transfection was conducted according to the instructions of Lipofectamine 2000 (Invitrogen, Carlsbad, CA, USA). In short, pBMSCs were plated in the 6-well plate, and allowed to grow to 90% confluence. The cells were incubated for 1 h with the Opti-MEM, and then transfected with 4 ug of the PCDH-*SPP1* vector, while 4 ug of the PCDH empty vector was used as a control by Lipofectamine 2000 (Invitrogen, Carlsbad, CA, USA). However, the transfection concentration of the *SPP1* gene wild type and mutant 3′UTR plasmid was 200 ng. Regarding RNA interference, pBMSCs were transfected with 100 nM siRNA when they grew to 90% confluency. The Opti-MEM was replaced by the complete DMEM, containing 10% FBS at five hours after transfection. Forty eight hours after transfection, the cells were harvested and detected by qPCR.

2.3. Testosterone Test

Our previous study found that 100 nM testosterone has a significant inhibitory effect on the adipogenic differentiation of pBMSCs. Therefore, we added 100 nM of testosterone during the differentiation of pBMSCs, and added the same volume of methanol and testosterone to the control group. Given that pBMSCs need 2 weeks to differentiate into mature adipocytes, we chose 18 days as the induction differentiation time of pBMSCs.

2.4. Oil Red O Staining

The Oil Red O was used to stain adipocytes, as previously described [33]. The cells were fixed by 10% formaldehyde for 30 min. Next, the formaldehyde-free cells were rinsed with 60% isopropanol for 10 s. After that, Oil Red O solution was used to stain the cells for 30 min. At last, the colored cells were inspected under a microscope.

2.5. Real-Time PCR

The total RNA was separated from cells using TRIzol® reagent and reverse transcription to synthesize cDNA. Quantitative PCR was carried out according to the instructions of Real-Time System. β-actin served as a housekeeping gene. Primer Premier 5.0 software was used to create the primer sequences for quantitative PCR (Table S2).

2.6. Dual Luciferase Reporter Assay

SPP1-3′UTR wild type (*SPP1*-3′UTR-WT) and mutant plasmid (*SPP1*-3′UTR-MUT) were synthesized by Beijing Genomics Institution (BGI, Beijing, China), and miRNA mimics were provided by GenePharma. In the present study, 100 nM miR-659 mimics or mimics control and 200 ng plasmids (*SPP1*-3′UTR-WT or *SPP1*-3′UTR-MUT) were co-transfected into 293T cells with Lipofectamine 2000. Then, 48 h after transfection, a Dual-Glo luciferase reporter gene was used to observe luciferase activities. Finally, the ratio of the Renilla fluorescence to the firefly fluorescence was calculated.

2.7. Statistical Analysis

All data were repeated at least three times in each experiment and showed as the mean ± SEM. Differences were analyzed by using t-test and one-way ANOVA associated with GraphPad Prism 8.0.1 software. * $p < 0\ 05$ was regarded as significant.

3. Result

3.1. Adipogenic Differentiation of pBMSCs

Bone marrow mesenchymal stem cells (BMSCs) were deemed to be good candidate cells for exploring de novo generation of adipocyte due to their easy availability. To validate the adipogenic differentiation ability of pBMSCs, lipid droplet accumulation of pBMSCs were visualized by Oil Red O staining at different time points. Results showed that there was no obvious droplet formation during the first 6 days, and multitudinous intracellular lipid droplets did not appear until 12 days later. Furthermore, the buildup of lipid droplets grew with the prolongation of differentiation time (Figure 1A). Additionally, the expression of adipogenic-specific genes *PPARγ* and *FABP4* were analyzed by qRT-PCR. The *PPARγ* was expressed at the early stage of pBMSCs differentiation while the expression of *FABP4* increased over time (Figure 1B), which was consistent with previous reports [38]. These results demonstrated the successful differentiation of pBMSCs into adipocytes.

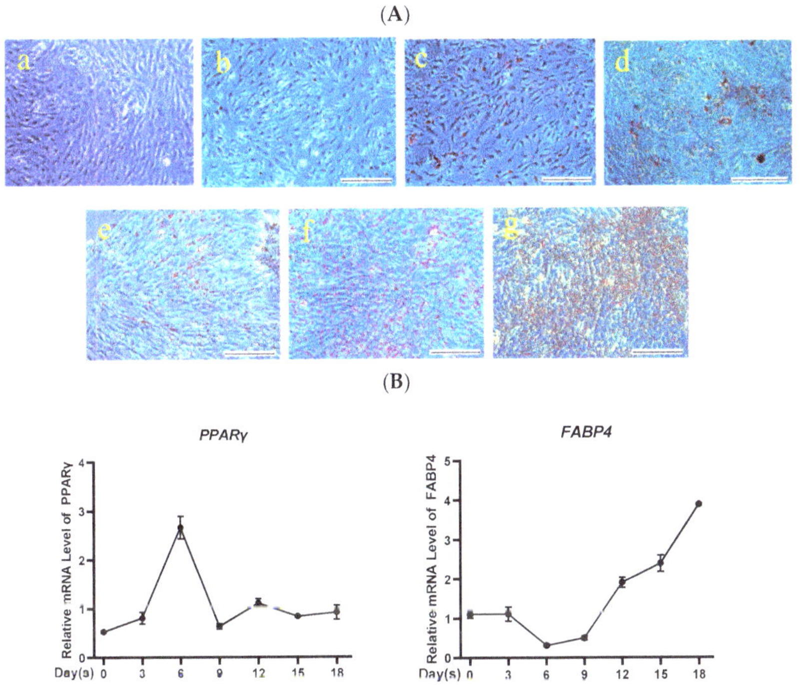

Figure 1. Adipogenic differentiation of pBMSCs. (**A**) Oil Red O was used to stain the pBMSCs, differentiated on day 0 (**A-a**), 3 (**A-b**), 6 (**A-c**), 9 (**A-d**), 12 (**A-e**), 15 (**A-f**), and 18 (**A-g**) (bar = 200 μm). (**B**) *PPARγ* and *FABP4* mRNA levels on the relevant days were examined by qRT-PCR.

3.2. *SPP1* Overexpression Suppressed the Adipogenic Differentiation of pBMSCs

Overexpression of the *SPP1* gene was performed by transfecting PCDH-*SPP1* into pBMSCs (Figure 2A). Then the Oil Red O staining assay and analysis of the expressions of adipogenesis marker factors were used for validating the functions of *SPP1* in pBMSCs. The results showed that after 14 days of induction differentiation, the pBMSCs containing the PCDH-*SPP1* vector displayed a significant decrease in accumulation of lipid droplets compared with the PCDH empty vector group (Figure 2B). Meanwhile, the expression of *PPARγ* and *FABP4* were also significantly lower in the PCDH-*SPP1* group than in the PCDH empty vector group (Figure 2C) ($p < 0.01$). These results demonstrated that the overexpression of *SPP1* inhibited the adipogenic differentiation of pBMSCs.

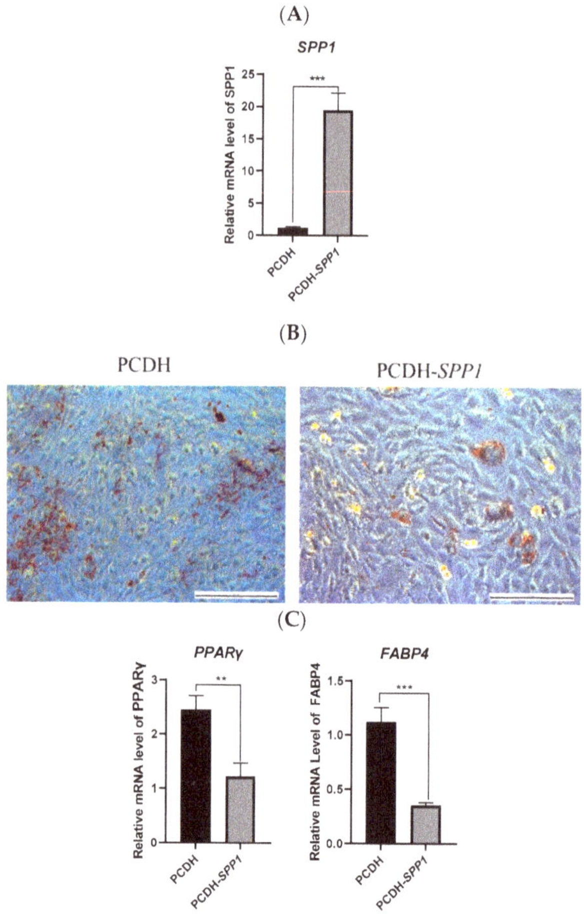

Figure 2. *SPP1* overexpression suppressed the adipogenic differentiation of pBMSCs. (**A**) pBMSCs were transfected with the PCDH empty vector (PCDH) or the *SPP1* overexpression vector (PCDH-*SPP1*), and *SPP1* expression was measured by qRT-PCR. (**B**) The transfected pBMSCs were induced to differentiate for 14 days and lipid droplets were observed by Oil Red O staining (bar = 200 μm). (**C**) Adipose marker factors *PPARγ* and *FABP4* mRNA expression of pBMSCs harboring PCDH or PCDH-*SPP1* were detected by qRT-PCR. ***, $p < 0.001$; **, $p < 0.01$.

3.3. Inhibition of SPP1 Promoted the Adipogenic Differentiation of pBMSCs

To further explore the role of *SPP1* on pBMSCs differentiation, three pairs of *SPP1* siRNA (si-212-*SPP1*, si-810-*SPP1*, and si-925-*SPP1*) were designed and transfected into pBMSCs. The quantitative PCR displayed that si-925-*SPP1* had the best inhibitory effect (Figure 3A). Oil Red O staining results revealed that treatment with the si-925-*SPP1* group accumulated great quantities of fat droplets, which was contrasted with the negative control (NC) group after 14 days of induction differentiation (Figure 3B). The mRNA levels of *PPARγ* and *FABP4* were significantly upregulated after si-925-*SPP1* transfection (Figure 3C). These results demonstrated that the inhibition of *SPP1* could promote the adipogenic differentiation of pBMSCs.

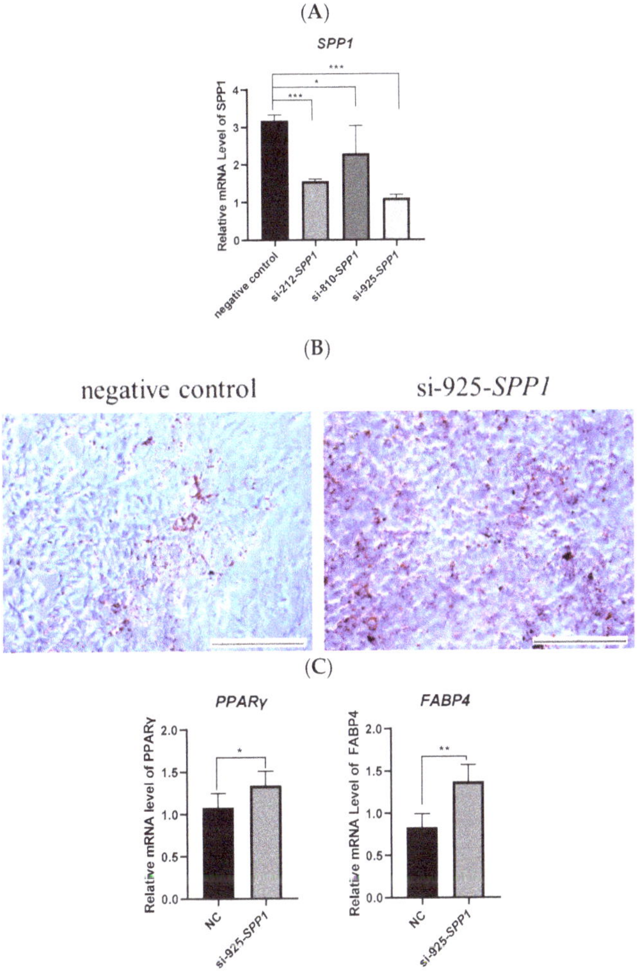

Figure 3. Inhibition of *SPP1* promoted the adipogenic differentiation of pBMSCs. (**A**) pBMSCs were transfected with siRNA (si-212-*SPP1*, si-810-*SPP1*, and si-925-*SPP1*) or negative control, and to assay-interfering efficiency by expression of *SPP1* two days after transfection. (**B**) Lipid droplet accumulation of pBMSCs transfected with si-925-SPP1 and negative control were verified by Oil Red O staining induction at 14 days (bar = 200 μm). (**C**) The mRNA levels of PPARγ and FABP4 in the si-925-SPP1 group and the negative control group were confirmed by qRT-PCR. ***, $p < 0.001$; **, $p < 0.01$; *, $p < 0.05$.

3.4. Testosterone Regulates the Expression of SPP1

This study tested whether the differential expression of *SPP1* in castrated and intact male pigs was caused by testosterone. In vitro, pBMSCs were treated with testosterone to simulate the androgen status of intact pigs. In contrast with the control group, the expression of *SPP1* in the testosterone group was slightly decreased at the early stage of pBMSCs differentiation. However, the *SPP1* expression levels in the testosterone group was markedly higher than those in the control group at the adipose deposition stage (after 12 days) (Figure 4), which was consistent with the function of *SPP1* in pBMSCs. The results indicated that the expression of *SPP1* was regulated by testosterone.

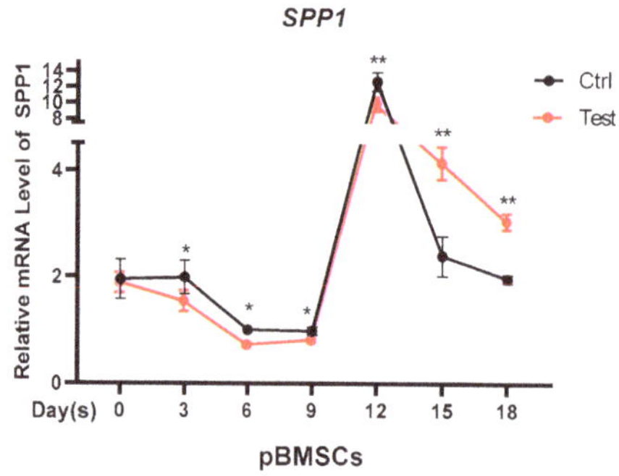

Figure 4. Testosterone regulates the expression of *SPP1*. The mRNA expression of *SPP1* was detected by PCR at a different stage of pBMSCs differentiation. Ctrl, control group. Test, testosterone group. **, $p < 0.01$; *, $p < 0.05$.

3.5. Novel-miR-659 Promoted Adipogenesis of pBMSCs by Targeting SPP1

Our previous study indicated that *SPP1* is a potential target of novel-miR-659 by TargetScan and miRDB software. To verify the targeting relationship between novel-miR-659 and *SPP1*, two plasmids containing either the WT or MUT 3′-UTR of *SPP1* were constructed (Figure 5A). Subsequently, cotransfection of novel-miR-659 mimics or mimics NC into 293T cells was performed. Notably, novel-miR-659 mimics markedly decreased the activity of WT *SPP1* compared with the mimics NC, but no significant changes were noted for the MUT (Figure 5B). Meanwhile, we determined the expression profile of novel-miR-659 in pBMSCs during adipogenesis induced by an adipogenic medium at different time points. This trend was the opposite of the expression trend of *SPP1* in adipogenesis of pBMSCs (Figure S1). These findings indicate that novel-miR-659 can directly target the 3′-UTR of *SPP1*.

To further investigate the function of novel-miR-659 on pBMSCs differentiation, novel-miR-659 mimics and inhibitors were transfected into pBMSCs. Transfection efficiency of novel-miR-659 and expression of *SPP1* were validated at day 14 of induction by qRT-PCR. The expression of *SPP1* decreased with novel-miR-659 upregulation and increased with novel-miR-659 knockdown, indicating that the expression trend of the *SPP1* gene was opposite to that of novel-miR-659 (Figure S2). After 14 days of induction Oil Red O staining, results showed that novel-miR-659 mimics promoted the formation of fat droplets while novel-miR-659 inhibitors inhibited the formation of fat droplets compared with mimics and inhibitors NC, respectively (Figure 5C). Quantitative PCR revealed that novel-miR-659 mimics significantly improved the expression of adipogenesis marker *PPARγ* and *FABP4*, whereas the novel-miR-659 inhibitor-treated group expressed lower amounts of these mRNAs (Figure 5D,E). Collectively, these results manifested that novel-miR-659 promoted adipogenesis of pBMSCs by repressing *SPP1*.

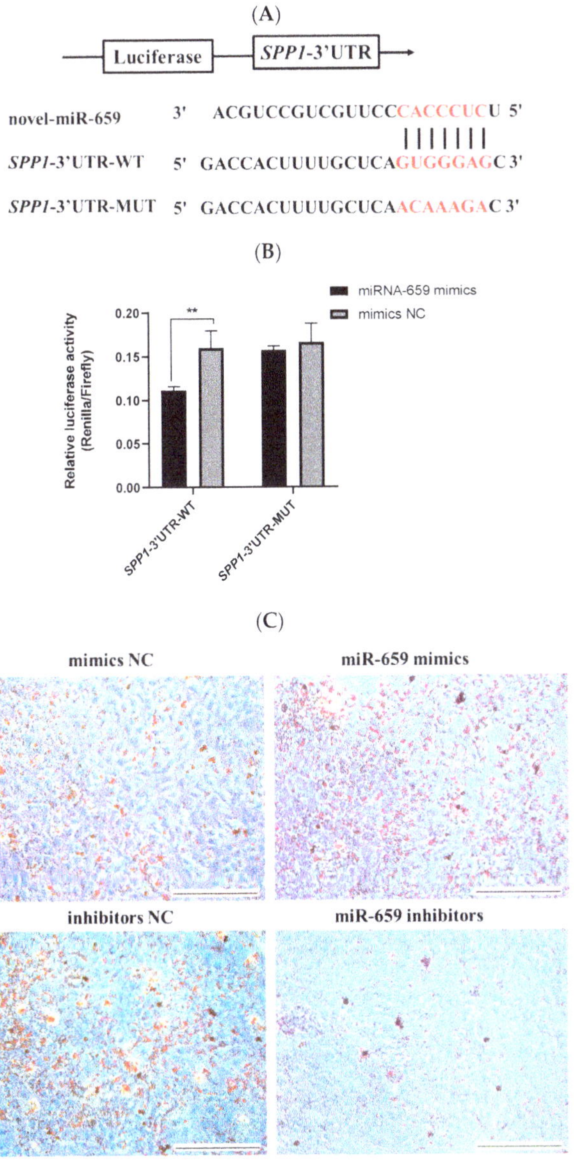

Figure 5. *Cont.*

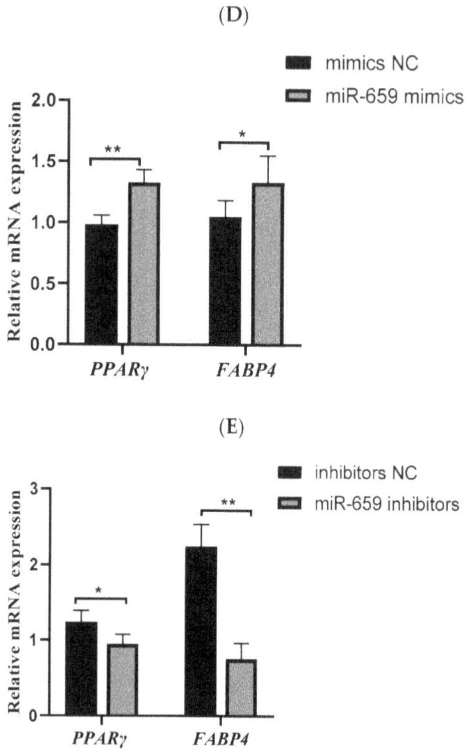

Figure 5. Novel-miR-659 promoted adipogenesis of pBMSCs by targeting *SPP1*. (**A**) The interaction binding site of novel-miR-659 and *SPP1* 3′UTR was predicted by TargetScan and miRDB. (**B**) Novel-miR-659 directly targeted the 3′UTR of *SPP1*, normalized Renilla fluorescence activity forty eight hours following cotransfection of novel-miR-659 mimics and mimics NC with *SPP1*-3′UTR-WT and *SPP1*-3′UTR-MUT. (**C**) Adipogenesis of pBMSCs transfected with mimics/inhibitors NC or miR-659 mimics/inhibitors were stained by Oil Red O at 14 days after differentiation (bar = 200 μm). (**D**,**E**) Expressions of *PPARγ* and *FABP4* in mimics NC, miR-659 mimics, inhibitors NC, and miR-659 inhibitors groups were detected by qPCR on day 14 after induction. **, $p < 0.01$; *, $p < 0.05$.

4. Discussion

Obesity is a global health issue, and androgen insufficiency has been linked to male obesity in animals. Our previous study identified differentially expressed mRNAs in backfat and abdominal fat tissues from intact and castrated full-sib male pigs, and found that *SPP1* was related to fat deposition. Some researchers have confirmed that blockage of *SPP1* expression facilitated MSCs adipogenic development but suppressed osteoclast formation [26]. Furthermore, Veronica et al. found that progenitor cells lacking *SPP1* were more efficient in adipogenic differentiation [27]. Additionally, Tang et al. displayed that *SPP1* prevents ASCs from becoming adipogenic by connecting to the external receptors' integrin αv/β1 and CD44 [28]. In line with these previously published findings, our studies of in vitro experiments revealed that pBMSCs from *SPP1* silence (or *SPP1* overexpression) were more (or less) effective in lipogenesis than NC controls, as identified by enhanced (or receded) fat-droplet accumulation (Figures 2 and 3), indicating that *SPP1* was a negative regulator of adipogenesis [31].

It has been reported that *SPP1* expression in epididymis main cells may be influenced by circulating testosterone [39]. Likewise, our results also revealed that the expression of *SPP1* was affected by testosterone, and the *SPP1* mRNA level was suppressed by the testosterone in the early stage of pBMSCs differentiation. However, testosterone promoted the *SPP1* expression during the later stages of pBMSCs differentiation (Figure 4), which was in keeping with the inhibition of *SPP1* on lipid deposition. The differentiation process from MSCs to mature adipocytes is generally considered to go through two stages. The first stage is called commitment, where MSCs are converted into preadipocytes. The preadipocyte is the intermediate stage between MSCs and adipocytes, and it can differentiate into adipocytes. The second stage, named terminal differentiation, is the transformation of preadipocytes into mature adipocytes. Preadipocytes are not morphologically distinguishable from adipocytes and lack specific markers to distinguish between the two stages [40]. Therefore, we believe that testosterone may play different roles in the control of *SPP1* expression at different stages of pBMSCs differentiation.

To learn more about the role of *SPP1* in the control of fat deposition. novel-miR-659 directly targeted the 3′UTR of *SPP1*, as shown in the dual-luciferase reporter assay (Figure 5B). Both novel-miR-659 overexpression and inhibition experiments demonstrated that novel-miR-659 promoted adipose deposition of pBMSCs by suppressing *SPP1* (Figure 5C–E). The novel-miR-659 is a novel miRNA in pigs, and its relationship with fat deposition is rarely studied. Recent studies have shown that miR-659 was obesity-specific, being expressed only in amnion from obese women than control women [41]. This is consistent with our findings that miR-659 promoted the accumulation of lipid droplets in pBMSCs.

5. Conclusions

In summary, our results show that SPP1 negatively regulated adipogenic differentiation of pBMSCs, and the SPP1 expression in adipose tissue of pigs is regulated by testosterone and novel-miR-659. The interaction between novel-miR-659 and SPP1 coregulate fat deposition in castrated male pigs. These data expand our mechanistic understanding of fat deposition in pigs caused by testosterone deficiency.

Supplementary Materials: The following supporting information can be downloaded at: https://www.mdpi.com/article/10.3390/ani12080944/s1, Table S1. SiRNA and miRNA sequences; Table S2. Primers used for real-time quantitative PCR; Figure S1. *SPP1* is a potential target of novel-miR-659; Figure S2. Transfection effect of novel-miR-659.

Author Contributions: Formal analysis, L.X. and Q.X.; Investigation, L.X., Q.X. and X.L.; Project administration, M.F.; Resources, X.L., S.C., Y.L. and S.H.; Writing—original draft, L.X., Q.X. and M.F. All authors have read and agreed to the published version of the manuscript.

Funding: This work was supported by the 2020 Research Program of Sanya Yazhou Bay Science and Technology City (Grant No. SKJC-2020-02-007); the National Natural Science Foundation of China (32072700); the National Key Research and Development Program of China (2021YFF1000603); the China Agriculture Research System of MOF and MARA(CARS-36); and the talent cultivation and developmental support program of China Agricultural University.

Institutional Review Board Statement: Not applicable.

Informed Consent Statement: Not applicable.

Data Availability Statement: The data are in the supplement.

Acknowledgments: The authors are grateful to Jinming Huang and Gang Liu for assistance with sample collection.

Conflicts of Interest: The authors declare no conflict of interest.

References

1. Kershaw, E.E.; Flier, J.S. Adipose tissue as an endocrine organ. *J. Clin. Endocrinol. Metab.* **2004**, *89*, 2548–2556. [CrossRef] [PubMed]
2. Lu, P.; Zhang, F.C.; Qian, S.W.; Li, X.; Cui, Z.M.; Dang, Y.J.; Tang, Q.Q. Artemisinin derivatives prevent obesity by inducing browning of WAT and enhancing BAT function. *Cell Res.* **2016**, *26*, 1169–1172. [CrossRef]
3. Berrington, D.G.A.; Hartge, P.; Cerhan, J.R.; Flint, A.J.; Hannan, L.; Macinnis, R.J.; Moore, S.C.; Tobias, G.S.; Anton-Culver, H.; Freeman, L.B.; et al. Body-mass index and mortality among 1.46 million white adults. *N. Engl. J. Med.* **2010**, *363*, 2211–2219. [CrossRef] [PubMed]
4. Zerradi, M.; Dereumetz, J.; Boulet, M.; Tchernof, A. Androgens, body fat Distribution and Adipogenesis. *Curr. Obes. Rep.* **2014**, *3*, 396–403. [CrossRef] [PubMed]
5. Wang, J.; Hua, L.; Chen, J.; Zhang, J.; Bai, X.; Gao, B.; Li, C.; Shi, Z.; Sheng, W.; Gao, Y.; et al. Identification and characterization of long non-coding RNAs in subcutaneous adipose tissue from castrated and intact full-sib pair Huainan male pigs. *BMC Genom.* **2017**, *18*, 542. [CrossRef] [PubMed]
6. Grossmann, M. Low Testosterone in Men with Type 2 Diabetes: Significance and Treatment. *J. Clin. Endocrinol. Metab.* **2011**, *96*, 2341–2353. [CrossRef] [PubMed]
7. Christoffersen, B.O.; Gade, L.P.; Golozoubova, V.; Svendsen, O.; Raun, K. Influence of castration-induced testosterone and estradiol deficiency on obesity and glucose metabolism in male Göttingen minipigs. *Steroids* **2010**, *75*, 676–684. [CrossRef]
8. Wang, K.; Li, W.; Bai, Y.; Yang, W.; Ling, Y.; Fang, M. ssc-miR-7134-3p regulates fat accumulation in castrated male pigs by targetingMARK4 gene. *Int. J. Biol. Sci.* **2017**, *13*, 189–197. [CrossRef] [PubMed]
9. Li, M.; Bjorntorp, P. Effects of testosterone on triglyceride uptake and mobilization in different adipose tissues in male rats in vivo. *Obes. Res.* **1995**, *3*, 113–119. [CrossRef]
10. Lundstrom, K.; Matthews, K.R.; Haugen, J.E. Pig meat quality from entire males. *Animal* **2009**, *3*, 1497–1507. [CrossRef] [PubMed]
11. Bonneau, M. Use of entire males for pig meat in the European Union. *Meat Sci.* **1998**, *49*, S257–S272. [CrossRef]
12. Poklukar, K.; Andek-Potokar, M.; Vrecl, M.; Batorek-Lukač, N.; Fazarinc, G.; Kress, K.; Weiler, U.; Stefanski, V.; Akrlep, M. The effect of immunocastration on adipose tissue deposition and composition in pigs. *Animal* **2021**, *15*, 100118. [CrossRef] [PubMed]
13. Gispert, M.; Angels, O.M.; Velarde, A.; Suarez, P.; Perez, J.; Font, I.F.M. Carcass and meat quality characteristics of immunocastrated male, surgically castrated male, entire male and female pigs. *Meat Sci.* **2010**, *85*, 664–670. [CrossRef] [PubMed]
14. Saintilan, R.; Sellier, P.; Billon, Y.; Gilbert, H. Genetic correlations between males, females and castrates for residual feed intake, feed conversion ratio, growth rate and carcass composition traits in Large White growing pigs. *J. Anim. Breed. Genet.* **2012**, *129*, 103–106. [CrossRef] [PubMed]
15. Wolf, E.; Braun-Reichhart, C.; Streckel, E.; Renner, S. Genetically engineered pig models for diabetes research. *Transgenic Res.* **2014**, *23*, 27–38. [CrossRef] [PubMed]
16. Friedenstein, A.J.; Gorskaja, J.F.; Kulagina, N.N. Fibroblast precursors in normal and irradiated mouse hematopoietic organs. *Exp. Hematol.* **1976**, *4*, 267–274. [PubMed]
17. Jürets, A.; Le Bras, M.; Staffler, G.; Stein, G.; Leitner, L.; Neuhofer, A.; Tardelli, M.; Turkof, E.; Zeyda, M.; Stulnig, T.M. Inhibition of Cellular Adhesion by Immunological Targeting of Osteopontin Neoepitopes Generated through Matrix Metalloproteinase and Thrombin Cleavage. *PLoS ONE* **2016**, *11*, e148333. [CrossRef] [PubMed]
18. Inoue, M.; Shinohara, M.L. Intracellular osteopontin (iOPN) and immunity. *Immunol. Res.* **2011**, *49*, 160–172. [CrossRef] [PubMed]
19. Wolak, T. Osteopontin–A multi-modal marker and mediator in atherosclerotic vascular disease. *Atherosclerosis* **2014**, *236*, 327–337. [CrossRef]
20. Bandopadhyay, M.; Bulbule, A.; Butti, R.; Chakraborty, G.; Ghorpade, P.; Ghosh, P.; Gorain, M.; Kale, S.; Kumar, D.; Kumar, S.; et al. Osteopontin as a therapeutic target for cancer. *Expert Opin. Ther. Tar.* **2014**, *18*, 883–895. [CrossRef] [PubMed]
21. Staines, K.A.; Macrae, V.E.; Farquharson, C. The importance of the SIBLING family of proteins on skeletal mineralisation and bone remodelling. *J. Endocrinol.* **2012**, *214*, 241–255. [CrossRef]
22. Lund, S.A.; Giachelli, C.M.; Scatena, M. The role of osteopontin in inflammatory processes. *J. Cell Commun. Signal.* **2009**, *3*, 311–322. [CrossRef] [PubMed]
23. Jang, J.; Kim, J. Improved Cellular Response of Osteoblast Cells Using Recombinant Human Osteopontin Protein Produced by Escherichia coli. *Biotechnol. Lett.* **2005**, *27*, 1767–1770. [CrossRef] [PubMed]
24. Denhardt, D.T.G.X. Osteopontin: A protein with diverse functions. *FASEB J.* **1993**, *15*, 1475–1482. [CrossRef]
25. Kiefer, F.W.; Zeyda, M.; Todoric, J.; Huber, J.; Geyeregger, R.; Weichhart, T.; Aszmann, O.; Ludvik, B.; Silberhumer, G.R.; Prager, G.; et al. Osteopontin Expression in Human and Murine Obesity: Extensive Local Up-Regulation in Adipose Tissue but Minimal Systemic Alterations. *Endocrinology* **2008**, *149*, 1350–1357. [CrossRef] [PubMed]
26. Gómez-Ambrosi, J.; Catalán, V.; Ramírez, B.; Rodríguez, A.; Colina, I.; Silva, C.; Rotellar, F.; Mugueta, C.; Gil, M.J.; Cienfuegos, J.A.; et al. Plasma Osteopontin Levels and Expression in Adipose Tissue Are Increased in Obesity. *J. Clin. Endocrinol. Metab.* **2007**, *92*, 3719–3727. [CrossRef] [PubMed]
27. Kahles, F.; Findeisen, H.M.; Bruemmer, D. Osteopontin: A novel regulator at the cross roads of inflammation, obesity and diabetes. *Mol. Metab.* **2014**, *3*, 384–393. [CrossRef] [PubMed]
28. Kiefer, F.W.; Neschen, S.; Pfau, B.; Legerer, B.; Neuhofer, A.; Kahle, M.; Hrabé De Angelis, M.; Schlederer, M.; Mair, M.; Kenner, L.; et al. Osteopontin deficiency protects against obesity-induced hepatic steatosis and attenuates glucose production in mice. *Diabetologia* **2011**, *54*, 2132–2142. [CrossRef] [PubMed]

29. Chapman, J.; Miles, P.D.; Ofrecio, J.M.; Neels, J.G.; Yu, J.G.; Resnik, J.L.; Wilkes, J.; Talukdar, S.; Thapar, D.; Johnson, K.; et al. Osteopontin is required for the early onset of high fat diet-induced insulin resistance in mice. *PLoS ONE* **2010**, *5*, e13959. [CrossRef] [PubMed]
30. Nomiyama, T.; Perez-Tilve, D.; Ogawa, D.; Gizard, F.; Zhao, Y.; Heywood, E.B.; Jones, K.L.; Kawamori, R.; Cassis, L.A.; Tschöp, M.H.; et al. Osteopontin mediates obesity-induced adipose tissue macrophage infiltration and insulin resistance in mice. *J. Clin. Investig.* **2007**, *117*, 2877–2888. [CrossRef]
31. Wang, M.; Guo, Y.; Zhou, Y.; Yuan, W.; Li, H.; Xiong, S.; Wang, K.; Zeng, G.; Huang, Q. Secreted-Osteopontin Contributes to Brown Adipogenesis In Vitro via a CD44-Dependent Pathway. *Horm. Metab. Res.* **2019**, *51*, 741. [CrossRef] [PubMed]
32. Lu, Y.; Xu, Y.; Yuan, W.; Wang, M.; Zhou, Y.; Chen, K.; Huang, Q. Downregulation of osteopontin inhibits browning of white adipose tissues through PI3K-AKT pathway in C57BL/6 mice. *Eur. J. Pharmacol.* **2020**, *866*, 172822. [CrossRef] [PubMed]
33. Zhong, X.; Shen, X.; Wen, J.; Kong, Y.; Chu, J.; Yan, G.; Li, T.; Liu, D.; Wu, M.; Zeng, G.; et al. Osteopontin-induced brown adipogenesis from white preadipocytes through a PI3K-AKT dependent signaling. *Biochem. Biophys. Res. Commun.* **2015**, *459*, 553–559. [CrossRef] [PubMed]
34. Chen, Q.; Shou, P.; Zhang, L.; Xu, C.; Zheng, C.; Han, Y.; Li, W.; Huang, Y.; Zhang, X.; Shao, C.; et al. An Osteopontin-Integrin Interaction Plays a Critical Role in Directing Adipogenesis and Osteogenesis by Mesenchymal Stem Cells. *Stem Cells* **2014**, *32*, 327–337. [CrossRef] [PubMed]
35. Moreno-Viedma, V.; Tardelli, M.; Zeyda, M.; Sibilia, M.; Burks, J.D.; Stulnig, T.M. Osteopontin-deficient progenitor cells display enhanced differentiation to adipocytes. *Obes. Res. Clin. Pract.* **2018**, *12*, 277–285. [CrossRef]
36. Fitter, S.; Zannettino, A.C.W. Osteopontin in the pathophysiology of obesity: Is Opn a fat cell foe? *Obes. Res. Clin. Pract.* **2018**, *12*, 249–250. [CrossRef]
37. Cai, Z.; Zhang, L.; Chen, M.; Jiang, X.; Xu, N. Castration-induced changes in microRNA expression profiles in subcutaneous adipose tissue of male pigs. *J. Appl. Genet.* **2014**, *55*, 259–266. [CrossRef]
38. Bai, C.; Chen, S.; Gao, Y.; Shan, Z.; Guan, W.; Ma, Y. Multi-lineage potential research of bone marrow mesenchymal stem cells from Bama miniature pig. *J. Exp. Zool. Part B Mol. Dev. Evol.* **2015**, *324*, 671–685. [CrossRef]
39. Shoelson, S.E. Inflammation and insulin resistance. *J. Clin. Investig.* **2006**, *116*, 1793–1801. [CrossRef]
40. Zuk, P.A.; Zhu, M.; Ashjian, P.; De Ugarte, D.A.; Huang, J.I.; Mizuno, H.; Alfonso, Z.C.; Fraser, J.K.; Benhaim, P.; Hedrick, M.H. Human Adipose Tissue Is a Source of Multipotent Stem Cells. *Mol. Biol. Cell* **2002**, *12*, 4279–4295. [CrossRef]
41. Nardelli, C.; Iaffaldano, L.; Ferrigno, M.; Labruna, G.; Maruotti, G.M.; Quaglia, F.; Capobianco, V.; Di Noto, R.; Del, V.L.; Martinelli, P.; et al. Characterization and predicted role of the microRNA expression profile in amnion from obese pregnant women. *Int. J. Obes.* **2014**, *38*, 466–469. [CrossRef] [PubMed]

animals

MDPI

Article

chi-miR-99b-3p Regulates the Proliferation of Goat Skeletal Muscle Satellite Cells In Vitro by Targeting *Caspase-3* and *NCOR1*

Rongrong Liao, Yuhua Lv, Jianjun Dai, Defu Zhang, Lihui Zhu * and Yuexia Lin *

Institute of Animal Husbandry and Veterinary Science, Shanghai Academy of Agricultural Sciences, Shanghai 201106, China
* Correspondence: zhulihui@saas.sh.cn (L.Z.); linyuexia@saas.sh.cn (Y.L.); Tel.: +86-216-220-5472 (L.Z.); +86-216-220-8651 (Y.L.)

Simple Summary: In this study, we investigated the role of chi-miR-99b-3p in goat skeletal muscle satellite cells (SMSCs). We confirmed that chi-miR-99b-3p promoted proliferation through targeting *Caspase-3* and nuclear receptor corepressor 1, and inhibiting the intrinsic apoptosis-related genes in SMSCs, whereas inhibition of chi-miR-99b-3p had the opposite effect. Furthermore, integrative transcriptomic analysis revealed that overexpression of chi-miR-99b-3p induced various differentially expressed (DE)-gene-associated processes. In addition, 47 dysregulated miRNAs, including 16 upregulated and 31 downregulated miRNAs, which were involved in the biological pathways associated with the DE genes, were identified. Our study demonstrated that chi-miR-99b-3p was an effective facilitator of goat SMSCs. This study is helpful for future studies in skeletal muscle development.

Abstract: We previously found that chi-miR-99b-3p was highly expressed in the skeletal muscle of 7-month-old (rapid growth period) goats and speculated that it may be associated with muscle development. To further investigate the role of chi-miR-99b-3p in goats, we found that chi-miR-99b-3p acted as a myogenic miRNA in the regulation of skeletal muscle development. Dual-luciferase reporter assays, qRT-PCR, and Western blot results confirmed that *Caspase-3* and nuclear receptor corepressor 1 were direct targets for chi-miR-99b-3p as their expression was inhibited by this miR. Cell proliferation and qRT-PCR assays showed that chi-miR-99b-3p promoted proliferation through relevant targets and intrinsic apoptosis-related genes in goat skeletal muscle satellite cells (SMSCs), whereas inhibition of chi-miR-99b-3p had the opposite effect. Furthermore, integrative transcriptomic analysis revealed that overexpression of chi-miR-99b-3p induced various differentially expressed (DE) genes mainly associated with the cell cycle, relaxin signaling pathway, DNA replication, and protein digestion and absorption. Notably, most of the cell-cycle-related genes were downregulated in SMSCs after miR-99b-3p upregulation, including the pro-apoptosis-related gene *BCL2*. In addition, 47 DE miRNAs (16 upregulated and 31 downregulated) were determined by Small RNA-sequencing in SMSCs after chi-miR-99b-3p overexpression. Based on the KEGG enrichment analysis, we found that these DE miRNAs were involved in the biological pathways associated with the DE genes. Our study demonstrated that chi-miR-99b-3p was an effective facilitator of goat SMSCs and provided new insights into the mechanisms by which miRNAs regulate skeletal muscle growth in goats.

Keywords: chi-miR-99b-3p; proliferation; apoptosis; goat; *Caspase-3*; *NCOR1*

1. Introduction

As an economically important agricultural animal, the goat provides humans with high-quality and low-cholesterol mutton. Muscle is an important tissue in goats, and constitutes the main form of mutton. Therefore, the identification of methods to improve goat muscle development and growth is an important economic problem. The formation of animal muscle tissue mainly depends upon the accumulation of proteins and the coordination of muscle cell proliferation, differentiation, and apoptosis, and includes three main stages: myotube formation, myotube differentiation into muscle fibers, and muscle fiber growth and maturation [1]. Skeletal muscle satellite cells (SMSCs) have stem-cell-like properties and are involved in postnatal skeletal muscle regeneration and growth [2–4]. Due to their myogenic and adipogenic differentiation capacity, SMSCs have become a suitable model for investigating postnatal muscle growth and intramuscular adipogenesis [5]. The development of SMSCs is a complex and delicate process that is regulated by many factors, including muscle cell migration, proliferation, differentiation, and apoptosis, and needs the coordinated control of multiple signal switches. For example, knockdown of *MUSTN1* inhibited the proliferation and differentiation of chicken SMSCs and promoted the apoptosis of SMSCs, while overexpression of *MUSTN1* showed the opposite effect [6], demonstrating the potential relationship among apoptosis, proliferation, and differentiation of SMSCs.

MicroRNAs (miRNAs) represent endogenous, small noncoding RNAs with a length of 20~25 bp, and are widely found in many species. They are highly evolutionarily conserved and play important roles in the complex and dynamic network that regulates development. miR-487b-3p downregulated *IRS1* expression to suppress goat myoblast cell proliferation and differentiation through the IRS1/PI3K/Akt pathway [4], miR-181a could suppress SMSC proliferation [2], and miR-223-3p was involved in muscle regeneration via the regulation of inflammatory cytokine secretion in the skeletal muscle microenvironment [7].

miR-99b-3p is evolutionarily conserved among different species. However, few studies have investigated the role of miR-99b-3p in goats, as most have been performed on human cancer cell proliferation. For example, evidence shows that miR-99b-3p can target protocadherin 19 to promote hepatocellular carcinoma metastasis and cell proliferation [8]. Furthermore, Hsa-miR-99b-3p is considered a novel prognostic marker for oral cancer, as its increased expression showed a trend toward the advanced tumor stage. In addition, high hsa-mir-99b-3p expression levels were related to better survival [9]. Here, miR-99b-3p exerted an antiproliferative effect and arrested gastric cancer cells in the S phase of the cell cycle by inhibiting cell viability and targeting *HoxD3*, which is upregulated in GC cell lines [10]. In addition, miR-99b-3p was reported to contribute to angiotensin-II-induced cardiac fibrosis in mice [11].

Studies have shown that miRNAs expressed in human skeletal muscle are associated with muscle size, strength, and mass. Muscle strength and cross-sectional area were positively correlated with miR-99b-5p expression in middle-aged men [12]. In our previous study, we found that chi-miR-99b-3p was highly expressed in the muscle tissue of Chongming white goats during the fast-growing period (7 months old), while it was downregulated in postweaning goats, which caused growth retardation in early weaned lambs [13,14]. Therefore, we speculated that chi-miR-99b-3p might be related to muscle development [13]. Here, we systematically investigated the potential role of chi-miR-99b-3p in goat SMSCs. Our findings help with understanding the function of miRNAs in skeletal muscle development and provide a theoretical basis for the molecular breeding of goats.

2. Materials and Methods

2.1. Cell Culture and Cell Transfection

Isolated goat SMSCs (separated from Anhui white goats) were a kind gift from Dr. Yinghui Ling (College of Animal Science and Technology, Anhui Agricultural University, Hefei, China). Detailed isolation steps for SMSCs can be found in a previous study [3]. The 9th to 12th generations of satellite cells were used in this study. The SMSCs were cultured in DMEM/F12 (Gibco, Grand Island, NY, USA) growth media containing 20% FBS (Gibco, Grand Island, NY, USA) and 1% penicillin/streptomycin (Gibco, Grand Island, NY, USA). The chi-miR-99b-3p mimic, negative control (NC mimic), chi-miR-99b-3p inhibitor were transfected into cells using Entranster^{TM}-R4000 reagent (Engreen, Beijing, China) according to the manufacturer's protocol. Cells were collected for further analysis at the indicated times. Oligonucleotides were designed and synthesized by Shanghai GenePharma Co., Ltd.

(Shanghai, China), and their sequences can be found in Table S1. For cell-differentiation-related analysis, SMSCs were first cultured in growth media until the cells grew to about 80% density. Differentiation medium, including 2% FBS (Gibco, Grand Island, NY, USA) and 1% penicillin/streptomycin (Gibco, Grand Island, NY, USA), was used to induce cell differentiation. Cells collected 1 d after the addition of differentiation medium were used as the control.

2.2. RNA Isolation and qRT-PCR

Total RNA was isolated using TRIzol reagent (Invitrogen, Waltham, MA, USA) and determined in an ND-2000 spectrophotometer (Thermo Scientific, Wilmington, DE, USA), and the qRT-PCR was performed as previously described [13]. Briefly, the expression levels of miRNAs and mRNA were determined using qRT-PCR performed on an ABI QuantStudio 5 PCR System (Applied Biosystems, Foster City, CA, USA) using SYBR Green PCR Master Mix (Fast Start Universal SYBR Green Master; Roche, Basel, Switzerland). Primer sequences are listed in Table S2. Glyceraldehyde 3-phosphate dehydrogenase was used as an endogenous control gene for mRNA, and U6 for miRNA. Relative quantitative levels were calculated based on the $2^{-\Delta\Delta Ct}$ method [4].

2.3. Western Blotting

Total protein was purified from satellite cells using radioimmunoprecipitation assay (RPMI) buffer (Beyotime, Shanghai, China), and electrophoresed on an 8−12% sodium dodecyl sulfate polyacrylamide gel, and transferred to polyvinylidene fluoride membranes (Millipore Sigma, Burlington, MA, USA). Next, the membranes were blocked with 5% (w/v) bovine serum albumin and incubated with primary antibodies against *Caspase-3* (GB11767c; Servicebio, Wuhan, China), *Caspase-7* (27155-1-ap; Proteintech, Chicago, IL, USA), *Caspase-9* (GB11053-1; Servicebio, Wuhan, Hubei, China), *BCL2* (BF9103; Affinity Biosciences, OH, USA), and β-actin (8227; Abcam, Cambridge, MA, USA), and visualized using an ECL Western blot detection kit (Thermo Scientific, Rockford, IL, USA) and ImageQuant LAS 4000 (GE Healthcare, Piscataway, NJ, USA).

2.4. Dual Luciferase Reporter Assay

Wild-type and mutated sequences of the 3′UTR of chi-miR-99b-3p were synthesized and cloned into psiCHECK-2 plasmid (Promega, Madison, WI, USA). Subsequently, the plasmids were cotransfected with the miRNA mimic or negative control miRNA into HEK293T cells. At 30 h post-transfection, luciferase activity was detected using a Dual-Glo® Luciferase Assay System (Promega) according to the manufacturer's instructions.

2.5. Cell Proliferation Assays

Cells were seeded into a 96-well plate and cell proliferation was measured using a cell counting kit-8 (CCK-8) reagent (Beyotime, Shanghai, China) 30 h after transfection according to the manufacturer's protocol.

2.6. Transcriptome Sequencing Analysis

Total RNA from the miRNA-treated cells was purified with beads containing oligo (dT) and then used to generate cDNA with adapter addition to generate cDNA fragments by PCR. The products were sequenced on a BGISEQ-500 system (BGI, Shenzhen, China). Clean reads were obtained by removing reads containing adapter, unknown ploy N, and low-quality reads from the raw data. Bowtie 2 (V2.2.5) (http://bowtie-bio.sourceforge.net/bowtie2/index.shtml, Baltimore, USA) (accessed on 2 March 2022) was used to align the clean reads to produce sets. Differentially expressed genes (DEGs) were quantified using the cutoff $|\log2 \text{ fold change}| > 1$, $p \leq 0.05$ by DESeq2 (https://bioconductor.org/packages/release/bioc/html/DESeq2.html) (accessed on 2 March 2022). The pathway enrichment analysis was performed by searching against the KEGG database. Significant terms with correction and $p \leq 0.05$ were considered statistically significant.

2.7. Small RNA Sequencing Analysis

Total RNA from the miRNA-treated cells was extracted using TRIzol reagent (Invitrogen, Carlsbad, CA, USA) and qualified using an Agilent 2100 Bioanalyzer (Thermo Fisher Scientific, Waltham, MA, USA). Libraries were prepared with a Small RNA Cloning Kit (TaKaRa, Dalian, China) and sequenced on a BGISEQ-500 platform (BGI, Shenzhen, China). Clean tags were mapped against miRBase for miRNA with Bowtie 2 (V2.2.5). MiRanda (v2.041) (https://www.cs.kent.ac.uk/people/staff/dat/miranda/downloads/cygwin/) (accessed on 2 March 2022), RNAhybrid (v2.1) (https://bibiserv.cebitec.uni-bielefeld.de/rnahybrid, Bielefeld, Germany) (accessed on 2 March 2022), and TargetScan (v5.0) (https://www.targetscan.org/vert_50/, Cambridge, USA) (accessed on 2 March 2022) were used to predict the target genes of miRNAs. Only overlapping genes were selected as candidate targets of the DE miRNAs. Then, the candidate targets were further screened by the DE genes above, and only genes recognized by the DE genes and predicted targets were considered as candidate genes targeted by the DE miRNAs. To annotate gene functions, all target genes were aligned against the Kyoto Encyclopedia of Genes (KEGG) database using phyper in R software (v3.3.1) (https://www.r-project.org/) (accessed on 2 March 2022). Differential expression analysis was performed using DEGseq (v1.4.5) (https://www.bioconductor.org/packages/release/bioc/html/DEGseq.html) (accessed on 2 March 2022) under the standard of $|\log2$ fold change$| \geq 1$ and Q value ≤ 0.001. Sequencing data were deposited into the NCBI SRA database with accession number PRJNA844520. The same total RNA from miRNA treated cells with triplicate biological replicates was used for transcriptome and small RNA-sequencing analysis.

2.8. DEmiRNA and DEG Interaction Analysis

The interaction network between the DE miRNAs and the DE genes was constructed using Cytoscape v.3.7.2 software (https://cytoscape.org/release_notes_3_7_2.html) (accessed on 2 March 2022) as previously described [13].

2.9. Statistical Analysis

One-way analysis of variance and Tukey's test were used for statistical analysis. Results are presented as the mean ± SEM of at least triplicate replicates. For statistically significant differences, $p < 0.05$ was required. All statistical analyses were carried out using SPSS 24.0 software (https://www.ibm.com/support/pages/downloading-ibm-spss-statistics-24) (accessed on 2 March 2022).

3. Results

3.1. Caspase-3 and NCOR1 Are Targeted by chi-miR-99b-3p

To determine the molecular function of chi-miR-99b-3p in muscle development, we first confirmed the mature sequence of chi-miR-99b-3p, and found that it was conserved in different species (Figure 1a); then, the potential targets of chi-miR-99b-3p were predicted by miRanda, Pita, and TargetScan software, as shown in Table S3. A total of 35 genes related to the development of skeletal muscle were predicted to be targeted by chi-miR-99b-3p. Among them, *Caspase-3*, histone deacetylase 9 (*HDAC9*), and nuclear receptor corepressor 1 (*NCOR1*) were the predicted targets of chi-miR-99b-3p. To further elucidate the relationship between chi-miR-99b-3p and the above three potential targets, we measured the expressions of chi-miR-99b-3p, *Caspase-3*, *HDAC9*, and *NCOR1* on different days post-culture of SMSCs during their differentiation (Figure 1b,c). The results showed that the endogenous expression of chi-miR-99b-3p increased, and the mRNA expressions of *Caspase-3*, *HDAC9*, and *NCOR1* decreased at 3 and 5 d of SMSC differentiation (Figure 1c). The expressions of *Caspase-3*, *HDAC9*, and *NCOR1* were significantly increased on day 7, with no significant difference in chi-miR-99b-3p expression between 1 d and 7 d of SMSC differentiation (Figure 1b).

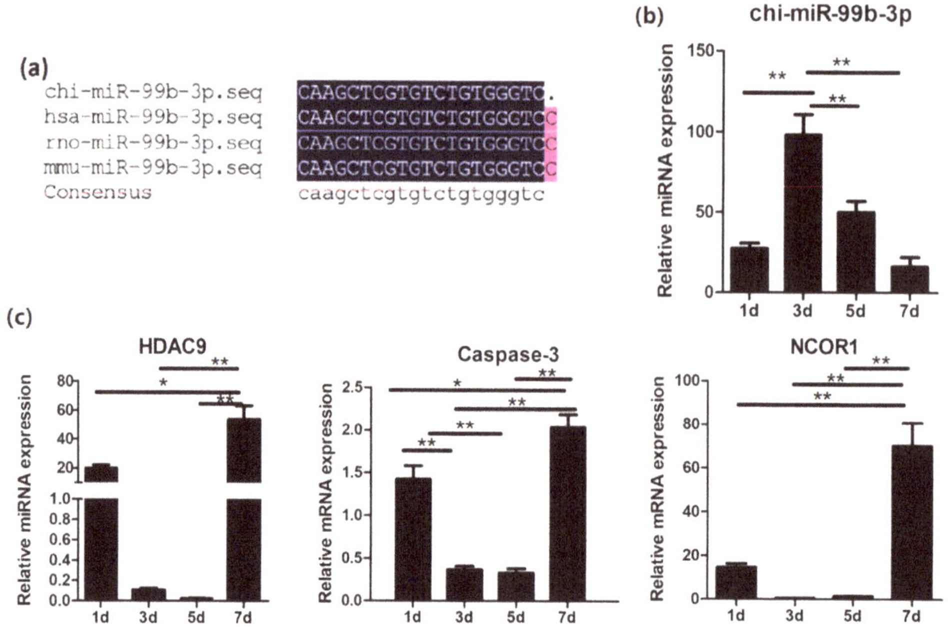

Figure 1. Relative expression of chi-miR-99b-3p and potential targets during the differentiation of SMSCs. (**a**) chi-miR-99b-3p was conserved across species. (**b**) Relative expression of chi-miR-99b-3p during the differentiation of SMSCs. (**c**) mRNA expressions of selected targets during the differentiation of SMSCs. All results are presented as means ± SEM. $n = 3$. * $p < 0.05$; ** $p < 0.01$.

We then used a vector psiCHECK-2 to construct 3′-UTR vectors and the corresponding mutant vectors for the potential target genes *NCOR1*, *HDAC9*, and *Caspase-3*. As shown in Figure 2a,b, enhanced chi-miR-99b-3p significantly inhibited the luciferase activity of NCOR1-3′-UTR-WT and CASP3-3′-UTR-WT ($p < 0.01$), but not of HDAC9 NCOR1-3′-UTR-WT (Figure S1), indicating that *NCOR1* and *CASP-3* were directly targeted by chi-miR-99b-3p. Furthermore, the levels of chi-miR-99b-3p and its target genes (*Caspase-3* and *NCOR1*) were measured at 30 h in goat skeletal muscle cells after transfection with a chi-miR-99b-3p mimic or inhibitor, and revealed that enhanced chi-miR-99b-3p expression repressed the mRNA expression of *Caspase-3* and *NCOR1*, whereas they were increased after the transfection with a chi-miR-99b-3p inhibitor (Figure 2c).

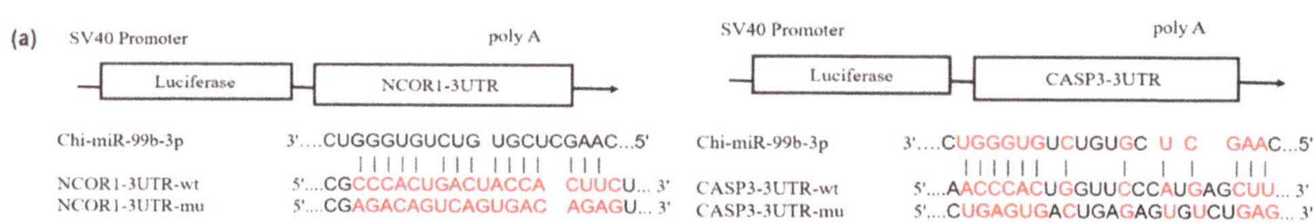

Figure 2. *Cont.*

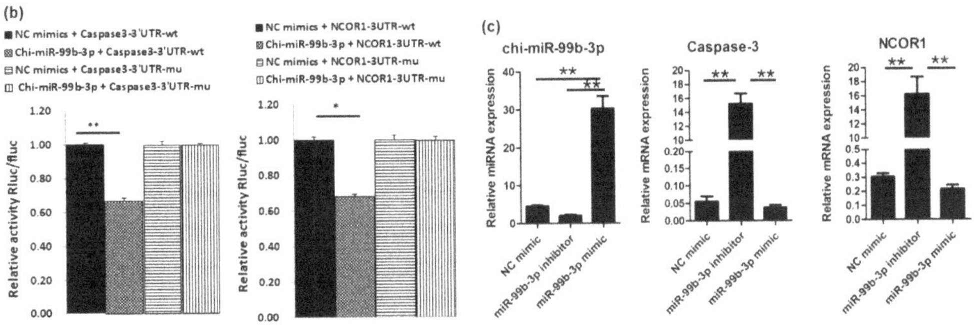

Figure 2. *Caspase-3* and *NCOR1* are the target genes of chi-miR-99b-3p. (**a**) Diagram depicting the binding sites of chi-miR-99b-3p on the 3′-UTR of *Caspase-3* and *NCOR1* (highlighted in red). (**b**) Detection of relative luciferase activity. wt, wild-type. mut, mutant-type. All results are presented as means ± SEM. $n = 3$. * $p < 0.05$; ** $p < 0.01$. (**c**) Overexpression or suppression of chi-miR-99b-3p impacted the expressions of *Caspase-3* and *NCOR1* at the mRNA level.

3.2. chi-miR-99b-3p Targets Caspase-3 and NCOR1 to Regulate Cell Proliferation and Differentiation

Compared with NC, the mRNA and protein expressions of the apoptosis-related genes *Caspase-9*, *7*, and *BCL2* were all successfully downregulated by the miR-99b-3p mimic, but were activated after transfection with the miR-99b-3p inhibitor (Figure 3a). The effects of cell viability and cell-differential-related gene expression at 30 h post-transfection were further analyzed (Figure 3b,c). Overexpression of miR-99b-3p significantly increased cell viability and inhibited the protein levels of *BCL2*, while the mRNA expression levels of apoptosis-related genes *Caspase-9*, *Caspase-7*, and *BCL2* were significantly increased in SMSCs after chi-miR-99b-3p downregulation compared with the control mimic-treated SMSCs (Figure 3a). Additionally, the mRNA expression levels of *MyoG* and *Pax7* were significantly increased in chi-miR-99b-3p-inhibited cells (Figure 3c).

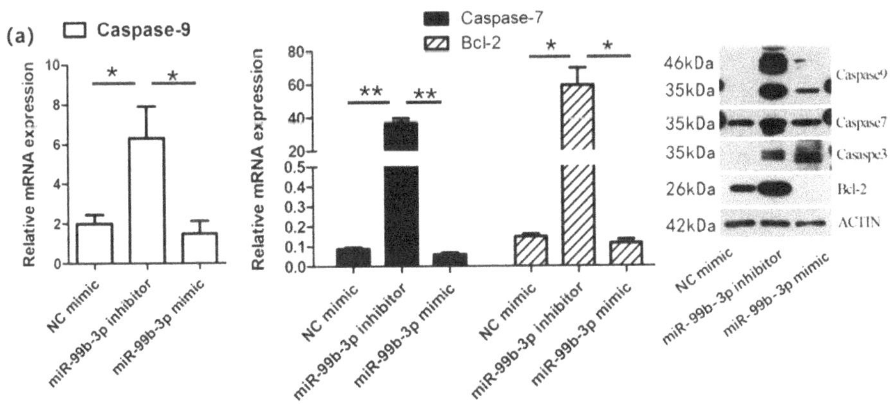

Figure 3. *Cont.*

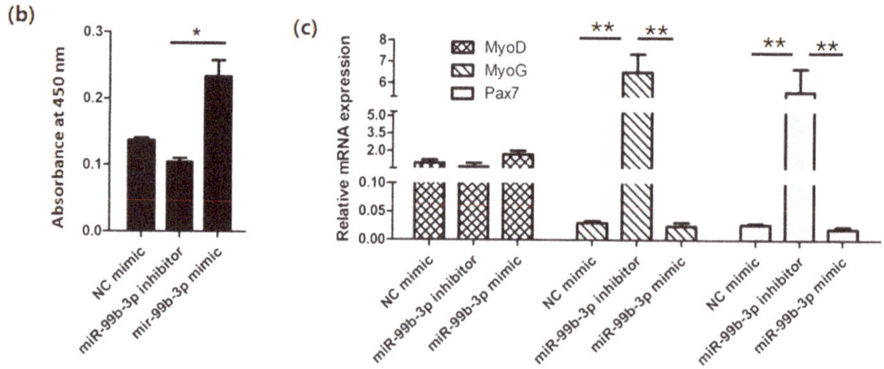

Figure 3. chi-miR-99b-3p regulates cell proliferation and differentiation via the intrinsic apoptotic pathway. (**a**) mRNA and protein levels of apoptosis-related genes after transfection with chi-miR-99b-3p mimic or inhibitor. (**b**) Cell counts detected by CCK-8 analysis. (**c**) mRNA levels of differentiation-related genes after transfection with chi-miR-99b-3p mimic or inhibitor. * $p < 0.05$; ** $p < 0.01$. Original Figure 3a in Figure S2.

3.3. mRNA Profile Analysis after Upregulation of chi-miR-99b-3p

Using mRNA sequencing analysis, 1490 DE genes were obtained |log2 fold change |≥1 and Q value ≤ 0.001 in miR-99b-3p mimic-transfected cells compared with the NC miRNA-treated cells (Figure 4a). Further functional analysis of these DE genes showed that after overexpression of chi-miR-99b-3p, the DEGs in the skeletal muscle cells were mainly involved in the cell cycle, relaxin signaling pathway, DNA replication, and protein digestion and absorption (Figure 4b), demonstrating that chi-miR-99b-3p was mainly involved in cell cycle regulation. In addition, the DE genes associated with the cell cycle process are listed in Table S4. Most of the cell-cycle-related genes were downregulated in SMSCs after chi-miR-99b-3p upregulation; among them, the pro-apoptosis-related gene *BCL2*, which is demonstrated to be downregulated in Figure 3a, was decreased in chi-miR-99b-3p upregulated cells based on sequencing data. However, we did not see downregulation of *Caspase-3* or *NCOR1* in the SMSCs after chi-miR-99b-3p upregulation, based on the filter condition of |log2 fold change | ≥ 1 and Q value ≤ 0.001. In addition, we also found 34 genes, including *SH2D4A*, *ABCB10*, *B3GNT6*, *OPCML*, and *SPATA13*, predicted to be candidate targets of chi-miR-99b-3p, which were downregulated in the cells after miR-99b-3p upregulation (Table S5).

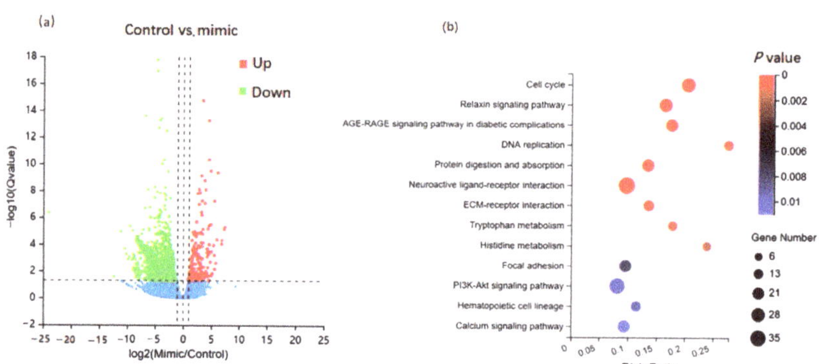

Figure 4. Differentially expressed mRNAs after chi-miR-99b-3p up-regulation. (**a**) A volcano plot representing DEGs after chi-miR-99b-3p overexpression (**b**) Functional enrichment analysis of DE genes.

3.4. miRNA Profile Analysis after chi-miR-99b-3p Upregulation

To evaluate the potential miRNAs influenced by chi-miR-99b-3p upregulation, small RNA-sequencing analysis was performed, and the DE miRNAs are shown in Table S6. Here, chi-miR-99b-3p upregulation induced 47 DE miRNAs, of which 16 were upregulated and 31 downregulated, including 30 known mature miRNAs (Figure 5a) and 17 novel miRNAs. Among them, chi-miR-199c-3p, chi-miR-199a-3p, chi-let-7c-3p, chi-let-7c-5p, chi-miR-10a-5p, chi-miR-125b-3p, and chi-miR-99a-5p represented the seven most highly expressed miRNAs in SMSCs after chi-miR-99b-3p upregulation. We next explored the potential targets of the DE miRNAs by screening the DE genes mentioned above, and overlapping genes were considered as candidate genes targeted for the DE miRNAs (Table S7), and were subsequently used for KEGG analysis. Further functional analysis revealed that these DE miRNAs were involved in the cell cycle, relaxin signaling pathway, DNA replication, and protein digestion and absorption (Figure 5b), and this was consistent with the results seen for the DE genes (Figure 4b).

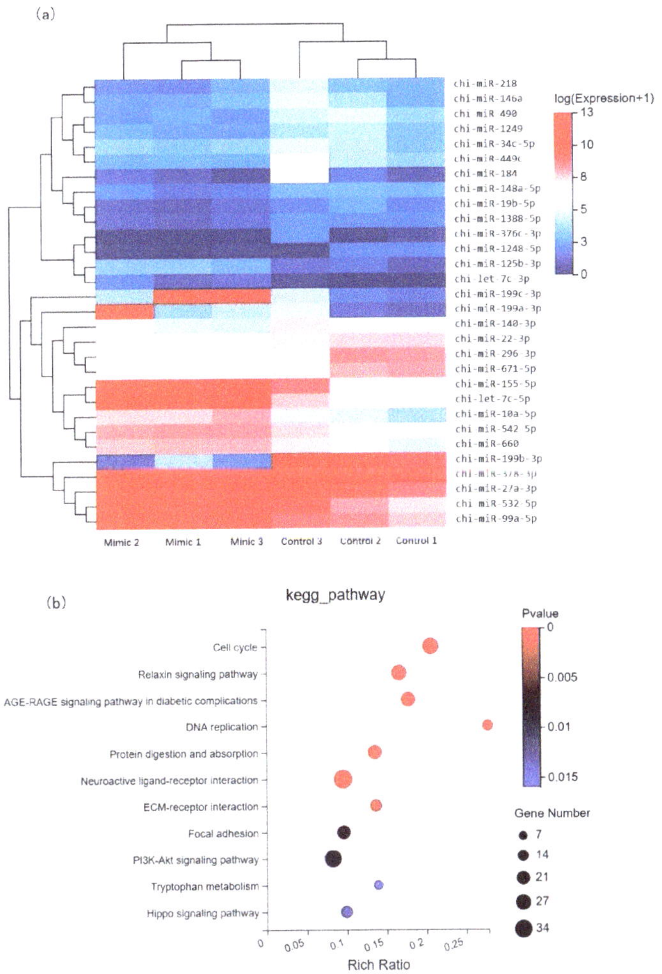

Figure 5. Differentially expressed miRNA expression in SMSCs after miR-99b-3p overexpression. (**a**) Heat map of known DE miRNAs in SMSCs after chi-miR-99b-3p overexpression. (**b**) KEGG functional analysis DE miRNA in SMSCs after chi-miR-99b-3p upregulation.

3.5. Integration Analysis of the DE miRNAs and DEGs

As shown in Figure 6, most miRNAs had multiple potential target genes, while different miRNAs could regulate the same target. For example, miR-671-5p was the regulator of *TTC4*, *CASQ2*, *ORC1*, *SLC6A17*, and *AQP7*; whereas miR-199a-3p, miR-199b-3p, and miR-199c-3p, which belong to the miR-199 cluster, could regulate the expressions of *ITGA6* and *TXNIP*. In addition, chi-miR-99b-3p could regulate *ORC1*, *SLC6A17*, and *TTC4*, which were downregulated in SMSCs after miR-99b-3p overexpression (Table S3).

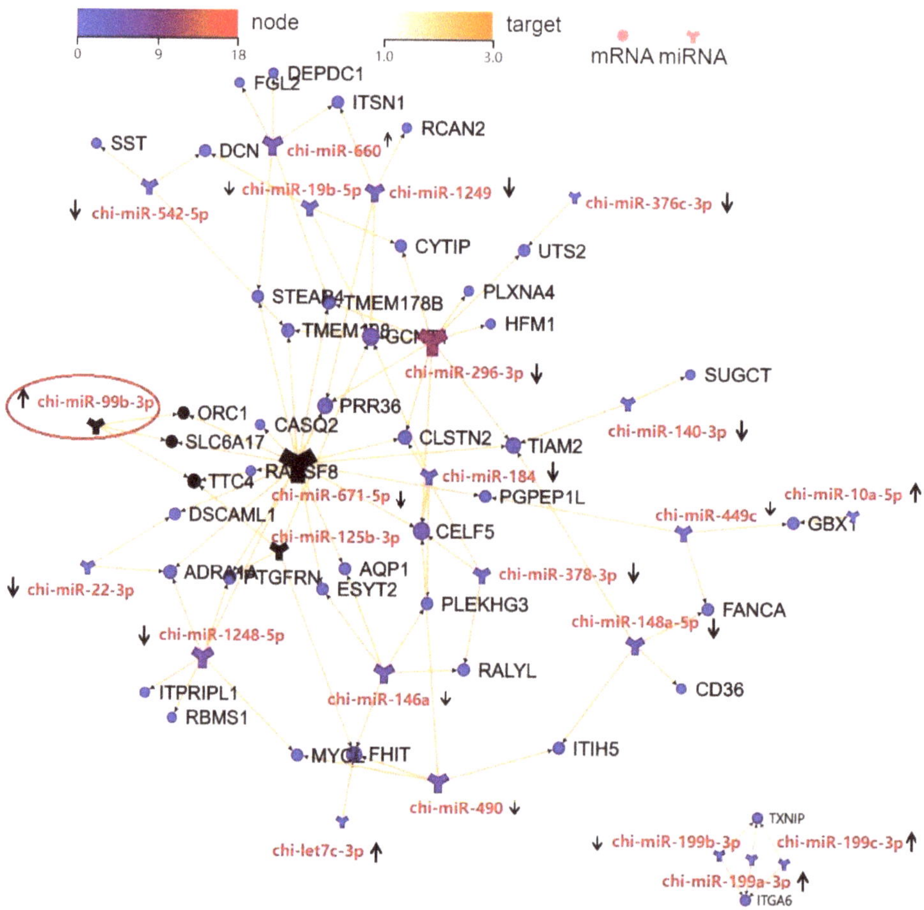

Figure 6. miRNA–mRNA interaction network in SMSCs after miR-99b-3p overexpression. DE miRNAs and mRNAs were considered.

4. Discussion

Skeletal muscle is the most abundant tissue in the body and possesses remarkable regenerative capacity. It consists of a diversity of cell types including multinucleated myofibers, SMSCs, endothelial cells, pericytes, mesenchymal progenitor cells, and immune cells [15]. SMSCs participate in postnatal muscle growth and regeneration in animals, which has been shown to be regulated by miRNAs [2,16]. For example, BMP2 is targeted by miR-378 to regulate the proliferation of sheep myoblast [17] and inhibit apoptosis in skeletal muscle [16]. miR-181a regulates the proliferation and differentiation of Hu sheep SMSCs by directly targeting YAP1 [2]. Here, we found that miR-99b-3p was able to promote cell proliferation and inhibit apoptosis in Anhui white goat SMSCs by blocking the intrinsic apoptosis pathway via targeting *caspase-3* and *NCOR1*. Moreover, integrated mRNA–miRNA analysis revealed that the overexpression of miR-99b-3p upregulated several cell-cycle-associated miRNAs that regulate the cell cycle in SMSCs. These results suggested that miR-99b-3p is an important component of skeletal muscle development.

Skeletal muscle development relies on the proliferation, differentiation, and regeneration of SMSCs to form new myofibers [15,18]. The myogenic regulatory factors, *MyoD* and *MyoG*, are the core components of the myogenic pathway, which play important roles during myogenesis and skeletal muscle differentiation [19]. *Pax7* is only expressed in satellite cells and is involved in regulating the self-renewal of adult skeletal muscle [20]. We found that the expression levels of *MyoG* and *Pax7* significantly increased after miR-99b-3p down-expression, while no significant difference was found with miR-99b-3p overexpression. Furthermore, we detected the expression level of miR-99b-3p during the differentiation process, and found that it first increased but then decreased, while the mRNA expression levels of *Caspase-3* and *NCOR1* were opposite to that of chi-miR-99b-3p. Additionally, the dual-luciferase reporter system confirmed the binding relationship between miR-99b-3p and Caspase-3, as well as *NCOR1*. These results indicated that miR-99b-3p might participate in the cell cycle regulation process of SMSCs. Additionally, we found that the expression levels of *MyoD* and *MyoG* gradually increased during the differentiation process (Figure S3), especially in the 7 d of SMSCs differentiation, when the expression of chi-miR-99b-3p decreased. However, whether chi-miR-99b-3p is related to the regulation of cell differentiation needs to be further explored.

Caspases are the main initiators of apoptosis and play crucial roles in mammals. *caspase-8*, *-9* and *-10* are initiator caspases, involved in the onset and amplification of apoptosis; *caspase-3*, *-6*, and *-7* are effector caspases, involved in the execution of apoptotic programs [21]. High expression of *caspase-3* was associated with augmented muscle protein catabolism and promoted apoptosis of SMSCs in hemodialysis patients [22]. Down-regulation of *Caspase-3* protects denervation-induced muscle atrophy through inhibiting apoptosis [23]. Furthermore, *NCOR1* is a master regulator of mitochondrial metabolism in muscle, which is inversely associated with key mitochondrial and MyoGenic genes [24]. These studies found that *NCOR1*-null mice had increased muscle mass, which was accompanied by increases in the number and activity of mitochondria in muscle cells, and significantly enhanced exercise capacity [25]. In addition, a previous study found that miR-99b-3p promoted the proliferation of hepatocellular carcinoma [8]. Here, we consistently found that miR-99b-3p inhibited the expressions of *caspase-3*, *-7*, *-9*, and *BCL2*, which promoted the intrinsic apoptotic pathway [26]. In addition, CCK-8 assays showed that the miR-99b-3p mimic promoted cell proliferation, while the miR-99b-3p inhibitor suppressed cell proliferation. *Caspase-3* and *NCOR1* are direct targets of miR-99b-3p. Thus, our findings indicated that miR-99b-3p might act as a "gatekeeper" for the intrinsic apoptotic pathway in skeletal muscle, which provides further insights into the mechanisms underlying the regulation of intrinsic skeletal muscle apoptosis.

To further investigate the detailed regulatory role of miR-99b-3p in skeletal muscle, RNA sequencing was used to assess the integrated miRNA–mRNA network changes after miR-99b-3p upregulation in SMSCs. We identified that the DE genes were enriched in the pathways associated with cell cycle, relaxin signaling, DNA replication, and protein digestion and absorption. Among them, the cell cycle pathway was the most enriched pathway, and the 48 DE genes were screened to participate in this pathway. Most of the DE genes were significantly downregulated in SMSCs after miR-99b-3p upregulation, including the pro-apoptosis-related gene *BCL2*, further suggesting that overexpression of miR-99b-3p can inhibit apoptosis in SMSCs. However, we did not find any downregulation of *Caspase-3* or *NCOR1* in SMSCs after miR-99b-3p overexpression by sequencing, which might have been due to the limitation of sequencing depth.

Additionally, we found that several miRNAs, including miR-199c-3p, miR-199a-3p, let-7c-3p, let-7c-5p, miR-10a-5p, miR-125b-3p, and miR-99a-5p, were highly expressed in SMSCs after miR-99b-3p upregulation, and these have been reported to enhance muscle regeneration [27–30] or promote cell proliferation in some cancers [31–33]. Furthermore, miR-199c-3p, miR-199a-3p, miR-99a-3p, and miR-99b-3p belong to the same miRNA cluster, suggesting that overexpression of miR-99b-3p induces the expression of a series of cell cycle regulation related miRNAs, further promoting cell proliferation and differentiation in SMSCs. Subsequently, KEGG enrichment analysis of these DE miRNAs revealed that most of the DE genes were predicted to be targeted by the DE miRNAs and were involved in the cell cycle process, further confirming a role for miR-99b-3p in SMSCs.

Although we identified potential miRNA–mRNA interactions regulated by miR-99b-3p and involved in goat skeletal muscle development, it is worth noting that further experimental studies are required to validate these miRNA–mRNA interactions in goat skeletal development.

5. Conclusions

The results of this preliminary study confirmed that chi-miR-99b-3p is an miRNA associated with the proliferation and differentiation of goat SMSCs. Overexpression of miR-99b-3p promoted SMSC proliferation by targeting *Caspase-3* and *NCOR1* and inhibiting the expression of apoptosis-related genes. Upregulation of miR-99b-3p promoted the expression of several miRNAs including miR-199c-3p, miR-199a-3p, let-7c-3p, let-7c-5p, miR-10a-5p, miR-125b-3p, and miR-99a-5p, which are closely related to differentiation, proliferation, and apoptosis. These results suggested that miR-99b-3p may regulate the differentiation, proliferation, and apoptosis of SMSCs through *Caspase-3* and *NCOR1* in goats.

Supplementary Materials: The following supporting information can be downloaded at: https://www.mdpi.com/article/10.3390/ani12182368/s1. Figure S1: *HDAC9* is not the target gene of chi-miR-99b-3p, (a) diagram depicting the binding sites of chi-miR-99b-3p on the 3′-UTR of *HDAC9*, (highlighted in red), (b) detection of relative luciferase activity; Figure S2: Original figures for figure 3a; Figure S3: mRNA expressions of selected targets during the differentiation of SMSCs; Table S1: Mimic/inhibitor sequence used in this study; Table S2: Details of primer sequence used in this study; Table S3: Potential targets of chi-miR-99b-3p involved in the regulation of skeletal muscle development; Table S4: DE genes associated with cell-cycle pathway; Table S5: Potential targets of miR-99b-3p screened by DE genes; Table S6: List of DE miRNA in SMSCs after chi-miR-99b-3p upregulation; Table S7: Overlapping genes screened by DE genes and predicted targeted of DE miRNAs.

Author Contributions: Conceptualization, R.L. and L.Z.; methodology, Y.L. (Yuexia Lin) and J.D.; software, L.Z.; validation, R.L. and Y.L. (Yuhua Lv), J.D. and D.Z.; formal analysis, R.L.; investigation, L.Z.; resources, Y.L. (Yuexia Lin); writing—original draft preparation, R.L.; writing—review and editing, R.L., D.Z. and Y.L. (Yuexia Lin); visualization, L.Z.; supervision, Y.L. (Yuexia Lin); funding acquisition, R.L. All authors have read and agreed to the published version of the manuscript.

Funding: This research was funded by the National Natural Science Foundation of China (Grant No. 31902149).

Institutional Review Board Statement: The research was conducted according to the approved protocols of the Ethics and Animal Welfare Committee of the Shanghai Academy of Agricultural Sciences (Ethic approval number: SAASPZ0521040).

Informed Consent Statement: Not applicable.

Data Availability Statement: The results from data analyses performed in this study are included in this article and the tables. The raw sequencing data were deposited into the NCBI SRA database with the free accession number PRJNA844520.

Conflicts of Interest: The authors declare no conflict of interest.

References

1. Mukund, K.; Subramaniam, S. Skeletal muscle: A review of molecular structure and function, in health and disease. *Wiley Interdiscip. Rev. Syst. Biol. Med.* **2020**, *12*, e1462. [CrossRef] [PubMed]
2. He, M.; Zhang, W.; Wang, S.; Ge, L.; Cao, X.; Wang, S.; Yuan, Z.; Lv, X.; Getachew, T.; Mwacharo, J.M.; et al. MicroRNA-181a regulates the proliferation and differentiation of Hu sheep skeletal muscle satellite cells and targets the YAP1 gene. *Genes* **2022**, *13*, 520. [CrossRef] [PubMed]
3. Sui, M.; Zheng, Q.; Wu, H.; Zhu, L.; Ling, Y.; Wang, L.; Fang, F.; Liu, Y.; Zhang, Z.; Chu, M.; et al. The expression and regulation of miR-1 in goat skeletal muscle and satellite cell during muscle growth and development. *Anim. Biotechnol.* **2020**, *31*, 455–462. [CrossRef] [PubMed]
4. Lyu, M.; Wang, X.; Meng, X.; Qian, H.; Li, Q.; Ma, B.; Zhang, Z.; Xu, K. Chi-miR-487b-3p inhibits goat myoblast proliferation and differentiation by targeting IRS1 through the IRS1/PI3K/Akt signaling pathway. *Int. J. Mol. Sci.* **2021**, *23*, 115. [CrossRef] [PubMed]
5. Zhuge, A.; Li, B.; Yuan, Y.; Lv, L.; Li, Y.; Wu, J.; Yang, L.; Bian, X.; Wang, K.; Wang, Q.; et al. Lactobacillus salivarius LI01 encapsulated in alginate-pectin microgels ameliorates D-galactosamine-induced acute liver injury in rats. *Appl. Microbiol. Biotechnol.* **2020**, *104*, 7437–7455. [CrossRef] [PubMed]
6. Hu, Z.; Xu, H.; Lu, Y.; He, Q.; Yan, C.; Zhao, X.; Tian, Y.; Yang, C.; Zhang, Z.; Qiu, M.; et al. MUSTN1 is an indispensable factor in the proliferation, differentiation and apoptosis of skeletal muscle satellite cells in chicken. *Exp. Cell Res.* **2021**, *407*, 112833. [CrossRef] [PubMed]
7. Cheng, N.; Liu, C.; Li, Y.; Gao, S.; Han, Y.C.; Wang, X.; Du, J.; Zhang, C. MicroRNA-223-3p promotes skeletal muscle regeneration by regulating inflammation in mice. *J. Biol. Chem.* **2020**, *295*, 10212–10223. [CrossRef]
8. Yao, X.; Zhang, H.; Liu, Y.; Liu, X.; Wang, X.; Sun, X.; Cheng, Y. MiR-99b-3p promotes hepatocellular carcinoma metastasis and proliferation by targeting protocadherin 19. *Gene* **2019**, *698*, 141–149. [CrossRef]
9. Jakob, M.; Mattes, L.M.; Küffer, S.; Unger, K.; Hess, J.; Bertlich, M.; Haubner, F.; Ihler, F.; Canis, M.; Weiss, B.G.; et al. MicroRNA expression patterns in oral squamous cell carcinoma: Hsa-miR-99b-3p and hsa-miR-100-5p as novel prognostic markers for oral cancer. *Head Neck* **2019**, *41*, 3499–3515. [CrossRef]
10. Chang, S.; Gao, Z.; Yang, Y.; He, K.; Wang, X.; Wang, L.; Gao, N.; Li, H.; He, X.; Huang, C. MiR-99b-3p is induced by vitamin D3 and contributes to its antiproliferative effects in gastric cancer cells by targeting HoxD3. *Biol. Chem.* **2019**, *400*, 1079–1086. [CrossRef]
11. Yu, Y.H.; Zhang, Y.H.; Ding, Y.Q.; Bi, X.Y.; Yuan, J.; Zhou, H.; Wang, P.X.; Zhang, L.L.; Ye, J.T. MicroRNA-99b-3p promotes angiotensin II-induced cardiac fibrosis in mice by targeting GSK-3β. *Acta Pharmacol. Sin.* **2021**, *42*, 715–725. [CrossRef]
12. Mitchell, C.J.; D'Souza, R.F.; Schierding, W.; Zeng, N.; Ramzan, F.; O'Sullivan, J.M.; Poppitt, S.D.; Cameron-Smith, D. Identification of human skeletal muscle miRNA related to strength by high-throughput sequencing. *Physiol. Genom.* **2018**, *50*, 416–424. [CrossRef]
13. Liao, R.; Lv, Y.; Zhu, L.; Lin, Y. Altered expression of miRNAs and mRNAs reveals the potential regulatory role of miRNAs in the developmental process of early weaned goats. *PLoS ONE* **2019**, *14*, e0220907. [CrossRef]
14. Wang, S.; Ma, T.; Zhao, G.; Zhang, N.; Tu, Y.; Li, F.; Cui, K.; Bi, Y.; Ding, H.; Diao, Q. Effect of age and weaning on growth performance, rumen fermentation, and serum parameters in lambs fed starter with limited ewe-lamb interaction. *Animals* **2019**, *9*, 825. [CrossRef]
15. Asakura, A.; Hirai, H.; Kablar, B.; Morita, S.; Ishibashi, J.; Piras, B.A.; Christ, A.J.; Verma, M.; Vineretsky, K.A.; Rudnicki, M.A. Increased survival of muscle stem cells lacking the MyoD gene after transplantation into regenerating skeletal muscle. *Proc. Natl. Acad. Sci. USA* **2007**, *104*, 16552–16557. [CrossRef]
16. Li, Y.; Jiang, J.; Liu, W.; Wang, H.; Zhao, L.; Liu, S.; Li, P.; Zhang, S.; Sun, C.; Wu, Y.; et al. MicroRNA-378 promotes autophagy and inhibits apoptosis in skeletal muscle. *Proc. Natl. Acad. Sci. USA* **2018**, *115*, E10849–E10858. [CrossRef]
17. Lu, Z.; Du, L.; Liu, R.; Di, R.; Zhang, L.; Ma, Y.; Li, Q.; Liu, E.; Chu, M.; Wei, C. MiR-378 and BMP-Smad can influence the proliferation of sheep myoblast. *Gene* **2018**, *674*, 143–150. [CrossRef]
18. Kowalski, K.; Kołodziejczyk, A.; Sikorska, M.; Płaczkiewicz, J.; Cichosz, P.; Kowalewska, M.; Stremińska, W.; Jańczyk-Ilach, K.; Koblowska, M.; Fogtman, A.; et al. Stem cells migration during skeletal muscle regeneration-the role of Sdf-1/Cxcr4 and Sdf-1/Cxcr7 axis. *Cell Adhes. Migr.* **2017**, *11*, 384–398. [CrossRef]
19. Zanou, N.; Gailly, P. Skeletal muscle hypertrophy and regeneration: Interplay between the myogenic regulatory factors (MRFs) and insulin-like growth factors (IGFs) pathways. *Cell. Mol. Life Sci.* **2013**, *70*, 4117–4130. [CrossRef]
20. Chen, Y.; Lin, G.; Slack, J.M. Control of muscle regeneration in the Xenopus tadpole tail by Pax7. *Development* **2006**, *133*, 2303–2313. [CrossRef]
21. Budihardjo, I.; Oliver, H.; Lutter, M.; Luo, X.; Wang, X. Biochemical pathways of caspase activation during apoptosis. *Annu. Rev. Cell Dev. Biol.* **1999**, *15*, 269–290. [CrossRef]
22. Boivin, M.A.; Battah, S.I.; Dominic, E.A.; Kalantar-Zadeh, K.; Ferrando, A.; Tzamaloukas, A.H.; Dwivedi, R.; Ma, T.A.; Moseley, P.; Raj, D.S. Activation of caspase-3 in the skeletal muscle during haemodialysis. *Eur. J. Clin. Investig.* **2010**, *40*, 903–910. [CrossRef]
23. Plant, P.J.; Bain, J.R.; Correa, J.E.; Woo, M.; Batt, J. Absence of caspase-3 protects against denervation-induced skeletal muscle atrophy. *J. Appl. Physiol. (1985)* **2009**, *107*, 224–234. [CrossRef]

24. Yamamoto, H.; Williams, E.G.; Mouchiroud, L.; Cantó, C.; Fan, W.; Downes, M.; Héligon, C.; Barish, G.D.; Desvergne, B.; Evans, R.M.; et al. NCoR1 is a conserved physiological modulator of muscle mass and oxidative function. *Cell* **2011**, *147*, 827–839. [CrossRef]
25. NCoR1 builds bigger muscles in mice. *Bonekey Rep.* **2012**, *1*, 15. [CrossRef]
26. Opdenbosch, V.N.; Lamkanfi, M. Caspases in Cell Death, Inflammation, and Disease. *Immunity* **2019**, *50*, 1352–1364. [CrossRef]
27. Fukuoka, M.; Fujita, H.; Numao, K.; Nakamura, Y.; Shimizu, H.; Sekiguchi, M.; Hohjoh, H. MiR-199-3p enhances muscle regeneration and ameliorates aged muscle and muscular dystrophy. *Commun. Biol.* **2021**, *4*, 427. [CrossRef] [PubMed]
28. Huang, Y.; Yu, M.; Kuma, A.; Klein, J.D.; Wang, Y.; Hassounah, F.; Cai, H.; Wang, X.H. Downregulation of let-7 by electrical acupuncture increases protein synthesis in mice. *Front. Physiol.* **2021**, *12*, 697139. [CrossRef] [PubMed]
29. Cao, X.; Tang, S.; Du, F.; Li, H.; Shen, X.; Li, D.; Wang, Y.; Zhang, Z.; Xia, L.; Zhu, Q.; et al. MiR-99a-5p regulates the proliferation and differentiation of skeletal muscle satellite cells by targeting MTMR3 in chicken. *Genes* **2020**, *11*, 369. [CrossRef] [PubMed]
30. Qiu, J.; Zhu, J.; Zhang, R.; Liang, W.; Ma, W.; Zhang, Q.; Huang, Z.; Ding, F.; Sun, H. MiR-125b-5p targeting TRAF6 relieves skeletal muscle atrophy induced by fasting or denervation. *Ann. Transl. Med.* **2019**, *7*, 456. [CrossRef] [PubMed]
31. Chen, Z.; Liu, J. MicroRNA-let-7 Targets HMGA2 to Regulate the Proliferation, Migration, and Invasion of Colon Cancer Cell HCT116. *Evid. Based Complement Alternat. Med.* **2021**, *2021*, 2134942. [CrossRef]
32. Wang, W.; Zhan, R.; Zhou, J.; Wang, J.; Chen, S. MiR-10 targets NgR to modulate the proliferation of microglial cells and the secretion of inflammatory cytokines. *Exp. Mol. Pathol.* **2018**, *105*, 357–363. [CrossRef]
33. Peng, T.; Peng, J.J.; Miao, G.Y.; Tan, Z.Q.; Liu, B.; Zhou, E. MiR-125/CDK2 axis in cochlear progenitor cell proliferation. *Mol. Med. Rep.* **2021**, *23*, 102. [CrossRef]

animals MDPI

Review

Candidate Genes and Their Expressions Involved in the Regulation of Milk and Meat Production and Quality in Goats (*Capra hircus*)

Jose Ignacio Salgado Pardo [1], Juan Vicente Delgado Bermejo [1], Antonio González Ariza [1], José Manuel León Jurado [2], Carmen Marín Navas [1], Carlos Iglesias Pastrana [1], María del Amparo Martínez Martínez [1] and Francisco Javier Navas González [1,3,*]

[1] Department of Genetics, Faculty of Veterinary Sciences, University of Córdoba, 14014 Córdoba, Spain; josalgadopardo@outlook.com (J.I.S.P.); juanviagr218@gmail.com (J.V.D.B.); angoarvet@outlook.es (A.G.A.); v32manac@uco.es (C.M.N.); ciglesiaspastrana@gmail.com (C.I.P.); ib2mamaa@uco.es (M.d.A.M.M.)
[2] Agropecuary Provincial Center of Córdoba, Provincial Council of Córdoba, 14014 Córdoba, Spain; jomalejur@yahoo.es
[3] Institute of Agricultural Research and Training (IFAPA), Alameda del Obispo, 14004 Córdoba, Spain
* Correspondence: fjng87@hotmail.com; Tel.: +34-63-853-5046 (ext. 621262)

Simple Summary: During the present decade, highly selected caprine farming has increased in popularity due to the hardiness and adaptability inherent to goats. Recent advances in genetics have enabled the improvement in goat selection efficiency. The present review explores how genetic technologies have been applied to the goat-farming sector in the last century. The main candidate genes related to economically relevant traits are reported. The major source of income in goat farming derives from the sale of milk and meat. Consequently, yield and quality must be specially considered. Meat-related traits were evaluated considering three functional groups (weight gain, carcass quality and fat profile). Milk traits were assessed in three additional functional groups (milk production, protein and fat content).

Abstract: Despite their pivotal position as relevant sources for high-quality proteins in particularly hard environmental contexts, the domestic goat has not benefited from the advances made in genomics compared to other livestock species. Genetic analysis based on the study of candidate genes is considered an appropriate approach to elucidate the physiological mechanisms involved in the regulation of the expression of functional traits. This is especially relevant when such functional traits are linked to economic interest. The knowledge of candidate genes, their location on the goat genetic map and the specific phenotypic outcomes that may arise due to the regulation of their expression act as a catalyzer for the efficiency and accuracy of goat-breeding policies, which in turn translates into a greater competitiveness and sustainable profit for goats worldwide. To this aim, this review presents a chronological comprehensive analysis of caprine genetics and genomics through the evaluation of the available literature regarding the main candidate genes involved in meat and milk production and quality in the domestic goat. Additionally, this review aims to serve as a guide for future research, given that the assessment, determination and characterization of the genes associated with desirable phenotypes may provide information that may, in turn, enhance the implementation of goat-breeding programs in future and ensure their sustainability.

Keywords: breeding; SNP; genomics; does; bucks; meat; milk

1. Introduction

Caprine farming has spread to almost every country in the world, due to the good prices and high value of goat-derived products (especially milk), attracting new farmers and investors [1]. The majority of the world caprine population is located in developing countries, occupying marginal territories under extreme climate conditions and held under elder farming systems [2]. This scene contrasts with that of Europe and North America, where otherwise, high-technological and intensive conditions rule the goat industry, which is highly focused on milk production and exploiting high-performance breeds subjected to genetic selection schemes [3]. In this context, the benefits that the domestic goat has obtained, as derived from the achievements made in the areas of genetics, nutrition and animal management, are rather limited in comparison to the level of integration that such techniques account for in other species.

The aforementioned framework evidences the secondary role to which caprine has been relegated within the scope of stockbreeding history [4]. This secondary position may be the result of two main conjoined facts; the traditional disregard of the caprine species as a destructive animal for pasture [5], and consumer preferences for other domestic species, which have conferred caprine-derived products with a low international market value, thus pushing caprine production to a marginal role in farming [6].

Recent archaeological findings indicate that the domestication of the goat took place more than 10,000 years ago in the 'Fertile Crescent' (Figure 1). This region is where the first settled agricultural communities of the Middle East and Mediterranean basin are thought to have originated, and would have covered the area from the Anatolian peninsula to the eastern territories of current Iran [7]. Wild bezoar (*Capra aegagrus*) and markhor (*Capra falconeri*) are thought to be the most likely ancestors of the domestic goat, according to phylogenetic studies implementing Y chromosome AMELY and ZFY sequences [8]. Additionally, other studies evaluating the major histocompatibility complex led to the possible inclusion of the Iberian mountain goat (*Capra pyrenaica*) and Alpine ibex (*Capra ibex*) in the history of goat domestication [3].

Figure 1. Relief Representation of Goatherd with Goat and Trees, ca. 1350–1333 BCE. New Kingdom, Amarna Period. Late Dynasty 18. Limestone, 8 1/4 × 16 3/4 × 2 1/2 in., 22.5 lb. (21 × 42.5 × 6.4 cm, 10.21 kg). Brooklyn Museum, Gift of the Ernest Erickson Foundation, Inc., New York, NY, USA, 86.226.30.

A higher tolerance to human handling and a better adaptability to the driven grazing may have been determining factors, which aimed to boost the popularity of certain animal populations [9]. Apart from the early breeding objectives that were sought following the domestication of the goat, the first civilizations became interested in the functional selection objectives linked to productive traits, such as the aptitude to captive breeding, prolificacy or body size [10].

The domestication and world dissemination of the species led to the first distinctive morphological traits of the original goat populations, such as the shape of the horns and ears [3]. This source of phenotypic variability could be the result of human/artificial selection, in addition to genetic drift and founder effects [11], which may explain the appearance of the characteristic traits in a lineage as a consequence of a narrow genetic base in its original population and its isolation. This may also be evidenced by other features, such as the presence of wattles, hair length or the wide variety of possible coat colours that have been developing as other distinctive traits in the first goat populations [3].

After centuries of their relationship with humans, natural selection for caprine adaptability to different environments and the artificial selection for productive, morphological and behaviour traits led to the appearance of 576 modern domestic goat breeds [10]. The Angora goat, whose presence in Phrygia and Cilicia (current Anatolia peninsula) was described 2400 years BC [12], was the first caprine breed in which a preliminary process of standardization was attempted. However, it would not be until 1890 and 1895 when the first caprine dairy goat breed standardization would take place in Switzerland, for the Saanen and Toggenburg goat breeds, respectively [13].

Although this event was a milestone and marked the beginning of caprine milk selection history [13], most of the significant advances would have to wait until the 1960s in France, when a bovine selection model would be applied to dairy goat production [14]. This turning point was promoted by the massive growth in the French cheese industry after the Second World War, which led to the development of intensive milking farm regions connected to the cheese factories. This improvement in animal production required farms to implement highly technical support, which brought about the routinization of progeny testing and artificial insemination [13]. It would take another decade for caprine meat breeding programmes (which have been dramatically scarcer than dairy goat breeding programs) to appear, with Boer goat breeding programmes being one of the few examples, appearing by the end of the 1970s in South Africa [15] (Figure 2).

The first documented register describing a caprine-selection-focused attempt dates back to 1962 [16], when the 'Universidad de Puerto Rico' studied the most profitable crosses between Puerto Rican local goat breeds and high-performance dairy goats. The study sought the most appropriate cross that would result in animals parallelly presenting the best adaptability, through the evaluation of goat kid survival rate, and the greatest productivity, through the evaluation of milk yield.

It was three years later (1965) when the first heritability estimations were performed in caprine [17]. This advance resulted from the integration of genealogical information as a compulsory step towards animal breeding estimations, which is a crucial step during the first stages of any breeding programme. The advances not only concerned the available genetic components or biostatistical tools, but also the revolution of phenotypic data collection. Contextually, the onset of 305-day lactation normalization in 1979 [18] made it possible to objectively compare the productivity of does from different breeds and controlled at different moments within lactation [19], which not only permitted farmers' taking directed decisions based on factual data, but also laid the grounds for genetic evaluations. Still, the control of environmental effects was challenging, and reliable estimations were not feasible.

After the implementation of Best Linear Unbiased Prediction methods (BLUP) in caprine in the mid-1980s [20], along with the Animal Model [21] for breeding value calculations, the design of more complex selection schemes arose. This permitted the complete and reliable integration of genealogical information into genetic evaluations, but also the evaluation of animals that were phenotypically controlled in broadly distinct environmental conditions. This means that BLUP allowed for the separate estimation of the genetic and non-genetic (environmental) factors; hence, the heritable fraction of functional traits could be isolated and more appropriately controlled.

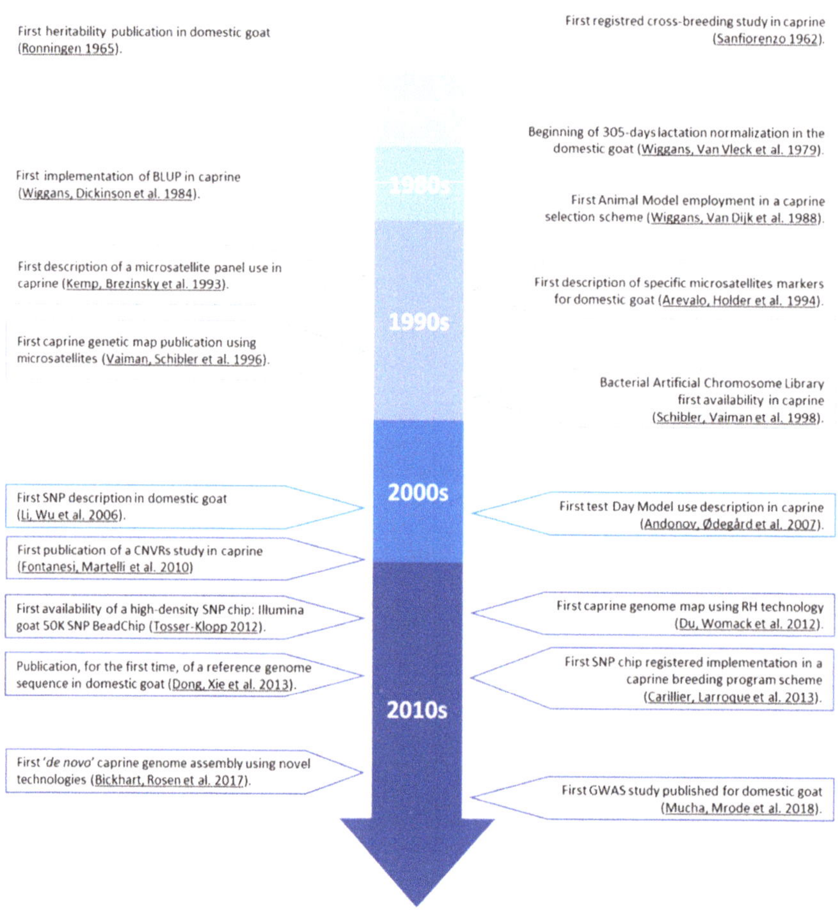

Figure 2. Timeline of caprine genetic advances milestones [16–18,20–33].

The 'Genomic Era' began with the discovery and utilization of microsatellite markers, which would be applied in the context of domestic goats as an extension of the use of bovine and ovine microsatellite panels in 1993 [22]. One year later, in 1994, the first description of caprine specific microsatellites was published [24]. The genomic information obtained from caprine microsatellite studies in these years permitted the development of studies based on the relationship between specific genomic regions, Quantitative Trait Loci (QTL), and desirable production traits [19].

QTLs are regions of the genome for which an association with the phenotypic variance of a certain trait has been proved [34]. Such an association may be supported by the fact that QTL regions may contain genes codifying for the specific regulation of the expression of a certain functional feature. For several years, many QTL were described using microsatellite genetic markers. However, even if they are still valid and preferrable when economic resources for research are scarce, the large size of some QTL [34] makes their mapping resolution and confidence intervals limited if the application of other, more efficient techniques is possible [35].

As previously mentioned, despite the fact that microsatellites offer a high degree of polymorphism for each marker, they are not as abundant as SNPs, and hence provide insufficient coverage of the genome [36]. In this regard, the first study of a caprine Single

Nucleotide Polymorphism (SNP) was published in 2006 [37]. SNP would progressively replace microsatellites as the preferred genetic marker.

Additional milestones were quickly reached in the following years. For instance, the first study using the Canadian Test-Day Model in caprine was published in 2007 [38]. The Canadian Test-Day Model is a 12-trait random regression animal model, in which the traits are milk, fat, and protein test-day yields, and somatic cell scores on test days within each of the first three lactations. Test-day records from later lactations are not used. Random regressions (genetic and permanent environmental) are based on Wilmink's three parameter function, which includes an intercept, regression on days in milk, and regression on an exponential function to the power –0.05 times days in milk (b0 + b1 $\times$ Exp(-0.05 $\times$ days in milk) + b2 $\times$ days in milk) [39]. This translates into genetic evaluations based on a better modelling of the lactation curve, providing more accurate results, which consequently enhances the selection progress.

These advances led to the publication of the first 'Copy Number Variations Regions' (CNVR) map for the domestic goat in 2010 [31]. Recent studies have shown that CNVR (intraspecific gains or losses of $\geq$1 kb of genomic DNA), represent important sources of variability of mammalian genomes (~0.4–25% of the genome). Their importance lies in the fact that CNVRs can change the gene structure and dosage, regulate gene expression and function and, hence, potentially have more effects than the most frequent single-nucleotide polymorphisms (SNPs) in determining phenotypic differences.

In 2012, the 'International Goat Genome Consortium' (IGGC) developed the first SNP chip for domestic goats; a high-density chip with 53,347 SNPs called '*Illumina 50K SNP BeadChip*' (Illumina Inc., San Diego, CA, USA) [30]. In 2014, a new version of the high density SNP chip was developed, which is the most advanced goat SNP chip at present, called the '*Illumina 52K SNP BeadChip*' a 60,000 SNP chip (Illumina Inc., San Diego, CA, USA), which is also carried out under the support of the IGGC [40]. Due to its robustness, low genotyping costs, automatic allele calling and the ability to interrogate the goat genome at high resolutions, these SNP chips were used to study the genetic diversity and population structure of native goats in various countries [41].

In 2013, after several attempts [23,25,42], the whole caprine genome was sequenced and optically mapped in China [29]. This advance made the reference genome sequence of the domestic goat available for the first time. The knowledge of the caprine genomic map was preceded by the first 'Bacterial Artificial Chromosome Library' published for caprine [43]. Recently, a new, 'de novo' genetic assembly (ARS1) has been developed. It allows for high-quality genomic mapping, thanks to the advanced sequencing methodology '*single-molecule PacBioRSII*' [27]. This methodology uses single-molecule, real-time (SMRT) sequencing technology, which takes advantage of an immobilized DNA polymerase/template complex nested in thousands of small wells (called zero-mode waveguides). The value of these new technologies lies in the fact that they are characterized by their high-throughput characteristics. Hence, they provide the opportunity to produce millions of reads with an inexpensive sequencing.

The use of SNP analysis in goat association studies marked a milestone with the first 'Genome Wide Association Study' (GWAS) in 2018 [26]. Thanks to GWAS tools, researchers could obtain more in-depth knowledge of the association between specific genome sequences and the phenotypic variation in traits of interest. GWAS appeared as an alternative to QTL and microsatellite studies, given that it allows for the identification of narrower genomic regions [34,44].

These new genomic tools are promising in terms of genetic selection within breeding programmes, because they allow scientists to locate the candidate genes that are directly involved in phenotypic traits of economic impact. Hence, they enable the acceleration of genetic progress. Therefore, the present review provides a chronological analysis of the advances made in caprine genetics and genomics and describes the main candidate genes for caprine meat and milk production and quality that have been reported in the literature to date.

2. Data Collection

A comprehensive bibliographic analysis was performed of the documents published over the last 10 years. Reviews, articles, short notes, Doctoral theses, and Master's theses were considered. The search for documents was performed using the 'Google Scholar' search engine (https://scholar.google.com/, accessed on 17 May 2021). This search engine was chosen as suggested by other papers developing bibliographic studies of a similar kind, as they comprise tools that enable data extraction for analysis [45–47]. The keywords chosen to perform the search of documents were "*caprine candidate genes, caprine meat genes, caprine milk genes*". Semantic fields were also considered.

The selection criteria for the documents included in the final dataset comprised: (1) date, looking for those that were published during the past 10 years (2) content, giving priority to those based on the study of candidate genes, and (3) species, including those primarily dealing with the caprine species, but also considering those involving the caprine species, in addition to other species, such as the ovine. In the case of data overlapping in different documents, the data published in the most recent publication were chosen. Only two reviews based on caprine candidate genes were found. The first was published in 2009 and the second in 2020. Based on the bibliographic references that these reviews provided, it was possible to expand the information and to include more candidate genes involved in caprine meat and milk production. A remarkable lack of articles dealing with candidate genes in the caprine species was observed during the last two years of this study. Particularly, references exclusively dealing with the caprine species were sparse. Gene data (karyotype location and information) were obtained using the database of the National Centre of Biotechnology Information NBCI (https://www.ncbi.nlm.nih.gov/gene/). This database was last accessed on 5 July 2021 [48].

3. Candidate Genes Regulating the Expression of Economically Relevant Traits in Caprine

3.1. Candidate Genes Regulating the Expression of Goat Meat Breeding Criteria and Traits

Growth performance and weight gain are two of the most relevant traits in caprine farming, given that they directly relate to the shortening of puberty age and the number of kilograms of meat to be sold, hence benefitting the farmer [49]. Contextually, body weight increases until 5–6 years of age, at which time it begins to decrease. The literature indicates the existence of sexual dimorphism, with males being heavier than females and a progressive decrease in weight gain as the number of kids born increases, with single-birth kids being heavier than double-birth kids, and these, in turn, being heavier than multiple-birth kids. Among other important factors, the season of birth, nutrition management, farming system, and the age and birth order of the doe have been reported to influence body-weight-related traits [50].

The current selection framework is characterized by a lack of reciprocity between caprine dairy and meat production. Hence, in practice, this is one of the main challenges that the sector needs to face; improving dairy characteristics, which negatively correlate with those related to butchering production [51], while taking advantage of the sturdiness and improved adaptability of caprine breeds [52].

The breeding objectives used in artificial selection to obtain a desirable dairy morphology are the opposite to those seeking the enhancement of meat/butcher traits. This may explain the marked differences that exist between the characteristic shape of goat dairy breeds (thin and triangular forms), and meat breeds (short, rectangular and excellent body conformation) [49].

Carcass dressing percentage in goats is approximately 50–55%, and is usually lower than in ovine. This is due to the greater bone proportion in caprine, along with a different carcass fat deposition, which, in goats, is perivisceral rather than subcutaneous or intramuscular [53]. In the same way, the caprine carcass has a lower fat coverage; hence, moisture losses (which can reach 8%) are greater than those in sheep. However, the lower intramuscular fat deposition in caprine makes chevon meat a healthier alternative to lamb [54].

In terms of meat quality characteristics, organoleptic traits include muscle appearance, juiciness, texture, hardness, flavour and aroma [55]. Many factors condition the expression of these traits, including nutrition, exercise/physical activity, age, and method of slaughter and bleeding [55]. Regarding nutritional quality, chevon muscle is high in amino acid diversity and is very lean, presenting a lower degree of fat interspersion than other species (but with more polyunsaturated fatty acids) [6].

Body weight and butcher performance, body growth, body weight gain, kid survival rate, feed conversion index and dressing percentage of the carcass are among the most frequently considered quantitative traits within the framework of caprine selection. Table 1 presents a summary of the most frequently addressed breeding criteria and traits concerning selection for meat production in goats. Figure 3 presents a scheme of caprine candidate gene background regulating the expression of meat production.

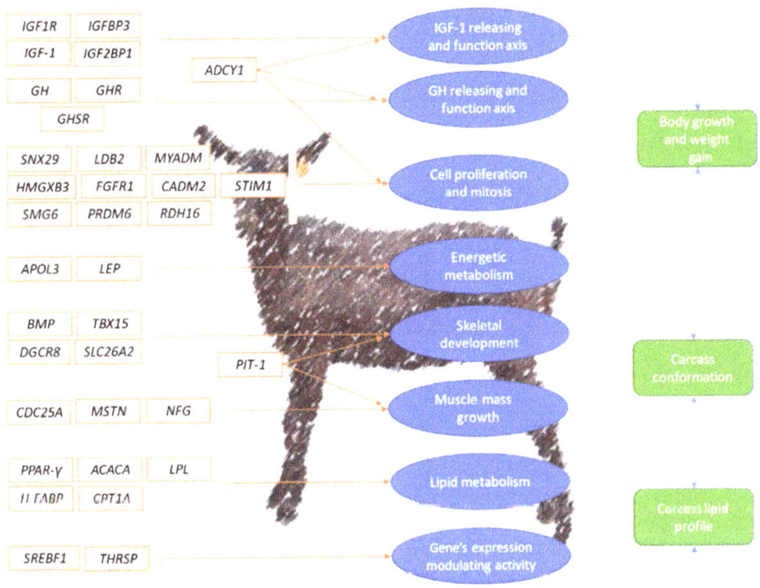

Figure 3. Summary scheme of caprine candidate genes' influence on meat production.

Table 1. Breeding criteria and traits for caprine meat production.

Breeding Criteria	Traits	Reference
Body weight	Birth weight, body weight at 7, 14, 21 and 28 days, monthly weighing until 18 months of age	[56,57]
Growth	Average daily gain (ADG) before weaning, from 3 to 6 months of age, from 6 to 12 months of age	[57]
Conformation/corporal structure	Body length, height at the withers, chest girth, shoulder width and pin-bone width	[58]
Carcass quality	Hot (immediately after slaughter) and cold (before chilling) carcass weight; weights of the head, skin, heart, and thoracic and abdominal viscera; carcass dressing percentage; leg length and carcass width; and carcass fat coverage and lean meat and bone yield percentages	[56,59]
Meat quality	Muscle pH measurements, colour, water retention capacity, nutritional composition (protein, fat, and collagen percentages), and fatty acid profile in samples of *longissimus thoracis et lumborum*, *triceps brachii* and *semimembranosus*	[56]

3.1.1. Actual and Potential Candidate Genes Regulating the Expression of Body Growth and Weight Gain

Among the genes for which a significant association has been reported with the regulation of the expression of body growth and weight gain, the following are the most relevant. First, the Adenylate Cyclase 1 gen (*ADCY1*), which encodes for a molecule involved in Cyclic Adenosine Monophosphate (cAMP) synthesis, is associated with chest width [60]. This gene regulates cellular division and mitosis [61], but also has been reported to hold regulatory control over Growth Hormone (GH) release [62].

Second, the Sorting Nexin 29 gene (*SNX29*), which has been described as a controller of the differentiation and proliferation of myocytes, has been directly ascribed to the complex of genes regulating the expression of muscle growth [63]. In parallel with this, its association with chest width and pin bone width, both relevant traits due to their association with dressing aptitude, has been recently reported [60]. Third, the LIM Domain Binding Factor 2 gene (*LDB2*), which encodes for a protein playing a crucial role in lymphocyte migration and atherosclerosis [64], has been shown to have a strong relationship with daily body weight gain in other species, such as poultry, as derived from GWAS studies [65].

Fourth, Myeloid-associated Differentiation Marker gene (*MYADM*) has been reported to be involved in the cytoplasmatic membrane synthesis of the myeloid-line cells and has impact on the erythrocyte morphology. Its association with post-weaning body weight in lambs has been described, which suggests it may be worth exploring in kids [66]. Fifth, the Insulin-like Growth Factor 1 Receptor gene (*IGF1R*) encodes for receptors for IGF-1 binding, modulating its blood concentration and activity, and has been shown to be associated with body growth and height in dogs [67].

Sixth, the Apolipoprotein L3 gene (*APOL3*) encodes for a protein involved in lipid transport and metabolism, and has been ascribed an interesting potential influence on milk and growth traits [68]. Seventh, and similar to the aforementioned, the Stromal Interaction Molecule 1 gene (*STIM1*) encodes for a membrane calcium transporter that is related to prolificacy traits [69], and has been suggested as a candidate gene for growth traits [70]. Eighth, the HMG-Box Containing 3 gene (*HMGXB3*), which has been described to be involved in cellular proliferation in neoplasia [71], has also been reported to be of great interest in meat production [70].

Ninth, among growth-hormone-linked genes, the Growth Hormone gene (*GH*), which encodes for the main hormone involved in the vertebrate corporal development, has been reported to be strongly related to body weight at birth and weight gain in goat kids [72]. In this regard, those genes linked to regulation or control of GH, such as the Growth Hormone Secretagogue Receptor gene (*GHSR*), may be relevant. For instance, *GHSR* encodes for a G protein-associated receptor that binds ghrelin, a GH-release-stimulating hormone [73]. Similarly, the Growth Hormone Receptor gene (*GHR*) encodes for the proteins to which *GH* binds, modulating the molecular mechanisms that this hormone sets in motion, which have a direct impact on corporal development [74].

Tenth, the insulin-like growth-factor-related genes, such as the insulin-like growth factor gene (*IGF-1*) play a pivotal role in the mammalian somatotrophic axis, participating in many metabolic functions involved in the regulation of the expression of growth, lactation, and reproduction traits. Indeed, its effects on the carcass conformation and fat distribution of cattle have been acknowledged in the research [75]. Additionally, the Insulin-like Growth Factor 2 Binding Protein 1 gene (*IGF2BP1*) encodes for a protein that takes part in the '*sonic hedgehog*' route, which is a metabolic chain impacting body/organ growth and development [76]. This is also extendible to the Insulin-like Growth Factor Binding Protein 3 gene (*IGFBP3*), which encodes for a protein that regulates the activity of an IGF-1 subfamily, but is also associated with growth, body size and prolificacy traits [77]. Another '*sonic hedgehog*' metabolic route participant is the Fibroblastic Growth Factor Receptor 1 (*FGFR1*), which unleashes a molecular-signals cascade associated with metabolism and organic development [78]. Eleventh, the Bone Morphogenic Protein gene (*BMP*) encodes for a transforming growth factor β (TGF-β) superfamily. Many of the genes in this complex, such as *BMP4* [79] and *BMP15* [80], have been reported to relate to growth traits.

Twelfth, the Nonsense-mediated mRNA Decay Factor gene (*SMG6*) has been suggested to affect growth due to its genome stabilization functions, given that genome instability inhibits mitosis and decreases cellular division and tissue growth [81]. Thirteenth, the Cell Adhesion Molecule 2 gene (*CADM2*) encodes for a synaptic signalling vector that is abundant in the brain and muscle, and has been described to be involved in weight at slaughter and corporal length [82]. Table 2 reports a summary of the candidate genes associated with body growth and weight gain that have been described in goat breeds, the chromosome location, exon counts, physiological function, and where to find them in the literature.

Table 2. Summary of the candidate genes associated with body growth and weight gain that have been described in goat breeds, their chromosome location, exon counts, physiological function and where to find them in the literature.

Acronym	Gene Name	Chromosome	Exon Count	Transcripts	Physiological Function	Goat Breeds in Which the Gene Has Been Described	Reference
ADCY1	Adenylate Cyclase 1	4	20	2	Involved in energetic metabolism, cellular mitosis and GH releasing.	Cameroon, West African Dwarf, Small East African and Landim goat	[60]
SNX29	Sorting Nexin 29	25	21	2	Regulates myoblast differentiation and proliferation.		
LDB2	LIM Domain Binding Factor 2	6	9	7	Modulates transendothelial leucocyte migration.	Nanjiang yellow goat	[83]
MYADM	Myeloid-associated Differentiation Marker	18	1	1	Involved in cellular membrane formation in a wide variety of cellular lines.	Bamu wild goat, Khonj wild goat, Australian feral Rangeland goats, Boer goats and Australian cashmere goat	[84]
IGF1R	Insulin-like Growth Factor 1 Receptor	21	21	1	Regulates IGF-1 activity.		
APOL3	Apolipoprotein L3	5	4	6	Involved in lipid blood transport and metabolism.	Leizhou goat	[70]
STIM1	Stromal Interaction Molecule 1	15	14	4	Associated with body weight gain.		
HMGXB3	HMG-Box Containing 3	7	21	3	Demonstrated relationship with cellular proliferation in neoplasia.		
GH	Growth Hormone	19	5	1	Related to corporal development.	Thai Native, Anglo-Nubian, Boer and Saanen goat	[85]
GHR	Growth Hormone Receptor	20	13	6			
IGFBP3	Insulin-like Growth Factor Binding Protein 3	4	5	1			
BMP4	Bone Morphogenic Protein 4	10	6	5		Boer goat, Chinese Xuhuai white goat and Chinese Haimen goat	[85]
BMP15	Bone Morphogenic Protein 15	10	2	1		Jining Grey goat	[85]
IGF-1	Insulin-like Growth Factor	5	7	9	Involved in corporal metabolism, growth and development.	Malabari and Black Attappady goat	[86]
IGF2BP1	Insulin-like Growth Factor Binding Protein 1	19	15	2	Involved in endocrine routes associated with body growth and development.	Shaanbei White Cashmere goat	[87]
FGFR1	Fibroblastic Growth Factor Receptor 1	27	20	13			
SMG6	Nonsense-mediated mRNA Decay Factor	19	20	2	Genome stabilization functions		
CADM2	Cell Adhesion Molecule 2	1	11	3	Involved in cellular migration and proliferation.	White and Black Guizhou, Nubiann, Boer and Huai goat.	[82]
GHSR	Growth Hormone Secretagogue Receptor	1	3	2	Regulates GH realise	Boer, Xuhuai and Haimen goat	[73]

3.1.2. Actual and Potential Candidate Genes Regulating the Expression of Body Size and Carcass Quality

The genes presented in this subsection have been reported to be associated with body size and carcass quality, either skeletal quality or muscle quality. First, the quest to find caprine specific genes related to the aforementioned traits is challenging, given the lack of existing species-specific literature. In this context, among the only examples found, the PR/SET Domain 6 gene (*PRDM6*) encodes for a histone methyltransferase, a molecule with an important role in corporal development and reproduction, and is, therefore, suspected to be related to body growth [88], and the Nerve Growth Factor gene (*NGF*) encodes for a crucial protein in the development and differentiation of the nervous system. However, it also plays an important role in muscle growth and prolificacy [89] in goats.

Interestingly, some of the genes for which an association with economically relevant traits has been reported have an additional relevance, as they coregulate the expression of dual-purpose traits; that is, they are associated with better carcass features while they regulate for better milk-quality traits. The Pituitary-specific Positive Transcription Factor 1 gene (*POU1F1 or PIT-1*) encodes for a positive regulator of many hormones, such as the Growth Hormone, Prolactin or Thyroid-stimulating Hormone, and has been shown to have a role in growth and lactation traits in caprine [90]. Furthermore, a certain degree of association is presumed with carcass conformation, owing to its relationship with body depth and leg size in bovine, alongside some milking morphological traits [91].

Additionally, regarding muscle development, the Leptin gene (*LEP*), which regulates the expression of a hormone that exerts its effects on daily voluntary food intake, energetic waste and corporal metabolic balance [92], has been proven to condition carcass weight, corporal fat yield and lean meat percentage in ovine [93]. The Myostatin gene (*MSTN*) is considered to be a highly relevant candidate gene for meat production, due to its widely studied association with body growth and muscle mass development in bovine [37].

Concerning bone quality and development, the T-box Transcription Factor 15 gene (*TBX15*) is highly expressed in mesenchymal precursor cells and chondrocytes, and has a direct impact on bone development [94]. This was also reported for the Drosha and DiGeorge Syndrome Critical Region gene 8 (*DGCR8*), which is involved in osteoclastic and bone-reabsorbing activity, and hence in the determination of skeletal development [95].

Regarding muscle development, the Cell Division Cycle 25 Homolog A gene (*CDC25A*) has been reported to take part in the cellular quiescence G1 phase and to be involved in myoblast differentiation in mice [96]. Additionally, the Solute Carrier Family 26 Member 2 gene (*SLC26A2*) encodes for a protein that seems to modify sulphate transporting residual activity, and is associated with short body size and skeletal dysplasia [97]. Table 3 summarizes the candidate genes associated with body size and carcass quality described in goat breeds to date, as well as their chromosome location, exon counts, physiological function and where to find them in the literature.

Table 3. Summary of the candidate genes associated with body size and carcass quality that have been described in goat breeds, their chromosome location, exon counts, physiological function and where to find them in the literature.

Acronym	Gene Name	Chromosome	Exon Count	Transcripts	Physiological Function	Goat Breeds in Which the Gene Has Been Described	Reference
TBX15	T-box Transcription Factor 15	3	9	2	Related to body size and chondrocytes and mesenchymal cell precursors regulator.	Gizhou Small goat	[98]
DGCR8	Drosha and DiGeorge Syndrome Critical Region microprocessor complex subunit gene 8	17	16	1	Associated with body size and involved in the osteoclastic development and remodelling bone activity.		

Table 3. *Cont.*

Acronym	Gene Name	Chromosome	Exon Count	Transcripts	Physiological Function	Goat Breeds in Which the Gene Has Been Described	Reference
CDC25A	Cell Division Cycle 25 Homolog A	22	15	1	Responsible for corporal development and involved in myoblast differentiation and G1 quiescence.		
LEP	Leptin	4	3	1	Related to corporal and muscle mass development.		
MSTN	Myostatin	2	3	1			
PRDM6	PR/SET Domain 6	7	8	1	Regulates skeletal development and body mass index.	Shaanbei White Cashmere goat	[88]
NGF	Nerve Growth Factor	3	3	2	Involved in muscle and nervous tissue development.	Black Attappady and Malabari goat	[89]
SLC26A2	Solute Carrier Family 26 Member 2	7	9	8	Involved in bone tissue development.	Leizhou goat	[70]
POU1F1	Pituitary-specific Positive Transcription Factor 1	1	6	2	Plays an important role in corporal metabolism, growth and corporal development.	Thai Native, Ango-nubian, Boer and Saanen goat	[85]

3.1.3. Actual and Potential Candidate Genes Regulating the Expression of Carcass Lipidic Profile

Among the genes regulating the expression of the lipidic profile of caprine carcasses we first find the Peroxisome Proliferator-activated Receptors Gamma gene (*PPAR*γ), which encodes for a protein that is directly related to carcass lipids, thanks to its regulatory activity over the expression of other genes involved in fat metabolism, promoting its mRNA synthesis and protein formation [99]. Second, the Lipoprotein Lipase gene (*LPL*), modulates fatty acid metabolism due to its implication in muscle lipid composition as reported in cattle [100].Third, following the aforementioned line, the Acetyl-Coenzyme A Carboxylase gene (*ACACA*) encodes for one of the most important enzymes involved in tissue fatty acid synthesis, and is associated with conformation traits and meat fat composition in bovine [101]. The Sterol Regulatory Element-binding Protein 1 (*SREBF1* or *SREBP-1c*) encodes for transcription factor family modulating lipid homeostasis [102] and seems to promote the expression of *ACACA* [103] and *PPAR*γ's transcription. Hence, its involvement in fatty acid metabolism is presumed [104]. Fourth, the Thyroid Hormone Responsive SPOT14 gene (*THRSP*) seems to mediate the expression of other genes that are involved in lipid synthesis, and is known to be related to the fatty acid profile in bovine muscle [105]. Fifth, the Carnitine Palmitoyl transferase 1A gene (*CPT1A*), which encodes for a mitochondrial enzyme responsible for acyl-carnitine synthesis, has been proven to promote fatty acid transport into the mitochondria [106]. Certain genes are associated with fatty acid and adipose tissue metabolism in other species. For instance, a study in pigs [107] showed that the Retinol deshidrogenase-16 gene (*RHD16*) plays a role in the metabolic reactions in adipose tissue [107], while the Heart Fatty Acid-Binding Protein gene (*H-FABP*) is reported to be involved in lipid metabolism and seems to influence the intramuscular fat composition in pigs [108]. However, their role in caprine has yet to be confirmed [59]. Table 4 presents a summary of the candidate genes associated with carcass lipidic profile that have been described in goat breeds, their chromosome location, exon counts, physiological function and where to find them in the literature.

Table 4. Summary of the candidate genes associated with carcass lipidic profile that have been described in goat breeds, their chromosome location, exon counts, physiological function and where to find them in the literature.

Acronym	Gene Name	Chromosome	Exon Count	Transcripts	Physiological Function	Goat Breeds in Which the Gene Has Been Described	Reference
PPAR-γ	Peroxisome Proliferator-activated Receptors Gamma	22	6	1	Regulates adipose differentiation and lipid metabolism.	White Yaoshan goat	[59]
LPL	Lipoprotein Lipase	8	10	2	Involved in triglycerides degradation, obtaining glycerol and fatty acids.		
H-FABP (FABP3)	Heart Fatty Acid-Binding Protein (fatty acid binding protein 3)	2	5	2	Related to lipid metabolism.		
SREBF1	Sterol Regulatory Element-binding Protein 1	19	8	1	Modulates expression of other lipid-metabolism-related genes.		
ACACA	Acetyl-Coenzyme A Carboxylase	19	60	7	Involved in liver fatty acid synthesis.		
THRSP (SPOT14)	Thyroid Hormone Responsive	29	2	1	Regulates expression of other lipid-metabolism-related genes.		
CPT1A	Carnitine Palmitoyl-transferase 1A	29	19	1	Responsible for acyl-carnitine obtention, product of fatty acid degradation.	Moroccan goat breeds	[109]
RDH16	Retinol deshidrogenase-16	5	4	1	Involved in energetic metabolism.	Gizhou goat	[98]

3.2. *Candidate Genes Regulating the Expression of Goat Dairy Breeding Criteria and Traits*

The international caprine milking sector comprised 215 million animals in 2019, which was one-quarter of the world caprine census [110]. Europe, with only 5% of the world caprine dairy census, produced 15% of the caprine worldwide milk production, which could mainly be ascribed to its high degree of specialization [1].

This high European productivity is supported on high-technology farms, on which goat breeds such as the Alpine, Saanen or Toggenburg have been bred under the scope of strict genetic selection schemes and breeding programmes [3]. However, this is only one side of the same coin and the opposite situation is found in developing countries. Specifically, in developing countries, genealogic control and productivity registers are barely available or are of poor quality [1]. The caprine European dairy sector is a well-regulated industry, where almost all the milk is processed into cheese [1]. Most of the farms' income derives from milk production, while the sale of chevon for the meat industry is a marginal source of income, due to the relatively low value of caprine meat within developed countries [6].

According to García, et al. [111], milk quality is defined as the milk's ability to tolerate the technological processes that lead to a market-demanded product in terms of nutritional value, food safety and sensorial parameters. This is especially relevant in the framework of the cheese industry, given that increasing milk protein and percentage (known as 'dry' or 'cheese extract') increases the chances to enhance farms' rentability and profitability, even more than milk volume production (in litres) [112]. Furthermore, traditional breeding criteria (milk yield in volume; milk lactose, fat and protein percentages; fatty acid profile and omega 3 (technologic quality); and somatic cell count (hygienic quality) [111]) have recently been joined by postprocessing technological traits related to *cheese-making* quality indicators, in order to improve cheese production efficiency. Among these traits we find, for instance, milk cheese yield (%CY) or the dairy cheese efficiency (dCY) [112], and the newly implemented term of 'recuperation' (%REC) for each milk component (proteins and fat) from the junket.

3.2.1. Actual and Potential Candidate Genes Regulating the Expression of Milk Production/Yield

Given the greater economic repercussion of goat dairy products against caprine-meat-derived products, it is understandable that the progress and knowledge on genes associated with milk yield and quality surpasses those for goat meat production-related genes. The evaluation of the literature suggested that a large fraction of the genes involved in meat quality- and/or yield-related traits correlatedly regulate the expression of milk yield- and content-related traits; hence, they indirectly correlate with the quality of milk and the products that derive from it, for example, cheese. For instance, the Pituitary-specific Positive Transcription Factor 1 gene (*POU1F1*) is associated with the expression of genes that encode for hormones regulating mammary gland development and milk production, such as prolactin, growth hormone, insulin-like growth factor 1, progesterone and oestrogens [113]. Among them, contextually, the Insulin-like Growth Factor 1 gene (*IGF-1*) is associated, as mentioned in the previous sections, with body growth and development or metabolism, but has also been reported to participate in the regulation of the expression of milk protein and fat [75], or the Vacuolar Protein Sorting 13 gene (*VPS13*) family, which is actively involved in milk production, as was proven in other mammalian domestic species [70]. For example, while the *VPS13C* gene seems to affect glucose homeostasis in high-productive milk cattle [114], the *VPS13B* gene was detected in a QTL related to leg length (which is associated with fertility, prolificacy and milk production) [115]. Among the genes which were found to be associated with dairy yield or quality, first, the Leptin gen (*LEP*) was reported as a candidate gene for milk yield and quality traits, due to its implications for the regulation of daily voluntary food intake, energetic distribution and metabolism in cattle [116]. Consequently, this brings about the implications of the Leptin Receptor gene (*LEPR*), which has also been reported to have an influence on milk traits, due to its function in leptin blood concentration and physiological activity [117]. Second, the ATP Binding Cassete Subfamily G Member 2 gen (*ABCG2*), besides its acknowledged membrane drug and xenobiotics' transporting activity, is presumably linked to a decrease in milk production and an increase in milk protein and fat [118]. Third, the Growth Hormone Receptor gene (*GHR*) encodes for the GH transmembrane receptors that unleash the cellular mechanisms set off by this hormone, and is directly associated with carbohydrate and lipid metabolism, and involved in the beginning and maintenance of lactation [119]. Fourth, it is common for genes regulating the expression of milk yield to have a negative or positive correlation with milk composition (protein, fat, lactose and somatic cell count, among others). In this context, the Prolactin gene (PRL), which encodes for a physiological multi-function hormone, whose specific actions occur at the reproductive and lactation levels, and, hence, influences milk yield and milk protein and fat percentages [120]. In the same way, the Prolactin Receptor gene (PRLR), which encodes for a receptor associated with a wide range of endocrine functions, such as mammalian lactation and body growth, is one of the most frequently addressed candidate genes for milk traits in the literature, given its regulation of the expression of milk yield and lactose, protein and fat percentages [121]. Fifth, the high expression of these genes in the mammary gland seems to be responsible for their implications for milk production/yield and quality. Contextually, these include the Ribosomal Protein L3 gene (*RPL3*), which has been described to have an especially high expression in the mammary gland, being involved in the regulation of energetic balance [122], or the Osteopontin gene (*OPN*), which encodes for a protein involved in many processes, such as cellular adhesion, chemotaxis and inflammation, cellular survival, and tissue remodelling, but is also potentially involved in milk production [123]. Furthermore, the Growth Hormone 1 gene (*GH1*) has been reported to likely be useful in stimulating the udder development in transgenic goats and, therefore, increasing their milk yield [48] as it has been reported for cattle milk production [124]. Sixth, the regulation of the expression of traits linked to hygienical quality of milk, measured through somatic cells counts, may be associated with the Lactoferrin gene (*LTF*), which is a very likely candidate gene for mastitis resistance. This occurs because, among the several functions in which it is involved, it has been reported to have a light antimicrobial effect in the mammary gland, which prevents the release of inflammatory mediators, such as the interleukins 1β and 6 or the α-tumoral necrosis factor [125]. This has also been reported for the Breast Cancer Type 1 gene (*BRCA1*), which encodes for a protein that is considered to be a DNA disorder-fixing molecule and to regulate the cell division cycle, as a potential early marker of disease, which has also been reported to increase somatic cell count in case of mastitis [126].

3.2.2. Actual and Potential Candidate Genes Regulating the Expression of Milk Content/Composition

First, casein genes conform to the most widely explored milk trait genes' expression regulation complex in the research. They attract special attention due to their implications for milk yield and quality determination [127]. Recent research advances in goat breeds such as Murciano-Granadina, Norwegian and Sarda goats have addressed the implication of the whole casein complex in milk production and contents (protein, fat, dry matter, lactose and somatic cells counts) [128–139]. The *CSN1S1*, *CSN1S2* and *CSN2* genes encode the αs1-casein, αs2-casein and β-casein, which are calcium-sensible caseins, while the *CSN3* gene encodes for κ-casein, which is involved in micelle stability [140]. Second, the α-Lactalbumin gene (*α-LA*) encodes for α-Lactalbumin, a highly important milk serum protein, which regulates the production of lactose in the milk of in all mammals. Contextually, this gene has been reported to increase lactose percentage in early-lactation milk from transgenics animals [141]. Third, the β-Lactoglobulin gene (*β-LG*) encodes for the β-Lactoglobulin, another essential milk serum protein, whose primary biologic functions seems to be related to phosphorus metabolism at the mammary gland, although its association with milk production and composition has not been comprehensively described in goat breeds [142]. Fourth, besides its acknowledged membrane drug and xenobiotics' transporting activity, the ATP Binding Cassette Subfamily G Member 2 gen (*ABCG2*), is presumably linked to a decrease in milk production and an increase in milk protein and fat [118]. This correlated action in genes, regulating the expression between milk traits and those of another nature, has been relatively frequently addressed in the literature. For instance, the Insulin-like Growth Factor 1 gene (*IGF-1*) is associated, as mentioned in the previous sections, with body growth and the development or metabolism, but has also been reported to participate in regulating the expression of milk protein and fat, respectively [75].

Even in genes specifically reported to participate in the regulation of the expression of milk fat contents, a multifunction character has often been suggested. For instance, the Diacylglycerol Acyltransferase gene (*DGAT1*) encodes for a acetyl-coenzyme A variant, which is involved in lipid and other biomolecules' synthesis, and was proved to associate with milk fat and protein contents [143]. The 1-acylglicerol-3-phosohate-O-acyltransferase gene (*AGPAT6*), which encodes for a crucial enzyme in glycolipids and triglycerides synthesis, the two main types of lipids present in milk, has been reported to present a high expression in the mammary gland epithelium; therefore, it has been suggested to have repercussions for milk fat percentage [144]. This was also noted for other genes, such as the Butyrophilin Subfamily 1 Member A1 gene (*BTNA1*), which was identified in a copy number variation analysis [84], and presents a widely acknowledged role in fat drops' secretion in the mammary gland [145]. The Fatty Acid Synthase gene (*FASN*), which encodes for a protein involved in fatty acid synthesis, and the Acetyl-CoA Carboxylase gene (*ACACA*), which encodes for a liver fatty acids synthesis mediator protein, have been proved to present a certain association with milk lipid profile and fat percentage in dairy cattle, respectively [146,147]. The Stearoyl-CoA Desaturase gene (*SCD*) which encodes for a protein with a strong association to monounsaturated fatty acids synthesis in adipose tissue and mammary gland, has a direct repercussion for milk fatty acid profile [148]. The action of particular genes may not be direct but derive from these genes' regulatory potential. Accordingly, the

Peroxisome Proliferator-activated Receptor Gamma gene (*PPARγ*) encodes for a protein with a high impact on other lipid-metabolism-related genes' transcription, such as *LPL*, *FASN*, *ACACA* y *SCD*, *PLIN2*, *PLIN3*, *FABP3*, *PNPLA2*, in same way as it regulates the expression of *NR1H3* y *SREBF1* genes [149]. The Adipophilin gene, also known as Perilipin 2 gene (*PLIN2*) [150] and the Perilipin 3 gene (PLIN3 ó TIP47) [151] take part in cytoplasmatic lipid transport and storage, and their implication for lipid secretion into milk has been widely reported. This also occurs at RNA levels, as was addressed for the Sterol Regulatory Element-binding Protein 1 (*SREBF1* or *SREBP-1c*). This encodes for a transcription factor *molecule*, which is a protein that regulates the expression of protein synthesis from mRNA, modulating the expression of those genes that encode for proteins with a fat milk percentage impact [152].

As those genes whose effects are highly expressed in the mammary gland should always be presumed to have an association with the expression of milk-related traits, either quantitative or qualitative. For example, the Patatin-like Phospholipase Domain Containing 2 gene (*PNPLA2*) encodes for a lipase that seems to be involved in intracellular triglycerides' degradation in adipocytes, and for which overexpression in cattle mammary gland was proved [149], or the Aldehyde Dehydrogenase 2 gene (*ALDH2*), which is thought to take part in triglyceride synthesis and mammary gland epithelium cell apoptosis in bovine, with repercussions for fat milk yield and somatic cell count [153]. Many of the genes involved in fatty acid metabolism or fat mobilization have also been reported to participate in the regulation of the expression of milk quality traits, especially those reported with the fat fraction of milk. he faTtty Acid Binding Protein 3 gene (*FABP3*), which encodes for a protein participating in long-chain fatty acid intracellular transport, as well as other lipids' cytoplasmatic movements, was proven to be associated with milk fat percentage [154]. Additionally, the Liver X Receptor-α (*NR1H3*) encodes for a transcription factor as well, ruling out the expression of other genes expression involved in fatty acids, cholesterol and carbohydrate metabolism, and was proved to contribute to milk lipid profile [155], as suggested for other genes, such as the Oxidized Low-Density Lipoprotein Receptor gene (*OLR1*), which encodes for a protein with fatty acid transport and oxidated-low density lipoprotein degradation functions, and has been reported to impact milk fat profile [156]. These genes not only directly affect the particular glands in which milk is produced, but may also be involved in the regulation of higher neurological routes. For example, the Brain-derived Neurotrophy Factor gene (*BDNF*), which encodes for a protein that regulates voluntary food intake and energetic distribution at the hypothalamic level, has been reported to be associated with fat milk percentage in cattle [157] or the Fat Mass and Obesity-associated Protein gene (*FTO*), which plays a central role in energetic homeostasis and waste control. This is, therefore, thought to affect the fat milk percentage [158]. The relationship across milk constituents may also be based on a correlated regulation of the function of certain genes, but this may affect other economically relevant traits, such as prolificacy. Therefore, the Acetyl-CoA Acetyltransferase gene (*ACAT*), which encodes for a cholesterol-metabolism regulating protein [159], has shown associations with other desirable productive traits (milk protein contents and fertility) in Holstein cattle [160] or the Long-chain Acyl-CoA Synthetase Isoform 1 gene (*ACSL1*), which encodes for a basic protein in triglycerides, phospholipids and cholesterol synthesis, making it a great candidate gene for milk-quality-related traits [161]. Table 5 presents a summary of the candidate genes associated with milk content that have been described in goat breeds, their chromosome location, exon counts, physiological function and where to find them in the literature, and Figure 4 presents a schematic representation of dairy goat candidate genes and their functions.

Table 5. Summary of the candidate genes associated with milk production/yield that have been described in goat breeds, their chromosome location, exon counts, physiological function and where to find them in the literature.

Acronym	Gene Name	Chromosome	Exon Count	Transcripts	Physiological Function	Goat Breeds in Which the Gene Has Been Described	Reference
LEP	Leptin	4	3	1	Regulates glycaemia, milk production and milk fat percentage.	Dairy cattle	[117]
LEPR	Leptin receptor	3	22	3			
BDNF	Brain-derived Neurotrophy Factor	15	6	5	Plays different roles in daily food intake and, in consequence, in nutrient and energy availability in the mammary gland.	Dairy cattle	[162]
FTO	Fat Mass and Obesity-associated Protein	18	9	1			
IGF-1	Insulin-like Growth Factor 1	5	7	9			
ABCG2	ATP Binding Cassete Subfamily G Member 2	6	22	8	Related to milk production and milk fat percentage.	Dairy cattle	[163]
GHR	Growth Hormone Receptor	20	13	6	Regulates cell growth, proliferation and apoptosis.	Dairy cattle	[117]
PRLR	Prolactin Receptor	20	11	7			[164]
PRL	Prolactin	23	5	1	Involved in mammary gland tissue preparation before lactation and regulates milk production.	Alpine goat	[165]
RPL3	Ribosomal Protein L3	5	10	1	Modulates energetic balance during the lactation yield peak.	Saanen goat	[70]
VPS13	Vacuolar Protein Sorting 13	8	73	3	Gene family associated with milk production.		
VPS13B	Vacuolar Protein Sorting 13 Homolog B	14	65	10	Related to leg's morphological traits associated with fertility and milk production.	Dairy cattle	[115]
VPS13C	Vacuolar Protein Sorting 13 Homolog C	10	85	1	Regulates glycaemia increasing milk production.	Dairy cattle	[114]
SPP1 (OPN)	Osteopontin	6	7	2	Involved in milk yield production and milk fat percentage.	Dairy cattle	[123]
GH1	Growth Hormone 1	19	5	1	Seems to stimulate udder development and is associated with milk dairy traits.	Dairy cattle	[124]
LTF	Lactoferrin	22	17	1	Associated with mastitis resistance somatic cell count.	Damascus goat	[166]
BRCA1	Breast Cancer Type 1	19	23	5			
POU1F1 (PIT1)	Pituitary-specific Positive Transcription Factor 1 (POU class 1 homeobox 1)	1	6	2	Modulates many milk-related hormones' action.		
CSN1S1	αs1-casein	6	19	8	Associated with total milk protein composition and milk yield production.	Murciano-Granadina and Norwegian goats	[128–138,167]
CSN1S2	αs2-casein	6	19	7	Encodes for most of the important milk proteins.	Sarda goat	[128–138,167]
CSN2	β-Casein	6	9	2			
CSN3	K-Casein	6	5	1		Norwegian, Saanen, Canaria, Malagueña, Murciano-Granadina and Payoya goat	
aLA LALBA (ALA)	α-lactalbumin	5	4	1	Encodes for most of the important milk serum proteins.	White Inner Mongolia Cashmere, Xinong, Guanzhong, Laoshan, Leizhou, White and Black Guizhou and Banjiao, Matou goat	[168]
β-LG LGB (PAEP)	β-lactoglobulin/progestagen-associated endometrial protein	11	8	4		Zarayby, Damascus, Albino and Balady Hybrid goat	[142]
DGAT1	Diacylglycerol Acyltransferase	14	18	2	Involved in triglycerides synthesis.	Xinong, Saanen and Guanzhong goat	[169]
AGPAT6 (AGPAT6)	1-acylglicerol-3-phosohate-O-acyltransferase	27	14	2			[144]

Table 5. *Cont.*

Acronym	Gene Name	Chromosome	Exon Count	Transcripts	Physiological Function	Goat Breeds in Which the Gene Has Been Described	Reference
BTN1A1	Butyrophilin Subfamily 1 Member A1	23	9	5	Essential in milk lipid micelles secretion in the mammary gland.	Bamu wild goat, Khonj wild goat, Australian feral rangeland goat, Boer and Australian Cashmere goat	[84]
FASN	Fatty Acid Synthase	19	42	1	Plays an important role in fatty acids synthesis.		[84]
ACACA	Acetyl-CoA Carboxylase	19	60	7	Involved in liver fatty acid synthesis.	Saanen and 'Grey local' goat	[147]
SCD	Stearoyl-CoA Desaturase	26	6	1	Catalyze unsaturated to mono-satured fatty acids transformation reaction.	Boer, Xuhuai White and Haimen goat	[170]
PPARγ (PPARG)	Peroxisome Proliferator-activated Receptor Gamma	22	6	1	Regulates lipogenesis and adipogenesis.	Damascus goat	[166]
PLIN3	Perilipin 3	7	8	2	Involved in tissue fatty acid synthesis.		
PLIN2	Perilipin 2	8	9	4			
FABP3	Fatty Acid Binding Protein 3	2	5	2	Involved in cytoplasmatic fatty acid storage and transport.	Not specified/Dairy goats	[149]
PNPLA2	Patatin-like Phospholipase Domain Containing 2	29	11	3	Encodes for one of the proteins involved in fat micelles formation.		
SREBF1	Sterol Regulatory Element-binding Protein 1	19	8	1	Transcription factors involved in lipid homeostasis.		
NR1H3	Liver X Receptor-α (nuclear receptor subfamily 1 group H member 3)	15	10	2			
OLR1	Oxidized Low Density Lipoprotein Receptor	5	6	2	Involved in lipid transport and oxided-LDL form degradation.	Sirohi goat	[171]
ALDH2	Aldehyde Dehydrogenase 2	17	13	1	Associated with triglycerides synthesis in mammary glandular tissue.		
ACAT1	Acetyl-CoA Acetyltransferase 1	15	12	1	Involved in cholesterol metabolism.	Xinong Saanen goat	[172]
ACAT2	Acetyl-CoA Acetyltransferase 2	9	9	1			
ACSL1	Long-chain Acyl-CoA Synthetase Isoform 1	27	24	5	Participates in triglycerides, phospholipids and cholesterol synthesis.		
BDNF	Brain-derived Neurotrophy Factor	15	6	5	Plays different roles in daily food intake and, in consequence, in nutrient and energy availability in the mammary gland.	Dairy cattle	[162]
FTO	Fat Mass and Obesity-associated Protein	18	9	1			
ABCG2	ATP Binding Cassete Subfamily G Member 2	6	22	8	Related to milk production and milk fat percentage.	Dairy cattle	[163]

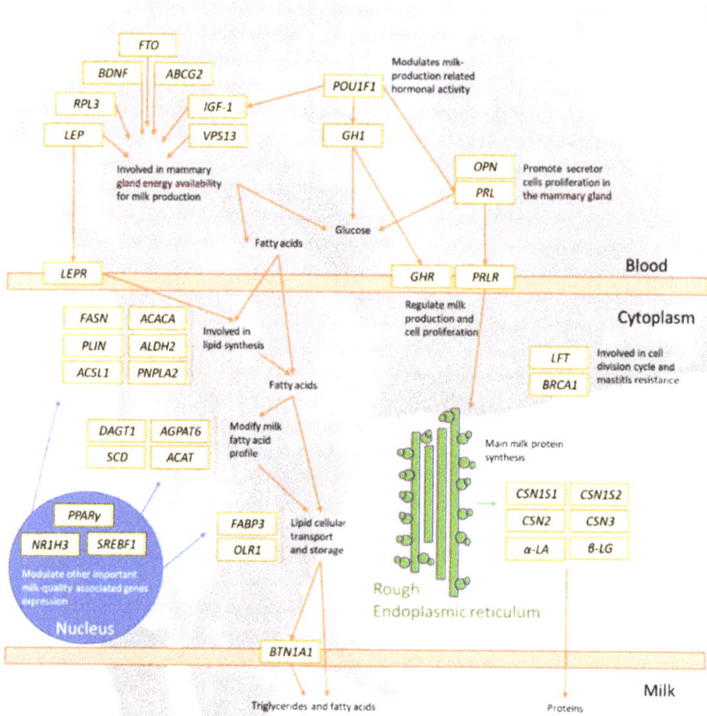

Figure 4. Dairy goat candidate genes interrelations' scheme.

4. Conclusions

The knowledge concerning the main candidate genes affecting caprine meat and milk's qualitative and quantitative production offers new opportunities to direct breeding practices towards the accurate and efficient selection of desirable traits. In this way, the latest genomic advances may allow to increase the response to selection, enabling a further genetic progress. There are pieces of evidence that suggest that the regulation of the expressions of very diverse traits (from body growth to milk composition) may somehow intertwine. This leads to the conclusion that gene regulation of the expression of caprine milk or meat very likely occurs in a multidimensional manner, taking place at different levels from central neurological control to the specific parts of the body, where, for instance, the meat and milk nutrients are produced. Consequently, knowing these functional regions of genomes may not only boost the efficiency, accuracy and progress of breeding schemes, but could also permit the reinforcement of local breeds' conservation strategies by enhancing the sustainability and profitability of their products.

Author Contributions: Conceptualization, F.J.N.G. and J.V.D.B.; data curation, J.M.L.J., F.J.N.G. and J.I.S.P.; formal analysis, A.G.A., F.J.N.G., C.M.N. and C.I.P.; funding acquisition, J.V.D.B.; investigation, J.I.S.P., F.J.N.G. and C.I.P.; methodology, F.J.N.G., A.G.A. and J.V.D.B.; project administration, J.V.D.B.; resources, J.I.S.P. and J.V.D.B.; software, J.M.L.J. and F.J.N.G.; supervision, F.J.N.G. and J.V.D.B.; validation, F.J.N.G. and J.V.D.B.; visualization, C.M.N.; writing—original draft, F.J.N.G. and J.I.S.P.; writing—review and editing, C.M.N., A.G.A., C.I.P., J.M.L.J., F.J.N.G., M.d.A.M.M. and J.V.D.B. All authors have read and agreed to the published version of the manuscript.

Funding: This research received no external funding.

Institutional Review Board Statement: Not applicable.

Informed Consent Statement: Not applicable.

Data Availability Statement: Not applicable.

Conflicts of Interest: The authors declare no conflict of interest.

References

1. Miller, B.A.; Lu, C.D. Current status of global dairy goat production: An overview. *Asian Australas. J. Anim. Sci.* **2019**, *32*, 1219. [CrossRef] [PubMed]
2. Gama, L.; Bressan, M. Biotechnology applications for the sustainable management of goat genetic resources. *Small Rumin. Res.* **2011**, *98*, 133–146. [CrossRef]
3. Amills, M.; Capote, J.; Tosser-Klopp, G. Goat domestication and breeding: A jigsaw of historical, biological and molecular data with missing pieces. *Anim. Genet.* **2017**, *48*, 631–644. [CrossRef] [PubMed]
4. Fahmy, M.H.; Shrestha, J. Genetics for the improvement of goat meat production. In Proceedings of the 7th International Conference on Goats, Tours, France, 15–21 May 2000; pp. 187–190.
5. Shrestha, J. Genetics and breeding of meat goats. In *Goat Meat Production and Quality*; Mahgoub, O., Kadim, I., Webb, E., Eds.; CAB International: London, UK, 2012; p. 52.
6. Mazhangara, I.R.; Chivandi, E.; Mupangwa, J.F.; Muchenje, V. The potential of goat meat in the red meat industry. *Sustainability* **2019**, *11*, 3671. [CrossRef]
7. Daly, K.G.; Delser, P.M.; Mullin, V.E.; Scheu, A.; Mattiangeli, V.; Teasdale, M.D.; Hare, A.J.; Burger, J.; Verdugo, M.P.; Collins, M.J. Ancient goat genomes reveal mosaic domestication in the Fertile Crescent. *Science* **2018**, *361*, 85–88. [CrossRef] [PubMed]
8. Pidancier, N.; Jordan, S.; Luikart, G.; Taberlet, P. Evolutionary history of the genus Capra (Mammalia, Artiodactyla): Discordance between mitochondrial DNA and Y-chromosome phylogenies. *Mol. Phylogenet. Evol.* **2006**, *40*, 739–749. [CrossRef] [PubMed]
9. Mirkena, T.; Duguma, G.; Haile, A.; Tibbo, M.; Okeyo, A.; Wurzinger, M.; Sölkner, J. Genetics of adaptation in domestic farm animals: A review. *Livest. Sci.* **2010**, *132*, 1–12. [CrossRef]
10. Gootwine, E. Genetics and breeding of sheep and goats. In *Animal Agriculture*; Elsevier: College Station, TX, USA, 2020; pp. 183–198.
11. Ferrando, A.; Manunza, A.; Jordana, J.; Capote, J.; Pons, A.; Pais, J.; Delgado, T.; Atoche, P.; Cabrera, B.; Martínez, A. A mitochondrial analysis reveals distinct founder effect signatures in Canarian and Balearic goats. *Anim. Genet.* **2015**, *46*, 452–456. [CrossRef]
12. Yalcin, B. *Sheep and Goats in Turkey*; FAO: Rome, Italy, 1986.
13. Lauvergne, J.; Renieri, C.; Machado, T.; Acharya, R. Goat genetic resources. In Proceedings of the 5th International Goat Conference, New Delhi, India, 2–8 March 1992.
14. Jenot, F. Goat unions, FNEC and technical institutes, an intersecting destiny: Trajectories and dynamics since 1950 in France. *Ethnozootechnie* **2011**, *91*, 123–133.
15. Casey, N.; Van Niekerk, W. The Boer goat. I. Origin, adaptability, performance testing, reproduction and milk production. *Small Rumin. Res.* **1988**, *1*, 291–302. [CrossRef]
16. Sanfiorenzo, J.H. Evaluation of six breeding lines of milk goats. *J. Agric. Univ. P. R.* **1962**, *46*, 205–212. [CrossRef]
17. Ronningen, K. Causes of variation in the flavour intensity of goat milk. *Acta Agric. Scand.* **1965**, *15*, 301–342. [CrossRef]
18. Wiggans, G.; Van Vleck, L.D.; Dickinson, F. Projection factors for goat lactation records. *J. Dairy Sci.* **1979**, *62*, 797–801. [CrossRef]
19. Gipson, T.A. Recent advances in breeding and genetics for dairy goats. *Asian Australas. J. Anim. Sci.* **2019**, *32*, 1275. [CrossRef] [PubMed]
20. Wiggans, G.; Dickinson, F.; King, G.; Weller, J. Genetic evaluation of dairy goat bucks for daughter milk and fat. *J. Dairy Sci.* **1984**, *67*, 201–207. [CrossRef]
21. Wiggans, G.; Van Dijk, J.; Misztal, I. Genetic evaluation of dairy goats for milk and fat yield with an animal model. *J. Dairy Sci.* **1988**, *71*, 1330–1337. [CrossRef]
22. Kemp, S.J.; Brezinsky, L.; Teale, A. A panel of bovine, ovine and caprine polymorphic microsatellites. *Anim. Genet.* **1993**, *24*, 363–365. [CrossRef]
23. Vaiman, D.; Schibler, L.; Bourgeois, F.; Oustry, A.; Amigues, Y.; Cribiu, E.P. A genetic linkage map of the male goat genome. *Genetics* **1996**, *144*, 279–305. [CrossRef]
24. Arevalo, E.; Holder, D.; Derr, J.; Bhebhe, E.; Linn, R. Caprine microsatellite dinucleotide repeat polymorphisms at the Sr-Crsp-1, Sr-Crsp-2, Sr-Crsp-3, Sr-Crsp-4 and Sr-Crsp-5 loci. *Anim. Genet.* **1994**, *25*, 202. [CrossRef]
25. Schibler, L.; Vaiman, D.; Oustry, A.; Guinec, N.; Dangy-Caye, A.L.; Billault, A.; Cribiu, E.P. Construction and extensive characterization of a goat bacterial artificial chromosome library with threefold genome coverage. *Mamm. Genome* **1998**, *9*, 119–124. [CrossRef]
26. Mucha, S.; Mrode, R.; Coffey, M.; Kizilaslan, M.; Desire, S.; Conington, J. Genome-wide association study of conformation and milk yield in mixed-breed dairy goats. *J. Dairy Sci.* **2018**, *101*, 2213–2225. [CrossRef] [PubMed]
27. Bickhart, D.M.; Rosen, B.D.; Koren, S.; Sayre, B.L.; Hastie, A.R.; Chan, S.; Lee, J.; Lam, E.T.; Liachko, I.; Sullivan, S.T. Single-molecule sequencing and chromatin conformation capture enable de novo reference assembly of the domestic goat genome. *Nat. Genet.* **2017**, *49*, 643–650. [CrossRef]

28. Carillier, C.; Larroque, H.; Palhière, I.; Clément, V.; Rupp, R.; Robert-Granié, C. A first step toward genomic selection in the multi-breed French dairy goat population. *J. Dairy Sci.* **2013**, *96*, 7294–7305. [CrossRef] [PubMed]
29. Dong, Y.; Xie, M.; Jiang, Y.; Xiao, N.; Du, X.; Zhang, W.; Tosser-Klopp, G.; Wang, J.; Yang, S.; Liang, J. Sequencing and automated whole-genome optical mapping of the genome of a domestic goat (*Capra hircus*). *Nat. Biotechnol.* **2013**, *31*, 135–141. [CrossRef] [PubMed]
30. Tosser-Klopp, G. Goat genome assembly, Availability of an international 50K SNP chip and RH panel: An update of the International Goat Genome Consortium projects. In Proceedings of the International Plant and Animal Genome XX Conference, Town and Country Hotel, San Diego, CA, USA, 14–18 January 2012.
31. Fontanesi, L.; Martelli, P.L.; Beretti, F.; Riggio, V.; Dall'Olio, S.; Colombo, M.; Casadio, R.; Russo, V.; Portolano, B. An initial comparative map of copy number variations in the goat (*Capra hircus*) genome. *BMC Genom.* **2010**, *11*, 639. [CrossRef]
32. Li, X.-L.; Wu, Z.-L.; Liu, Z.-Z.; Gong, Y.-F.; Zhou, R.-Y.; Zheng, G.-R. SNP identification and analysis in part of intron 2 of goat MSTN gene and variation within and among species. *J. Hered.* **2006**, *97*, 285–289. [CrossRef]
33. Andonov, S.; Ødegård, J.; Svendsen, M.; Ådnøy, T.; Vegara, M.; Klemetsdal, G. Comparison of random regression and repeatability models to predict breeding values from test-day records of Norwegian goats. *J. Dairy Sci.* **2013**, *96*, 1834–1843. [CrossRef]
34. Gebreselassie, G.; Berihulay, H.; Jiang, L.; Ma, Y. Review on genomic regions and candidate genes associated with economically important production and reproduction traits in sheep (*Ovies aries*). *Animals* **2020**, *10*, 33. [CrossRef]
35. Zhang, H.; Wang, Z.; Wang, S.; Li, H. Progress of genome wide association study in domestic animals. *J. Anim. Sci. Biotechnol.* **2012**, *3*, 26. [CrossRef] [PubMed]
36. Lashmar, S.; Visser, C.; van Marle-Köster, E. Validation of the 50k Illumina goat SNP chip in the South African Angora goat. *S. Afr. J. Anim. Sci.* **2015**, *45*, 56–59. [CrossRef]
37. Li, X.; Wu, Z.L.; Gong, Y.; Liu, Y.; Liu, Z.; Wang, X.; Xin, T.; Ji, Q. Single-nucleotide polymorphism identification in the caprine myostatin gene. *J. Anim. Breed.* **2006**, *123*, 141–144. [CrossRef] [PubMed]
38. Andonov, S.; Ødegård, J.; Boman, I.; Svendsen, M.; Holme, I.; Ådnøy, T.; Vukovic, V.; Klemetsdal, G. Validation of test-day models for genetic evaluation of dairy goats in Norway. *J. Dairy Sci.* **2007**, *90*, 4863–4871. [CrossRef] [PubMed]
39. Pizarro Inostroza, M.G.; Navas González, F.J.; Landi, V.; León Jurado, J.M.; Delgado Bermejo, J.V.; Fernández Álvarez, J.; Martínez Martínez, M.d.A. Software-Automatized Individual Lactation Model Fitting, Peak and Persistence and Bayesian Criteria Comparison for Milk Yield Genetic Studies in Murciano-Granadina Goats. *Mathematics* **2020**, *8*, 1505. [CrossRef]
40. Tosser-Klopp, G.; Bardou, P.; Bouchez, O.; Cabau, C.; Crooijmans, R.; Dong, Y.; Donnadieu-Tonon, C.; Eggen, A.; Heuven, H.C.; Jamli, S. Design and characterization of a 52K SNP chip for goats. *PLoS ONE* **2014**, *9*, e86227. [CrossRef] [PubMed]
41. Onzima, R.; Upadhyay, M.; Mukiibi, R.; Kanis, E.; Groenen, M.; Crooijmans, R. Genome-wide population structure and admixture analysis reveals weak differentiation among Ugandan goat breeds. *Anim. Genet.* **2018**, *49*, 59–70. [CrossRef]
42. Schibler, L.; Vaiman, D.; Oustry, A.; Giraud-Delville, C.; Cribiu, E.P. Comparative gene mapping: A fine-scale survey of chromosome rearrangements between ruminants and humans. *Genome Res.* **1998**, *8*, 901–915. [CrossRef]
43. Du, X.; Womack, J.; Owens, K.; Elliott, J.; Sayre, B.; Bottcher, P.; Milan, D.; Podesta, M.G.; Zhao, S.; Malek, M. A whole-genome radiation hybrid panel for goat. *Small Rumin. Res.* **2012**, *105*, 114–116. [CrossRef]
44. Hirschhorn, J.N.; Daly, M.J. Genome-wide association studies for common diseases and complex traits. *Nat. Rev. Genet.* **2005**, *6*, 95–108. [CrossRef]
45. González Ariza, A.; Arando Arbulu, A.; Navas González, F.J.; Nogales Baena, S.; Delgado Bermejo, J.V.; Camacho Vallejo, M.E. The Study of Growth and Performance in Local Chicken Breeds and Varieties: A Review of Methods and Scientific Transference. *Animals* **2021**, *11*, 2492. [CrossRef]
46. McLean, A.K.; Navas Gonzalez, F.J. Can scientists influence donkey welfare? Historical perspective and a contemporary view. *J. Equine Vet. Sci.* **2018**, *65*, 25–32. [CrossRef]
47. Iglesias Pastrana, C.; Navas González, F.J.; Ciani, E.; Barba Capote, C.J.; Delgado Bermejo, J.V. Effect of research impact on emerging camel husbandry, welfare and social-related awareness. *Animals* **2020**, *10*, 780. [CrossRef] [PubMed]
48. Zonaed Siddiki, A.; Miah, G.; Islam, M.; Kumkum, M.; Rumi, M.H.; Baten, A.; Hossain, M.A. Goat Genomic Resources: The Search for Genes Associated with Its Economic Traits. *Int. J. Genom.* **2020**, *2020*, 5940205. [CrossRef] [PubMed]
49. Webb, E.; Casey, N.; Simela, L. Growth, development and growth manipulation in goats. In *Goat Meat Production and Quality*; Mahgoub, O., Kadim, I.T., Cottington Webb, E., Ovid Technologies, Inc., Eds.; CAB International: Cambridge, MA, USA, 2012; pp. 196–199.
50. Shrestha, J.; Fahmy, M. Breeding goats for meat production: 3. Selection and breeding strategies. *Small Rumin. Res.* **2007**, *67*, 113–125. [CrossRef]
51. Shrestha, J.; Fahmy, M. Breeding goats for meat production: 2. Crossbreeding and formation of composite population. *Small Rumin. Res.* **2007**, *67*, 93–112. [CrossRef]
52. Fernández Álvarez, J.; León Jurado, J.; Navas González, F.J.; Iglesias Pastrana, C.; Delgado Bermejo, J.V. CAPRIGRAN Linear Appraisal Evidences Dairy Selection Signs in Murciano-Granadina Goats and Bucks: Presentation of the New Linear Appraisal Scale. *Arch. Zootecn.* **2021**, *70*, 240–245. [CrossRef]
53. Webb, E.; Casey, N.; Simela, L. Goat meat quality. *Small Rumin. Res.* **2005**, *60*, 153–166. [CrossRef]
54. Webb, E. Goat meat production, composition, and quality. *Anim. Front.* **2014**, *4*, 33–37. [CrossRef]

55. Raes, K.; Balcaen, A.; Dirinck, P.; De Winne, A.; Claeys, E.; Demeyer, D.; De Smet, S. Meat quality, fatty acid composition and flavour analysis in Belgian retail beef. *Meat Sci.* **2003**, *65*, 1237–1246. [CrossRef]
56. Herrera, P.Z.; Bermejo, J.V.D.; Henríquez, A.A.; Vallejo, M.E.C.; Costa, R.G. Effects of extensive system versus semi-intensive and intensive systems on growth and carcass quality of dairy kids. *Rev. Bras. Zootec.* **2011**, *40*, 2613–2620. [CrossRef]
57. Jembere, T.; Dessie, T.; Rischkowsky, B.; Kebede, K.; Okeyo, A.; Haile, A. Meta-analysis of average estimates of genetic parameters for growth, reproduction and milk production traits in goats. *Small Rumin. Res.* **2017**, *153*, 71–80. [CrossRef]
58. Huson, H.; Sonstegard, T.; Silverstein, J.; Woodward Greene, J.; Masiga, C.; Muchadeyi, F.; Rees, J.; Sayre, B.; Elbeltagy, A.; Rothschild, M.F. Genetic and phenotypic characterization of African goat populations to prioritize conservation and production efforts for small-holder farmers in Sub-Saharan africa. In Proceedings of the 10th World Congress on Genetics Applied to Livestock Production, Vancouver, BC, Canada, 17–22 August 2014; American Society of Animal Science: Champaign, IL, USA, 2014.
59. Zhang, N.; Teng, Z.; Qi, Q.; Hu, G.; Lian, H.; Gao, T. Carcass traits, meat quality characteristics, and lipid metabolism-related gene expression pattern of Yaoshan white goats raised in traditional extensive production system: Effects of slaughter age and meat cuts. *Small Rumin. Res.* **2020**, *182*, 29–36. [CrossRef]
60. Liu, M.; Woodward-Greene, J.; Kang, X.; Pan, M.G.; Rosen, B.; Van Tassell, C.P.; Chen, H.; Liu, G.E. Genome-wide CNV analysis revealed variants associated with growth traits in African indigenous goats. *Genomics* **2020**, *112*, 1477–1480. [CrossRef] [PubMed]
61. Yarwood, S.J.; Kilgour, E.; Anderson, N.G. Cyclic AMP potentiates growth hormone-dependent differentiation of 3T3-F442A preadipocytes: Possible involvement of the transcription factor CREB. *Mol. Cell. Endocrinol.* **1998**, *138*, 41–50. [CrossRef]
62. Harwood, J.P.; Grewe, C.; Aguilera, G. Actions of growth hormone-releasing factor and somatostatin on adenylate cyclase and growth hormone release in rat anterior pituitary. *Mol. Cell. Endocrinol.* **1984**, *37*, 277–284. [CrossRef]
63. Bi, Y.; Chen, Y.; Xin, D.; Liu, T.; He, L.; Kang, Y.; Pan, C.; Shen, W.; Lan, X.; Liu, M. Effect of indel variants within the sorting nexin 29 (SNX29) gene on growth traits of goats. *Anim. Biotechnol.* **2020**, *19*, 1–6. [CrossRef]
64. Hägg, S.; Skogsberg, J.; Lundström, J.; Noori, P.; Nilsson, R.; Zhong, H.; Maleki, S.; Shang, M.M.; Brinne, B.; Bradshaw, M. Multi-organ expression profiling uncovers a gene module in coronary artery disease involving transendothelial migration of leukocytes and LIM domain binding 2: The Stockholm Atherosclerosis Gene Expression (STAGE) study. *PLoS Genet.* **2009**, *5*, e1000754. [CrossRef]
65. Liu, R.; Sun, Y.; Zhao, G.; Wang, F.; Wu, D.; Zheng, M.; Chen, J.; Zhang, L.; Hu, Y.; Wen, J. Genome-wide association study identifies loci and candidate genes for body composition and meat quality traits in Beijing-You chickens. *PLoS ONE* **2013**, *8*, e61172. [CrossRef]
66. Gonzalez, M.V.; Mousel, M.R.; Herndon, D.R.; Jiang, Y.; Dalrymple, B.P.; Reynolds, J.O.; Johnson, W.C.; Herrmann-Hoesing, L.M.; White, S.N. A divergent Artiodactyl MYADM-like repeat is associated with erythrocyte traits and weight of lamb weaned in domestic sheep. *PLoS ONE* **2013**, *8*, e74700. [CrossRef]
67. Hoopes, B.C.; Rimbault, M.; Liebers, D.; Ostrander, E.A.; Sutter, N.B. The insulin-like growth factor 1 receptor (IGF1R) contributes to reduced size in dogs. *Mamm. Genome* **2012**, *23*, 780–790. [CrossRef]
68. Bickhart, D.M.; Hou, Y.; Schroeder, S.G.; Alkan, C.; Cardone, M.F.; Matukumalli, L.K.; Song, J.; Schnabel, R.D.; Ventura, M.; Taylor, J.F. Copy number variation of individual cattle genomes using next-generation sequencing. *Genome Res.* **2012**, *22*, 778–790. [CrossRef]
69. Wu, J.; Peng, X.; Li, F.; Qiao, M.; Wu, H.; Mei, S. Association of porcine STIM1 gene with litter size traits. *Anim. Sci. Pap. Rep.* **2016**, *34*, 165–173.
70. Zhang, B.; Chang, L.; Lan, X.; Asif, N.; Guan, F.; Fu, D.; Li, B.; Yan, C.; Zhang, H.; Zhang, X. Genome-wide definition of selective sweeps reveals molecular evidence of trait-driven domestication among elite goat (*Capra species*) breeds for the production of dairy, cashmere, and meat. *GigaScience* **2018**, *7*, giy105. [CrossRef] [PubMed]
71. Granados-Principal, S.; Liu, Y.; Rodriguez-Aguayo, C.; Sood, A.K.; Chang, J.C. HMG Box Domain Containing 3 (HMGXB3), a novel TGFβ-induced cancer gene that influences cancer stem cells and metastasis in triple negative breast cancer. *Cancer Res.* **2015**, *75*, 4086.
72. Li, M.Y.; Min, L.J.; Sun, G.Q.; Pan, Q.J.; Shen, W.; Wang, G.L. Polymorphism analysis of the goat growth hormone gene in the 5′regulatory sequence. *Hereditas* **2004**, *26*, 831–835. [PubMed]
73. Jin, Q.; Fang, X.T.; Yang, L.; Zhang, C.L.; Sun, J.J.; Chen, D.X.; Shi, X.; Du, Y.; Lei, C.; Chen, H. Novel SNPs of the caprine growth hormone secretagogue receptor (GHSR) gene and their association with growth traits in goats. *Biochem. Genet.* **2010**, *48*, 847–856. [CrossRef] [PubMed]
74. Lucy, M.; Johnson, G.; Shibuya, H.; Boyd, C.; Herring, W. Rapid communication: Polymorphic (GT) n microsatellite in the bovine somatotropin gene promoter. *J. Anim. Sci.* **1998**, *76*, 2209. [CrossRef]
75. Mullen, M.P.; Berry, D.P.; Howard, D.J.; Diskin, M.G.; Lynch, C.O.; Giblin, L.; Kenny, D.A.; Magee, D.A.; Meade, K.G.; Waters, S.M. Single nucleotide polymorphisms in the insulin-like growth factor 1 (IGF-1) gene are associated with performance in Holstein-Friesian dairy cattle. *Front. Genet.* **2011**, *2*, 3. [CrossRef]
76. Van Laere, A.-S.; Nguyen, M.; Braunschweig, M.; Nezer, C.; Collette, C.; Moreau, L.; Archibald, A.L.; Haley, C.S.; Buys, N.; Tally, M. A regulatory mutation in IGF2 causes a major QTL effect on muscle growth in the pig. *Nature* **2003**, *425*, 832–836. [CrossRef]
77. Bale, L.K.; Conover, C.A. Regulation of insulin-like growth factor binding protein-3 messenger ribonucleic acid expression by insulin-like growth factor I. *Endocrinology* **1992**, *131*, 608–614.

78. Ji, W.; Yu, Y.; Li, Z.; Wang, G.; Li, F.; Xia, W.; Lu, S. FGFR1 promotes the stem cell-like phenotype of FGFR1-amplified non-small cell lung cancer cells through the Hedgehog pathway. *Oncotarget* **2016**, *7*, 15118. [CrossRef]
79. Fang, X.; Xu, H.; Zhang, C.; Chen, H.; Hu, X.; Gao, X.; Gu, C.; Yue, W. Polymorphism in BMP4 gene and its association with growth traits in goats. *Mol. Biol. Rep.* **2009**, *36*, 1339–1344. [CrossRef] [PubMed]
80. Chu, M.; Jiao, C.; He, Y.; Wang, J.; Liu, Z.; Chen, G. Association between PCR-SSCP of bone morphogenetic protein 15 gene and prolificacy in Jining Grey goats. *Anim. Biotechnol.* **2007**, *18*, 263–274. [CrossRef] [PubMed]
81. García-de Teresa, B.; Hernández-Gómez, M.; Frías, S. DNA damage as a driver for growth delay: Chromosome instability syndromes with intrauterine growth retardation. *BioMed Res. Int.* **2017**, *2017*, 8193892. [CrossRef] [PubMed]
82. Xu, Z.; Wang, X.; Zhang, Z.; An, Q.; Wen, Y.; Wang, D.; Liu, X.; Li, Z.; Lyu, S.; Li, L. Copy number variation of CADM2 gene revealed its association with growth traits across Chinese *Capra hircus* (goat) populations. *Gene* **2020**, *741*, 144519. [CrossRef] [PubMed]
83. Guo, J.; Tao, H.; Li, P.; Li, L.; Zhong, T.; Wang, L.; Ma, J.; Chen, X.; Song, T.; Zhang, H. Whole-genome sequencing reveals selection signatures associated with important traits in six goat breeds. *Sci. Rep.* **2018**, *8*, 10405. [CrossRef]
84. Dong, Y.; Zhang, X.; Xie, M.; Arefnezhad, B.; Wang, Z.; Wang, W.; Feng, S.; Huang, G.; Guan, R.; Shen, W. Reference genome of wild goat (*Capra aegagrus*) and sequencing of goat breeds provide insight into genic basis of goat domestication. *BMC Genom.* **2015**, *16*, 431. [CrossRef]
85. Supakorn, C. The important candidate genes in goats—A review. *Walailak J. Sci. Technol.* **2009**, *6*, 17–36.
86. Naicy, T.; Venkatachalapathy, R.; Aravindakshan, T.; Kurian, E. Association of a Cac8I polymorphism in the IGF1 gene with growth traits in Indian goats. *J. Genet. Eng. Biotechnol.* **2017**, *15*, 7–11. [CrossRef]
87. Wang, Z.; Zhang, X.; Jiang, E.; Yan, H.; Zhu, H.; Chen, H.; Liu, J.; Qu, L.; Pan, C.; Lan, X. InDels within caprine IGF 2 BP 1 intron 2 and the 3′-untranslated regions are associated with goat growth traits. *Anim. Genet.* **2020**, *51*, 117–121. [CrossRef]
88. Wang, Z.; Wang, C.; Guo, Y.; She, S.; Wang, B.; Jiang, Y.; Bai, Y.; Song, X.; Li, L.; Shi, L. Screening of deletion variants within the goat PRDM6 gene and its effects on growth traits. *Animals* **2020**, *10*, 208. [CrossRef]
89. Naicy, T.; Venkatachalapathy, T.; Aravindakshan, T.; Bosewell, A.; Silpa, M. Association of a SacII polymorphism in the Nerve Growth Factor (NGF) gene exon 3 with growth traits in Indian goats. *Small Rumin. Res.* **2018**, *158*, 19–21. [CrossRef]
90. Lan, X.; Pan, C.; Chen, H.; Zhang, C.; Li, J.; Zhao, M.; Lei, C.; Zhang, A.; Zhang, L. An AluI PCR-RFLP detecting a silent allele at the goat POU1F1 locus and its association with production traits. *Small Rumin. Res.* **2007**, *73*, 8–12. [CrossRef]
91. Renaville, R.; Gengler, N.; Vrech, E.; Prandi, A.; Massart, S.; Corradini, C.; Bertozzi, C.; Mortiaux, F.; Burny, A.; Portetelle, D. Pit-1 gene polymorphism, milk yield, and conformation traits for Italian Holstein-Friesian bulls. *J. Dairy Sci.* **1997**, *80*, 3431–3438. [CrossRef]
92. Houseknecht, K.L.; Baile, C.A.; Matteri, R.L.; Spurlock, M.E. The biology of leptin: A review. *J. Anim. Sci.* **1998**, *76*, 1405–1420. [CrossRef] [PubMed]
93. Barzekar, R.; Salehi, A.; Mahjoubi, F. Polymorphisms of the ovine leptin gene and its association with growth and carcass traits in three Iranian sheep breeds. *Iran. J. Biotechnol.* **2009**, *7*, 241–246.
94. Singh, M.K.; Petry, M.; Haenig, B.; Lescher, B.; Leitges, M.; Kispert, A. The T-box transcription factor Tbx15 is required for skeletal development. *Mech. Dev.* **2005**, *122*, 131–144. [CrossRef]
95. Sugatani, T.; Hildreth III, B.E.; Toribio, R.E.; Malluche, H.H.; Hruska, K.A. Expression of DGCR8-dependent microRNAs is indispensable for osteoclastic development and bone-resorbing activity. *J. Cell. Biochem.* **2014**, *115*, 1043–1047. [CrossRef]
96. Sarkar, S.; Dey, B.K.; Dutta, A. MiR-322/424 and-503 are induced during muscle differentiation and promote cell cycle quiescence and differentiation by down-regulation of Cdc25A. *Mol. Biol. Cell.* **2010**, *21*, 2138–2149. [CrossRef]
97. Barreda-Bonis, A.C.; Barraza-García, J.; Parrón, M.; Pastor, I.; Heath, K.E.; González-Casado, I. Multiple SLC26A2 mutations occurring in a three-generational family. *Eur. J. Med. Genet.* **2018**, *61*, 24–28. [CrossRef]
98. Wang, X.; Liu, J.; Zhou, G.; Guo, J.; Yan, H.; Niu, Y.; Li, Y.; Yuan, C.; Geng, R.; Lan, X. Whole-genome sequencing of eight goat populations for the detection of selection signatures underlying production and adaptive traits. *Sci. Rep.* **2016**, *6*, 38932. [CrossRef]
99. Liu, S.H.; Chiu, C.Y.; Wang, L.P.; Chiang, M.T. Omega-3 fatty acids-enriched fish oil activates AMPK/PGC-1α signaling and prevents obesity-related skeletal muscle wasting. *Mar. Drugs* **2019**, *17*, 380. [CrossRef] [PubMed]
100. Oh, D.; La, B.; Lee, Y.; Byun, Y.; Lee, J.; Yeo, G.; Yeo, J. Identification of novel single nucleotide polymorphisms (SNPs) of the lipoprotein lipase (LPL) gene associated with fatty acid composition in Korean cattle. *Mol. Biol. Rep.* **2013**, *40*, 3155–3163. [CrossRef] [PubMed]
101. Shin, S.C.; Heo, J.P.; Chung, E.R. Effect of Single Nucleotide Polymorphisms of Acetyl-CoA Carboxylase α (ACACA) gene on carcass traits in Hanwoo (Korean cattle). *Asian Australas. J. Anim. Sci.* **2011**, *24*, 744–751. [CrossRef]
102. Eberlé, D.; Hegarty, B.; Bossard, P.; Ferré, P.; Foufelle, F. SREBP transcription factors: Master regulators of lipid homeostasis. *Biochimie* **2004**, *86*, 839–848. [CrossRef]
103. Yang, C.; Liu, J.; Wu, X.; Bao, P.; Long, R.; Guo, X.; Ding, X.; Yan, P. The response of gene expression associated with lipid metabolism, fat deposition and fatty acid profile in the *longissimus dorsi* muscle of Gannan yaks to different energy levels of diets. *PLoS ONE* **2017**, *12*, e0187604. [CrossRef] [PubMed]
104. Shimano, H. Sterol regulatory element-binding proteins (SREBPs): Transcriptional regulators of lipid synthetic genes. *Prog. Lipid Res.* **2001**, *40*, 439–452. [CrossRef]

105. Oh, D.Y.; Lee, Y.S.; La, B.M.; Lee, J.Y.; Park, Y.S.; Lee, J.H.; Ha, J.J.; Yi, J.K.; Kim, B.K.; Yeo, J.S. Identification of exonic nucleotide variants of the thyroid hormone responsive protein gene associated with carcass traits and Fatty Acid composition in korean cattle. *Asian-Australas. J. Anim. Sci.* **2014**, *27*, 1373. [CrossRef]
106. van der Leij, F.R.; Huijkman, N.C.; Boomsma, C.; Kuipers, J.R.; Bartelds, B. Genomics of the human carnitine acyltransferase genes. *Mol. Genet. Metab.* **2000**, *71*, 139–153. [CrossRef]
107. Sodhi, S.S.; Park, W.C.; Ghosh, M.; Kim, J.N.; Sharma, N.; Shin, K.Y.; Cho, I.C.; Ryu, Y.C.; Oh, S.J.; Kim, S.H. Comparative transcriptomic analysis to identify differentially expressed genes in fat tissue of adult Berkshire and Jeju Native Pig using RNA-seq. *Mol. Biol. Rep.* **2014**, *41*, 6305–6315. [CrossRef]
108. Wei-Jun, P.; Liang, B.; Gong-She, Y. Relationship among H-FABP gene polymorphism, intramuscular fat content, and adipocyte lipid droplet content in main pig breeds with different genotypes in western China. *Acta Genet. Sin.* **2006**, *33*, 515–524.
109. Benjelloun, B.; Alberto, F.J.; Streeter, I.; Boyer, F.; Coissac, E.; Stucki, S.; BenBati, M.; Ibnelbachyr, M.; Chentouf, M.; Bechchari, A. Characterizing neutral genomic diversity and selection signatures in indigenous populations of Moroccan goats (*Capra hircus*) using WGS data. *Front. Genet.* **2015**, *6*, 107. [CrossRef] [PubMed]
110. FAO. FAOSTAT Statistical Database. 2019. Available online: http://www.fao.org/faostat/en/ (accessed on 17 July 2021).
111. García, V.; Rovira, S.; Boutoial, K.; López, M. Improvements in goat milk quality: A review. *Small Rumin. Res.* **2014**, *121*, 51–57. [CrossRef]
112. Pazzola, M.; Stocco, G.; Dettori, M.L.; Bittante, G.; Vacca, G.M. Effect of goat milk composition on cheesemaking traits and daily cheese production. *J. Dairy Sci.* **2019**, *102*, 3947–3955. [CrossRef] [PubMed]
113. Lin, J.; Yu, Q.; Zhang, Q.; Yang, Q. Construction of mammary gland specific expression plasmid pIN and its expression in vitro and in vivo. *Afr. J. Biotechnol.* **2012**, *11*, 6946–6955.
114. Lemley, C.; Butler, S.; Butler, W.; Wilson, M. Insulin alters hepatic progesterone catabolic enzymes cytochrome P450 2C and 3A in dairy cows. *J. Dairy Sci.* **2008**, *91*, 641–645. [CrossRef] [PubMed]
115. Capitan, A.; Michot, P.; Baur, A.; Saintilan, R.; Hoze, C.; Valour, D.; Guillaume, F.; Boichon, D.; Barbat, A.; Boichard, D. Genetic tools to improve reproduction traits in dairy cattle. *Reprod. Fert. Dev.* **2015**, *27*, 14–21. [CrossRef] [PubMed]
116. Liefers, S.; Te Pas, M.; Veerkamp, R.; Van Der Lende, T. Associations between leptin gene polymorphisms and production, live weight, energy balance, feed intake, and fertility in Holstein heifers. *J. Dairy Sci.* **2002**, *85*, 1633–1638. [CrossRef]
117. Banos, G.; Woolliams, J.; Woodward, B.; Forbes, A.; Coffey, M. Impact of single nucleotide polymorphisms in leptin, leptin receptor, growth hormone receptor, and diacylglycerol acyltransferase (DGAT1) gene loci on milk production, feed, and body energy traits of UK dairy cows. *J. Dairy Sci.* **2008**, *91*, 3190–3200. [CrossRef] [PubMed]
118. Ron, M.; Cohen-Zinder, M.; Peter, C.; Weller, J.; Erhardt, G. A polymorphism in ABCG2 in Bos indicus and Bos taurus cattle breeds. *J. Dairy Sci.* **2006**, *89*, 4921–4923. [CrossRef]
119. Rahmatalla, S.A.; Müller, U.; Strucken, E.M.; Reissmann, M.; Brockmann, G.A. The F279Y polymorphism of the GHR gene and its relation to milk production and somatic cell score in German Holstein dairy cattle. *J. Appl. Genet.* **2011**, *52*, 459–465. [CrossRef]
120. Dybus, A. Associations of growth hormone (GH) and prolactin (PRL) genes polymorphisms with milk production traits in Polish Black-and-White cattle. *Anim. Sci. Pap. Rep.* **2002**, *20*, 203–212.
121. Lu, A.; Hu, X.; Chen, H.; Dong, Y.; Zhang, Y.; Wang, X. Novel SNPs of the bovine PRLR gene associated with milk production traits. *Biochem. Genet.* **2011**, *49*, 177–189. [CrossRef] [PubMed]
122. Wesolowski, S.R.; Allan, M.F.; Nielsen, M.K.; Pomp, D. Evaluation of hypothalamic gene expression in mice divergently selected for heat loss. *Physiol. Genom.* **2003**, *13*, 129–137. [CrossRef] [PubMed]
123. Leonard, S.; Khatib, H.; Schutzkus, V.; Chang, Y.; Maltecca, C. Effects of the osteopontin gene variants on milk production traits in dairy cattle. *J. Dairy Sci.* **2005**, *88*, 4083–4086. [CrossRef]
124. Mullen, M.P.; Lynch, C.; Waters, S.; Howard, D.J.; O'boyle, P.; Kenny, D.; Buckley, F.; Horan, B.; Diskin, M. Single nucleotide polymorphisms in the growth hormone and insulin-like growth factor-1 genes are associated with milk production, body condition score and fertility traits in dairy cows. *Genet. Mol. Res.* **2011**, *26*, 1819–1830. [CrossRef]
125. Ward, P.; Paz, E.; Conneely, O. Lactoferrin. *Cell. Mol. Life Sci.* **2005**, *62*, 2540–2548. [CrossRef]
126. Yuan, Z.; Chu, G.; Dan, Y.; Li, J.; Zhang, L.; Gao, X.; Gao, H.; Li, J.; Xu, S.; Liu, Z. BRCA1: A new candidate gene for bovine mastitis and its association analysis between single nucleotide polymorphisms and milk somatic cell score. *Mol. Biol. Rep.* **2012**, *39*, 6625–6631. [CrossRef]
127. Moioli, B.; D'Andrea, M.; Pilla, F. Candidate genes affecting sheep and goat milk quality. *Small Rumin. Res.* **2007**, *68*, 179–192. [CrossRef]
128. Dagnachew, B.S.; Ådnøy, T. Additive and dominance effects of casein haplotypes on milk composition and quality in Norwegian dairy goats. *Small Rumin. Res.* **2014**, *122*, 59–69. [CrossRef]
129. Dagnachew, B.S.; Thaller, G.; Lien, S.; Ådnøy, T. Casein SNP in Norwegian goats: Additive and dominance effects on milk composition and quality. *Genet. Sel. Evol.* **2011**, *43*, 31. [CrossRef]
130. Devold, T.G.; Nordbø, R.; Langsrud, T.; Svenning, C.; Brovold, M.J.; Sørensen, E.S.; Christensen, B.; Ådnøy, T.; Vegarud, G.E. Extreme frequencies of the αs1-casein "null" variant in milk from Norwegian dairy goats—Implications for milk composition, micellar size and renneting properties. *Dairy Sci. Technol.* **2011**, *91*, 39–51. [CrossRef]
131. Hayes, B.; Hagesæther, N.; Adnøy, T.; Pellerud, G.; Berg, P.R.; Lien, S. Effects on production traits of haplotypes among casein genes in Norwegian goats and evidence for a site of preferential recombination. *Genetics* **2006**, *174*, 455–464. [CrossRef] [PubMed]

132. Pizarro Inostroza, M.G.; Navas González, F.J.; Landi, V.; León Jurado, J.M.; Delgado Bermejo, J.V.; Fernández Álvarez, J.; Martínez Martínez, M.d.A. Bayesian Analysis of the Association between Casein Complex Haplotype Variants and Milk Yield, Composition, and Curve Shape Parameters in Murciano-Granadina Goats. *Animals* **2020**, *10*, 1845. [CrossRef] [PubMed]
133. Nilsen, H.; Olsen, H.G.; Hayes, B.; Sehested, E.; Svendsen, M.; Nome, T.; Meuwissen, T.; Lien, S. Casein haplotypes and their association with milk production traits in Norwegian Red cattle. *Genet. Sel. Evol.* **2009**, *41*, 24. [CrossRef]
134. Pizarro Inostroza, M.G.; Landi, V.; Navas González, F.J.; León Jurado, J.M.; Delgado Bermejo, J.V.; Fernández Álvarez, J.; Martínez Martínez, M.d.A. Integrating casein complex SNPs additive, dominance and epistatic effects on genetic parameters and breeding values estimation for murciano-granadina goat milk yield and components. *Genes* **2020**, *11*, 309. [CrossRef]
135. Pizarro Inostroza, M.G.; Landi, V.; Navas González, F.J.; León Jurado, J.M.; Martínez Martínez, A.; Fernández Álvarez, J.; Delgado Bermejo, J.V. Does the acknowledgement of αS1-casein genotype affect the estimation of genetic parameters and prediction of breeding values for milk yield and composition quality-related traits in Murciano-Granadina? *Animals* **2019**, *9*, 679. [CrossRef]
136. Pizarro Inostroza, M.G.; Landi, V.; Navas Gonzalez, F.J.; León Jurado, J.M.; Martínez Martínez, M.d.A.; Fernández Álvarez, J.; Delgado Bermejo, J.V. Non-parametric association analysis of additive and dominance effects of casein complex SNPs on milk content and quality in Murciano-Granadina goats. *J. Anim. Breed. Genet.* **2020**, *137*, 407–422. [CrossRef]
137. Pizarro, M.; Landi, V.; Navas, F.; León, J.; Martínez, A.; Fernández, J.; Delgado, J. Nonparametric analysis of casein complex genes' epistasis and their effects on phenotypic expression of milk yield and composition in Murciano-Granadina goats. *J. Dairy Sci.* **2020**, *103*, 8274–8291. [CrossRef]
138. Pizarro, M.; Landi, V.; Navas, F.; León, J.; Martínez, A.; Fernández, J.; Delgado, J. Non-parametric analysis of the effects of nongenetic factors on milk yield, fat, protein, lactose, dry matter content and somatic cell count in Murciano-Granadina goats. *Ital. J. Anim. Sci.* **2020**, *19*, 960–973. [CrossRef]
139. Vacca, G.M.; Dettori, M.L.; Piras, G.; Manca, F.; Paschino, P.; Pazzola, M. Goat casein genotypes are associated with milk production traits in the Sarda breed. *Anim. Genet.* **2014**, *45*, 723–731. [CrossRef]
140. Rijnkels, M.; Elnitski, L.; Miller, W.; Rosen, J.M. Multispecies comparative analysis of a mammalian-specific genomic domain encoding secretory proteins. *Genomics* **2003**, *82*, 417–432. [CrossRef]
141. Bleck, G.T.; White, B.R.; Miller, D.J.; Wheeler, M.B. Production of bovine α-lactalbumin in the milk of transgenic pigs. *J. Anim. Sci.* **1998**, *76*, 3072–3078. [CrossRef] [PubMed]
142. Kahilo, K.; El-Shazly, S.; El-Khadrawy, A.; Fattouh, I. Genetic polymorphism in β-lactoglobulin gene of some goat breeds in Egypt and its influence on milk yield. *Life Sci. J.* **2014**, *11*, 232–238.
143. Strucken, E.M.; Rahmatalla, S.; Koning, D.-J.D.; Brockmann, G.A. Haplotype analysis and linkage disequilibrium for DGAT1. *Arch. Anim. Breed.* **2010**, *53*, 247–255. [CrossRef]
144. He, C.; Wang, C.; Chang, Z.; Guo, B.; Li, R.; Yue, X.; Lan, X.; Chen, H.; Lei, C. AGPAT6 polymorphism and its association with milk traits of dairy goats. *Genet. Mol. Res.* **2011**, *10*, 2747–2756. [CrossRef]
145. Ogg, S.L.; Weldon, A.K.; Dobbie, L.; Smith, A.J.; Mather, I.H. Expression of butyrophilin (Btn1a1) in lactating mammary gland is essential for the regulated secretion of milk–lipid droplets. *Proc. Natl. Acad. Sci. USA* **2004**, *101*, 10084–10089. [CrossRef]
146. Morris, C.A.; Cullen, N.G.; Glass, B.C.; Hyndman, D.L.; Manley, T.R.; Hickey, S.M.; McEwan, J.C.; Pitchford, W.S.; Bottema, C.D.; Lee, M.A. Fatty acid synthase effects on bovine adipose fat and milk fat. *Mamm. Genome* **2007**, *18*, 64–74. [CrossRef]
147. Federica, S.; Francesco, N.; Giovanna, D.M.; Carmela, S.M.; Gennaro, C.; Carmela, T.; Bianca, M. Identification of novel single nucleotide polymorphisms in promoter III of the acetyl-CoA carboxylase-α gene in goats affecting milk production traits. *J. Hered.* **2009**, *100*, 386–389. [CrossRef]
148. Tudisco, R.; Calabrò, S.; Cutrignelli, M.; Moniello, G.; Grossi, M.; Gonzalez, O.; Piccolo, V.; Infascelli, F. Influence of organic systems on Stearoyl-CoA desaturase gene expression in goat milk. *Small Rumin. Res.* **2012**, *106*, S37–S42. [CrossRef]
149. Shi, H.; Luo, J.; Zhu, J.; Li, J.; Sun, Y.; Lin, X.; Zhang, L.; Yao, D.; Shi, H. PPARγ regulates genes involved in triacylglycerol synthesis and secretion in mammary gland epithelial cells of dairy goats. *PPAR Res.* **2013**, *2013*, 310948. [CrossRef]
150. Chong, B.M.; Reigan, P.; Mayle-Combs, K.D.; Orlicky, D.J.; McManaman, J.L. Determinants of adipophilin function in milk lipid formation and secretion. *Trends Endocrinol. Metab.* **2011**, *22*, 211–217. [CrossRef] [PubMed]
151. Wolins, N.E.; Rubin, B.; Brasaemle, D.L. TIP47 associates with lipid droplets. *J. Biol. Chem.* **2001**, *276*, 5101–5108. [CrossRef]
152. Xu, H.; Luo, J.; Zhao, W.; Yang, Y.; Tian, H.; Shi, H.; Bionaz, M. Overexpression of SREBP1 (sterol regulatory element binding protein 1) promotes de novo fatty acid synthesis and triacylglycerol accumulation in goat mammary epithelial cells. *J. Dairy Sci.* **2016**, *99*, 783–795. [CrossRef] [PubMed]
153. Shen, B.; Pan, Q.; Yang, Y.; Gao, Y.; Liu, X.; Li, W.; Han, Y.; Yuan, X.; Qu, Y.; Zhao, Z. miR-224 affects mammary epithelial cell apoptosis and triglyceride production by downregulating ACADM and ALDH2 genes. *DNA Cell Biol.* **2017**, *36*, 26–33. [CrossRef]
154. Dubey, P.K.; Goyal, S.; Mishra, S.K.; Arora, R.; Mukesh, M.; Niranjan, S.K.; Kathiravan, P.; Kataria, R.S. Identification of polymorphism in fatty acid binding protein 3 (FABP3) gene and its association with milk fat traits in riverine buffalo (*Bubalus bubalis*). *Trop. Anim. Health. Prod.* **2016**, *48*, 849–853. [CrossRef]
155. Yao, D.; Luo, J.; He, Q.; Xu, H.; Li, J.; Shi, H.; Wang, H.; Chen, Z.; Loor, J. Liver X receptor α promotes the synthesis of monounsaturated fatty acids in goat mammary epithelial cells via the control of stearoyl-coenzyme A desaturase 1 in an SREBP-1-dependent manner. *J. Dairy Sci.* **2016**, *99*, 6391–6402. [CrossRef]
156. Khatib, H.; Leonard, S.; Schutzkus, V.; Luo, W.; Chang, Y. Association of the OLR1 gene with milk composition in Holstein dairy cattle. *J. Dairy Sci.* **2006**, *89*, 1753–1760. [CrossRef]

157. Zielke, L.G.; Bortfeldt, R.H.; Tetens, J.; Brockmann, G.A. BDNF contributes to the genetic variance of milk fat yield in German Holstein cattle. *Front. Genet.* **2011**, *2*, 16. [CrossRef]
158. Zielke, L.G.; Bortfeldt, R.H.; Reissmann, M.; Tetens, J.; Thaller, G.; Brockmann, G.A. Impact of variation at the FTO locus on milk fat yield in Holstein dairy cattle. *PLoS ONE* **2013**, *8*, e63406.
159. Cochran, S.D.; Cole, J.B.; Null, D.J.; Hansen, P.J. Discovery of single nucleotide polymorphisms in candidate genes associated with fertility and production traits in Holstein cattle. *BMC Genet.* **2013**, *14*, 49. [CrossRef]
160. Miltiadou, D.; Hager-Theodorides, A.; Symeou, S.; Constantinou, C.; Psifidi, A.; Banos, G.; Tzamaloukas, O. Variants in the 3′UTR of the ovine Acetyl-Coenzyme A Acyltransferase 2 gene are associated with dairy traits and exhibit differential allelic expression. *J. Dairy Sci.* **2017**, *100*, 6285–6297. [CrossRef] [PubMed]
161. Yao, D.; Luo, J.; He, Q.; Shi, H.; Li, J.; Wang, H.; Xu, H.; Chen, Z.; Yi, Y.; Loor, J.J. SCD1 alters long-chain fatty acid (LCFA) composition and its expression is directly regulated by SREBP-1 and PPARγ1 in dairy goat mammary cells. *J. Cell. Physiol.* **2017**, *232*, 635–649. [CrossRef] [PubMed]
162. Strucken, E.M.; Laurenson, Y.C.; Brockmann, G.A. Go with the flow—Biology and genetics of the lactation cycle. *Front. Genet.* **2015**, *6*, 118. [CrossRef] [PubMed]
163. Hou, J.; An, X.; Song, Y.; Wang, J.; Ma, T.; Han, P.; Fang, F.; Cao, B. Combined effects of four SNPs within goat PRLR gene on milk production traits. *Gene* **2013**, *529*, 276–281. [CrossRef] [PubMed]
164. Cohen-Zinder, M.; Seroussi, E.; Larkin, D.M.; Loor, J.J.; Everts-Van Der Wind, A.; Lee, J.-H.; Drackley, J.K.; Band, M.R.; Hernandez, A.; Shani, M. Identification of a missense mutation in the bovine ABCG2 gene with a major effect on the QTL on chromosome 6 affecting milk yield and composition in Holstein cattle. *Genome Res.* **2005**, *15*, 936–944. [CrossRef] [PubMed]
165. Le Provost, F.; Leroux, C.; Martin, P.; Gaye, P.; Djiane, J. Prolactin gene expression in ovine and caprine mammary gland. *Neuroendocrinology* **1994**, *60*, 305–313. [CrossRef]
166. Yakan, A.; Özkan, H.; ŞAKAR, A.E.; Ünal, N.; Özbeyaz, C. Gene expression levels in some candidate genes for mastitis resistance, milk yield, and milk quality of goats reared under different feeding systems. *Turk. J. Vet. Anim. Sci.* **2018**, *42*, 18–28. [CrossRef]
167. Pizarro, M.G.; Landi, V.; Navas González, F.J.; León, J.M.; Delgado, J.V. Non-parametric analysis of the effects of αS1-casein genotype and parturition non-genetic factors on milk yield and composition in Murciano-Granadina goats. *Ital. J. Anim. Sci.* **2019**, *18*, 1021–1034. [CrossRef]
168. Lan, X.; Pan, C.; Chen, H.; Zhang, C.; Zhang, A.; Zhang, L.; Li, J.; Lei, C. An MspI PCR-RFLP detecting a single nucleotide polymorphism at alpha-lactalbumin gene in goat. *Czech J. Anim. Sci.* **2007**, *52*, 138. [CrossRef]
169. Winter, A.; Krämer, W.; Werner, F.A.; Kollers, S.; Kata, S.; Durstewitz, G.; Buitkamp, J.; Womack, J.E.; Thaller, G.; Fries, R. Association of a lysine-232/alanine polymorphism in a bovine gene encoding acyl-CoA: Diacylglycerol acyltransferase (DGAT1) with variation at a quantitative trait locus for milk fat content. *Proc. Natl. Acad. Sci. USA* **2002**, *99*, 9300–9305. [CrossRef]
170. Zhang, C.-L.; Gao, X.-Y.; Shao, R.-Y.; Wang, Y.-H.; Fang, X.-T.; Chen, H. Stearoyl-CoA desaturase (SCD) gene polymorphism in goat breeds. *Biochem. Genet.* **2010**, *48*, 822–828. [CrossRef] [PubMed]
171. Dixit, S.; Sivalingam, J.; Tyagi, A.; Saroha, V.; Sharma, A.; Nagda, R. Association of novel SNPs in the candidate genes affecting caprine milk fatty acids related to human health. *Meta Gene* **2015**, *4*, 45–56. [CrossRef] [PubMed]
172. Shi, H.; Zhu, J.; Luo, J.; Cao, W.; Shi, H.; Yao, D.; Li, J.; Sun, Y.; Xu, H.; Yu, K. Genes regulating lipid and protein metabolism are highly expressed in mammary gland of lactating dairy goats. *Funct. Integr. Genom.* **2015**, *15*, 309–321. [CrossRef] [PubMed]

animals

MDPI

Article

Estimates of Genetic Parameters for Milk, the Occurrence of and Susceptibility to Clinical Lameness and Claw Disorders in Dairy Goats

Natasha Jaques [1,*], Sally-Anne Turner [2], Emilie Vallée [3], Cord Heuer [3] and Nicolas Lopez-Villalobos [1]

1 School of Agriculture and Environment, Massey University, Palmerston North 4442, New Zealand
2 Dairy Goat Co-Operative (NZ) Ltd., 18 Gallagher Drive, Melville, Hamilton 3206, New Zealand
3 EpiCentre, School of Veterinary Science, Private Bag 11-222, Palmerston North 4442, New Zealand
* Correspondence: njaques@massey.ac.nz

Simple Summary: Lameness and claw disorders are important animal welfare issues in dairy goats, and knowledge is limited. Estimates of genetic parameters for occurrence and susceptibility to clinical lameness and claw disorders and the genetic associations with milk production traits in dairy goats were obtained in this study. These results indicate that a selection index can be developed to select animals resistant or tolerant to lameness and claw disorders.

Abstract: The New Zealand goat industry accesses niche markets for high-value products, mainly formula for infants and young children. This study aimed to estimate the genetic parameters of occurrence and susceptibility of clinical lameness and selected claw disorders and establish their genetic associations with milk production traits. Information on pedigree, lameness, claw disorders, and milk production was collected on three farms between June 2019 and July 2020. The dataset contained 1637 does from 174 sires and 1231 dams. Estimates of genetic and residual (co)variances, heritabilities, and genetic and phenotypic correlations were obtained with uni- and bi-variate animal models. The models included the fixed effects of farm and parity, deviation from the median kidding date as a covariate, and the random effects of animal and residual error. The heritability (h^2) estimates for lameness occurrence and susceptibility were 0.07 and 0.13, respectively. The h^2 estimates for claw disorder susceptibilities ranged from 0.02 to 0.23. The genotypic correlations ranged from weak to very strong between lameness and milk production traits (−0.94 to 0.84) and weak to moderate (0.23 to 0.84) between claw disorder and milk production traits.

Keywords: goat; heritability; genetic correlation; genetic evaluation; lameness; claw disorder; susceptibility; occurrence; milk production

1. Introduction

Dairy goats produced 2% of the total milk produced in New Zealand in 2017 [1]. Under the management of the Dairy Goat Co-operative (NZ) Ltd. (Hamilton, New Zealand), the dairy goat industry in New Zealand became the first commercial producer of dairy goat infant formula and has continued to dominate this niche market internationally [1]. Lameness has been identified as a problem in some dairy goat herds.

Given the potential economic impact, lameness prevention, and control could contribute to farm efficiency, sustainability, and productivity by improving goat welfare. Two New Zealand studies have investigated lameness [2,3]. The first study followed a small cohort of goats during their first two years of life [2]. Deeming et al. [2] reported a small percentage (0 to 8.9%) of lame goats within the group across four locomotion-scoring events after parturition. The second study recorded lameness prevalence between 6.7 and 25.5% in four dairy goat herds [3]. It is unknown what causes lameness in dairy goat herds and whether claw disorders are the predominant reason for lameness. Due to such large variability of lameness prevalence within and between past studies, it is worth exploring risk factors associated with it, including genetics.

Lameness is the impairment of normal locomotion due to pain caused by an injury, disease, or claw disorder [4]. Its etiology can be infectious or non-infectious [5]. Infectious claw diseases are caused by pathogenic bacteria or viruses that colonize on or within their host animal and cause injury to that animal when the conditions are favorable. Non-infectious factors, such as farm management, environment, or genetics, may predispose the goat to lameness.

Selective breeding has been investigated as a long-term option to reduce the prevalence of lameness and claw disorders in dairy cattle and sheep [6–8]. Heringstad et al. [8] extensively reviewed the genetics of claw health (infectious and non-infectious claw disorders) in dairy cattle. Overall, claw disorders had low to moderate (0.01 to 0.39) heritabilities across studies. Grouping the claw disorders as one claw health trait or grouping based on etiology resulted in higher estimates for heritabilities. Heritabilities for lameness and claw disorders were low, 0.02 to 0.34. Higher heritability estimates were obtained when lameness was treated as an ordinal trait (1–5 score) than a binary trait (0.08 to 0.15 vs. 0.02 to 0.04, respectively).

A few studies have reported the heritability of claw disorders and lameness in sheep. Studies investigating footrot in 8 to 9 months old lambs reported heritabilities between 0.03 and 0.41 when an occurrence (binary) recording system was used [9–11]. Conington et al. [6] studied the prevalence of white line disease and reported heritabilities between 0.09 and 0.33. O'Brien et al. [12] analyzed lameness in various breeds of ewes and lambs and reported an overall heritability for lameness between 0.06 and 0.12. Despite heritability studies in sheep [9–11], none have reported genetic or phenotypic correlations between different claw disorder traits and lameness traits.

After an extensive literature search using Web of Science, Scopus, and Google Scholar, only two studies were found that referenced genetics in the occurrence of lameness and claw disorders in dairy goats. Hill et al. [13] reported that the incidence of horn separation was random and, therefore, most likely caused by external factors rather than genetic factors. A second, more recent study reported a moderate to high repeatability of limb problems (arthritis and overgrown claws) and lameness in Damascus, a dairy goat breed in Greece. Repeatability includes the genetic and environmental effects, and low repeatability suggests a greater influence of the temporary environment on the trait of interest [14]. The evidence in sheep and dairy cows suggests that the heritabilities of lameness and claw disorders need to be investigated further in dairy goats.

Exploration of selection against the occurrence and susceptibility of lameness and claw disorders in dairy goats requires estimating genetic parameters, variance and covariances, heritabilities, and genetic and phenotypic correlations between production traits, and occurrence of lameness and claw disorders. Currently, the Dairy Goat Co-operative (NZ) Ltd. (Hamilton, New Zealand). manages the yearly genetic evaluation of dairy goats for lactation yields of milk, fat, protein, lactose, and average somatic cell score (SCS) [15]. Therefore, the objective of this study was to estimate the genetic parameters (genetic variances and covariances, heritability, and genetic and phenotypic correlations) for the occurrence and susceptibility of lameness, claw disorders, and milk production and composition traits in dairy goats.

2. Materials and Methods

The animal study protocol was approved by the Ethics Committee of Massey University, New Zealand (MUAEC Protocol 19/51, 29/05/2019).

2.1. Data

Data were collected from three dairy goat farms in Waikato, New Zealand, for one production year, from June 2019 to June 2020. All farms volunteered to participate in this study and were shareholders in the Dairy Goat Co-operative (NZ) Ltd. (Hamilton, New

Zealand). These farmers had witnessed lameness within their herds and were interested in reducing its prevalence on their farms. Two selected farms had herd sizes around the national average of 750 milking does [16], while one farm was significantly larger, with over 1600 milking does.

The final dataset of seasonal lactation goats contained observations from 1637 does. The pedigree included 174 sires and 1231 dams spanning two generations. Of the does, 83% had records of the sire and the dam, 14% had only information on the dam, and 3% had only information on the sire.

There were a few reference sires between farms, where farm A shared one sire with farm B and three sires with farm C. There was one sire shared between farms B and C. The goat breeds on the three farms consisted mainly of Saanen crosses with other breeds, such as Toggenburg, Alpine, and Nubian [16]. Despite many crosses within the herd, heterosis was not estimated as the data input for the breed information of the animal, sire, and dam and was not precisely recorded within the pedigree.

2.2. Milk Trait Characteristics

An orthogonal polynomial model of order 3 was used to predict the total yields of milk, protein, fat, and lactose to 270 days [17]. Aside from calculating total production yields, milk concentrations and ratios were also calculated. The lactation fat, protein, and lactose percentages were calculated by taking the respective lactation yields and dividing these by the total milk yield and then multiplying them by 100. The protein-to-fat ratio (P:F), protein-to-lactose ratio (P:L), and lactose-to-fat plus protein ratio (L:(F + P)) were calculated from the total yields for the respective milk characteristics.

2.3. Lameness and Claw Disorder Trait Definitions

Locomotion and claw disorder scoring events were carried out five times for farm A and four times for farms B and C across one lactation from July 2019 to June 2020. Lameness was scored using a 5-point locomotion scale developed by Deeming et al. [18] with minor modifications (Table S1). Briefly, the scores were defined as 0—normal, 1—uneven, 2—mildly lame, 3—moderately lame, and 4—severely lame. A goat was classified as clinically lame if its locomotion was scored as a three or a four. The clinical lameness traits were susceptibility (continuous variable) and occurrence (binomial variable). Clinical lameness susceptibility was the number of times the goat was clinically lame (scored 3 or 4) over the total number of times they were scored between July 2019 and June 2020. The clinical lameness occurrence trait was the diagnosis of clinical lameness (presence or absence) recorded during at least one locomotion scoring event across the production year.

Three claw disorders were selected to focus on; horn separation, rot, and granulomas. Horn separation, a form of white line disease, was a combination of minor and severe hoof horn separation. Minor horn separation was defined as less than 50% of the hoof horn being removed. Severe horn separation was defined as 50% or more of the hoof wall being removed. Severe horn separation was included as a separate trait because it was also associated with clinical lameness (unpublished). Rot was defined as the combination of digital dermatitis and footrot. Granulomas were cherry-like tissue growths protruding from the claw. Each claw disorder was classified as present or absent at each hoof-trimming event. The five claw disorder traits estimated were horn separation susceptibility, severe horn separation susceptibility, rot susceptibility, granuloma susceptibility, and claw disorder occurrence. Claw disorder (horn separation, rot, and granuloma) susceptibility was calculated by the number of times the goat was diagnosed with one of these disorders over the total number of times the hoof trimmer assessed the goat between July 2019 and June 2020. Claw disorder occurrence was the binomial score given (presence or absence) if the goat had at least one case of horn separation or rot or granuloma.

2.4. Models to Estimate Genetic Parameters

The genetic parameters (heritability, genetic and phenotypic variances, and covariances) were estimated by Restricted Maximum Likelihood procedures using ASReml version 4 [19]. The milk production traits, lameness, and claw disorder susceptibilities were treated as continuous variables, while lameness and claw disorder occurrence traits were treated as binary variables. This study classified the occurrence as a binary trait and susceptibility as a continuous trait. Both classifications of the traits were assumed to have a normal distribution for both animal and residual variances. In matrix notation, the univariate animal model can be represented as:

$$\mathbf{y} = \mathbf{Xb} + \mathbf{Za} + \mathbf{e}$$

where $\mathbf{y}$ is the vector of phenotypic observations from measured goats based on one trait, $\mathbf{b}$ is the vector of fixed effects for goats, $\mathbf{a}$ is the vector of random animal effects, $\mathbf{e}$ is the vector of random residual errors, $\mathbf{X}$ is the incidence matrix for the fixed effects, and $\mathbf{Z}$ is the incidence matrix combining observations with the random animal additive genetic effects.

Fixed effects in vector $\mathbf{b}$ include farm, parity, and deviation from the median kidding date as a covariate. The following expectations (E) of the variables were assumed $E(\mathbf{y}) = \mathbf{Xb}$, $E(\mathbf{a}) = \mathbf{0}$, and $E(\mathbf{e}) = \mathbf{0}$. The variances of random effects were assumed as follows: $var(\mathbf{a}) = \mathbf{A}\sigma_a^2$ and $var(\mathbf{e}) = I\sigma_e^2$, where σ_a^2 is the additive genetic variance, σ_e^2 is the random residual variance, $\mathbf{A}$ the numerator relationship matrix between all animals considered in the data set, and $\mathbf{I}$ is an identity matrix of order equal to the number of animals with records.

Heritabilities for various traits were estimated with the following equation using variances obtained from the previous univariate analysis:

$$h^2 = \frac{\sigma_a^2}{\sigma_p^2}$$

where σ_p^2 is the phenotypic variance calculated as the sum of genetic and environmental variances. Heritability estimates ranging 0–0.10, 0.10–0.60, and 0.60–1.00 were considered as low, moderate, and high, respectively [20]. The bivariate animal model was as follows:

$$\begin{bmatrix} y_1 \\ y_2 \end{bmatrix} = \begin{bmatrix} X_1 & 0 \\ 0 & X_2 \end{bmatrix} \begin{bmatrix} b_1 \\ b_2 \end{bmatrix} + \begin{bmatrix} Z_1 & 0 \\ 0 & Z_2 \end{bmatrix} \begin{bmatrix} a_1 \\ a_2 \end{bmatrix} + \begin{bmatrix} e_1 \\ e_2 \end{bmatrix},$$

where $\mathbf{y_1}$ and $\mathbf{y_2}$ are the vectors of phenotypic observations from measured goats based on the two traits of interest, $\mathbf{b_1}$ and $\mathbf{b_2}$ are the vectors of fixed effects, $\mathbf{a_1}$ and $\mathbf{a_2}$ are the vectors of random animal effects, $\mathbf{e_1}$ and $\mathbf{e_2}$ are the vectors of random residual errors, $\mathbf{X_1}$ and $\mathbf{X_2}$ are the incidence matrices for the fixed effects, and $\mathbf{Z_1}$ and $\mathbf{Z_2}$ are the incidence matrices for the random animal effects. The distribution properties of the variance–covariance elements for traits used in the model for the random animal additive genetic effects and the residual effect are as follows:

$$var \begin{bmatrix} a_1 \\ a_2 \end{bmatrix} = \begin{bmatrix} \mathbf{A}\sigma_{a1}^2 & \mathbf{A}\sigma_{a12} \\ \mathbf{A}\sigma_{a21} & \mathbf{A}\sigma_{a2}^2 \end{bmatrix},$$

and

$$var \begin{bmatrix} e_1 \\ e_2 \end{bmatrix} = \begin{bmatrix} \mathbf{I}\sigma_{e1}^2 & \mathbf{I}\sigma_{e12} \\ \mathbf{I}\sigma_{e12} & \mathbf{I}\sigma_{e2}^2 \end{bmatrix},$$

where σ_{a1}^2, σ_{a2}^2, and σ_{a12} are the additive genetic (co)variance components for the traits of interest, and σ_{e1}^2, σ_{e2}^2, and σ_{e12} are the error (co)variance components for the traits of interest.

The following equations defined phenotypic and genotypic correlations:

Phenotypic correlation (r_P)

$$r_P = \frac{\sigma_{p12}}{(\sigma_{p1} \times \sigma_{p2})}$$

Genotypic correlation (r_G)

$$r_G = \frac{\sigma_{a12}}{(\sigma_{a1} \times \sigma_{a2})},$$

where σ_{p12} and σ_{a12} are the phenotypic and additive genetic covariance between two traits, σ_{p1} and σ_{a1} are the phenotypic and additive genetic standard deviations for one trait, and σ_{p2} and σ_{a2} are the phenotypic and additive genetic standard deviations for the second trait.

3. Results

3.1. Descriptive Statistics

Descriptive statistics are presented in Table 1. The coefficient of variation (CV) was high for all traits. The CV for production traits ranged from 11.8 to 23.5, while the CV for milk concentration traits ranged from 3.93 to 11.4. The CV for lameness and claw disorders were the highest out of all the traits, ranging from 50 to 338.

Table 1. Descriptive statistics [1] for milk production, lameness, and claw disorder traits in dairy goats from three farms within New Zealand, collected between July 2019 and June 2020.

Trait	N	Average	SD	Min	Max	CV
Lactation length (days)	1637	257	67.5	33.0	350	26.2
Total yields (kg; 270 days)						
Milk	1637	978	228	442	1718	23.3
Fat	1637	31.0	6.99	13.9	54.3	22.6
Protein	1637	30.9	6.76	13.7	53.9	21.9
Lactose	1637	44.4	10.5	19.8	79.4	23.5
Somatic cell score [2]	1637	9.48	1.12	5.98	13.1	11.8
Milk percentages (%)						
Fat	1637	3.19	0.36	1.91	4.37	11.4
Protein	1637	3.18	0.23	2.47	4.08	7.31
Lactose	1637	4.54	0.18	3.62	5.14	3.93
Milk ratios [3]						
Protein:fat	1637	1.01	0.11	0.74	1.49	10.8
Protein:lactose	1637	0.70	0.05	0.53	0.96	7.46
Lactose:fat + protein	1637	0.72	0.06	0.56	0.93	7.83
Clinical lameness						
Susceptibility	1625	0.09	0.18	0.00	1.00	208
Occurrence	1637	0.24	0.42	0.00	1.00	180
Claw disorders						
Horn sep. sus. ($\times 10^3$) [4]	1637	0.43	0.32	0.00	1.00	75.3
Severe horn sep. sus. ($\times 10^3$) [4]	1637	0.18	0.23	0.00	1.00	127
Rot susceptibility	1637	0.05	0.13	0.00	1.00	244
Granuloma susceptibility	1637	0.04	0.15	0.00	1.00	338
Occurrence [5]	1637	0.80	0.40	0.00	1.00	50.0

[1] N = number of records, SD = standard deviation, Min = minimum value, Max = maximum value, CV = coefficient of variation. [2] Somatic cell score = average $\log_2$ (somatic cell count/1000). [3] Ratios were calculated using percentages. [4] Horn sep. sus. = horn separation susceptibility; Severe horn sep. sus. = Severe horn separation susceptibility. [5] Occcurence is the presence of at least one claw disorder or clinical lameness event across the year.

3.2. Genetic Parameters

Estimates of variance components and heritabilities for the different traits are presented in Table 2. The heritabilities for milk production traits ranged from 0.24 to 0.69.

Meanwhile, heritabilities for lameness and claw disorder traits ranged from 0.07 to 0.13 and 0.02 to 0.23, respectively.

Table 2. Estimates of additive genetic and residual variances, heritability, and the corresponding standard errors in parenthesis for milk production, lameness, and claw disorder traits in dairy goats from three farms within New Zealand collected between July 2019 and June 2020 production year.

Traits	Additive Genetic Variance	Residual Variance	Heritability
Total yields (kg; 270 days)			
Milk	8508 (2035)	24,206 (1914)	0.26 (0.06)
Fat	8.93 (1.95)	23.9 (1.84)	0.27 (0.06)
Protein	6.24 (1.53)	19.6 (1.47)	0.24 (0.06)
Lactose	19.3 (4.40)	50.3 (4.10)	0.28 (0.06)
Somatic cell score [1]	0.28 (0.06)	0.31 (0.07)	0.80 (0.07)
Milk percentages (%)			
Fat ($\times 10^4$)	0.05 (0.01)	0.07 (0.01)	0.56 (0.06)
Protein ($\times 10^4$)	0.03 (0.00)	0.02 (0.00)	0.69 (0.06)
Lactose ($\times 10^4$)	0.02 (0.00)	0.01 (0.00)	0.63 (0.06)
Milk ratios [2]			
Protein:fat ($\times 10^3$)	3.92 (0.63)	5.57 (0.53)	0.56 (0.06)
Protein:lactose ($\times 10^3$)	1.77 (0.20)	0.86 (0.15)	0.67 (0.06)
Lactose:fat + protein ($\times 10^3$)	1.82 (0.22)	1.20 (0.17)	0.60 (0.06)
Clinical lameness			
Susceptibility ($\times 10^3$)	4.03 (1.76)	26.4 (1.84)	0.13 (0.06)
Occurrence ($\times 10^3$)	0.01 (0.01)	0.15 (0.01)	0.07 (0.05)
Claw disorders			
Horn sep. sus. ($\times 10^3$) [3]	0.02 (0.00)	0.05 (0.00)	0.23 (0.06)
Severe horn sep. sus. ($\times 10^3$) [3]	0.01 (0.00)	0.03 (0.00)	0.22 (0.06)
Rot susceptibility ($\times 10^3$)	1.84 (0.71)	12.1 (0.77)	0.13 (0.05)
Granuloma susceptibility ($\times 10^3$)	0.45 (0.82)	21.1 (1.08)	0.02 (0.04)
Occurrence ($\times 10^3$)	0.01 (0.01)	0.11 (0.01)	0.11 (0.05)

[1] Somatic cell score = average $\log_2$(somatic cell count/1000). [2] Ratios were calculated using percentages. [3] Horn sep. sus. = horn separation susceptibility; Severe horn sep. sus. = Severe horn separation susceptibility.

Table 3 present the estimates for the genetic and phenotypic correlations. Generally, if there was a relationship, it was due to moderate to very strong positive genetic correlations among the milking traits (0.62 to 0.97). There were a few exceptions. Firstly, a weak positive correlation was observed between somatic cell score and P:F. Secondly, there were strong negative correlations of L:(F + P) with protein percentage and P:L. Lastly, there were moderate negative correlations between P:L and fat percentage, between P:L and milk yield, and between fat yield and P:F.

There were five strong to very strong positive genetic correlations between claw disorder traits and between lameness and claw disorder traits. The correlations were between horn susceptibility and claw disorder occurrence, granuloma susceptibility and rot susceptibility, granuloma susceptibility and lameness occurrence, rot susceptibility and lameness susceptibility, and rot susceptibility and lameness occurrence.

Generally, the correlations between lameness and claw disorder traits and milk traits ranged from near zero to weak, with some exceptions. There was a very strong positive correlation between lameness occurrence and fat percentage. A very strong negative correlation was also found between lameness occurrence and lactose percentage. Rot and granulomas had weak to very strong correlations with milk production yield traits. Horn separation susceptibility and severe horn separation susceptibility had weak and moderate correlations with milk concentration traits, respectively. There were also weak correlations between lactose percentage and granuloma susceptibility and between lactose percentage and claw disorder occurrence.

Table 3. Estimates of phenotypic (above diagonal) and genetic (below diagonal) correlations [1] and standard errors (in parentheses) for milk production, lameness, and claw disorder traits in dairy goats from three farms within New Zealand collected between July 2019 and June 2020 production year.

Trait	1	2	3	4	5	6	7	8	9
1. Milk yield		0.81 (0.01)	0.91 (0.00)	0.98 (0.00)	−0.18 (0.03)	0.15 (0.08)	−0.20 (0.05)	0.21 (0.19)	0.00 (0.03)
2. Fat yield	0.62 (0.09)		0.86 (0.01)	0.83 (0.01)	−0.23 (0.02)	0.31 (0.02)	−0.06 (0.05)	−0.03 (0.05)	−0.46 (0.20)
3. Protein yield	0.79 (0.05)	0.74 (0.06)		0.92 (0.00)	−0.14 (0.03)	−0.06 (0.05)	0.09 (0.05)	0.16 (0.04)	0.04 (0.26)
4. Lactose yield	0.94 (0.01)	0.70 (0.07)	-		−0.23 (0.02)	−0.15 (0.05)	−0.18 (0.05)	−0.03 (0.05)	−0.05 (0.03)
5. Somatic cell score [2]	0.26 (0.17)	0.17 (−0.01)	−0.01 (0.15)	0.15 (0.33)		−0.10 (0.04)	0.14 (0.26)	0.03 (0.05)	0.21 (0.02)
6. Fat percentage	-	0.43 (0.11)	−0.05 (0.20)	−0.17 (0.20)	−0.49 (0.24)		−0.12 (0.32)	0.32 (0.06)	−0.14 (0.09)
7. Protein percentage	−0.21 (0.22)	−0.03 (0.20)	−0.06 (0.25)	−0.18 (0.22)	0.07 (0.11)	0.26 (0.44)		−0.02 (0.08)	0.08 (0.04)
8. Lactose percentage	-	0.00 (0.11)	0.44 (0.13)	0.00 (0.11)	−0.06 (0.29)	0.28 (0.18)	0.08 (0.18)		−0.23 (0.03)
9. Protein:fat ratio [3]	0.07 (0.13)	−0.56 (0.10)	0.13 (0.13)	−0.03 (0.13)	0.35 (0.12)	−0.22 (0.44)	−0.23 (0.12)	−0.31 (0.09)	
10. Protein:lactose ratio [3]	−0.53 (0.09)	−0.19 (0.11)	-	−0.42 (0.23)	0.23 (0.10)	−0.59 (0.23)	0.87 (0.02)	−0.27 (0.07)	0.25 (0.09)
11. Lactose:fat + protein ratio [3]	0.55 (0.09)	−0.12 (0.12)	-	0.56 (0.08)	−0.05 (0.12)	−0.44 (3.83)	−0.77 (0.03)	−0.01 (0.16)	0.28 (0.09)
12. Lameness susceptibility	0.09 (0.21)	−0.11 (0.20)	0.16 (0.21)	0.09 (0.21)	0.26 (0.20)	-	0.11 (0.14)	-	0.32 (0.17)
13. Lameness occurrence	0.07 (0.26)	−0.07 (0.24)	0.14 (0.26)	0.10 (0.25)	0.43 (0.24)	0.75 (0.12)	-	−0.94 (0.03)	0.22 (0.20)
14. Horn sep. sus. [4]	0.05 (0.03)	0.10 (0.18)	0.08 (0.18)	0.10 (0.18)	0.07 (0.17)	−0.04 (0.13)	0.37 (0.12)	0.15 (0.13)	0.00 (0.14)
15. Severe horn sep. sus. [4]	0.07 (0.18)	0.06 (0.17)	−0.05 (0.18)	0.10 (0.18)	0.06 (0.17)	−0.40 (0.15)	0.34 (0.09)	-	−0.12 (0.15)
16. Rot susceptibility	0.31 (0.22)	0.14 (0.20)	0.41 (0.22)	0.31 (0.22)	−0.02 (0.20)	0.20 (0.08)	0.08 (0.15)	0.07 (0.15)	0.17 (0.17)
17. Granuloma susceptibility	0.52 (0.42)	0.63 (0.48)	0.84 (0.49)	0.63 (0.46)	−0.07 (0.35)	0.00 (0.38)	−0.25 (0.27)	0.27 (0.24)	0.07 (0.30)
18. Claw disorder occurrence	−0.08 (0.23)	0.07 (0.23)	−0.10 (0.23)	0.02 (0.24)	−0.19 (0.22)	0.05 (0.17)	0.18 (0.09)	0.06 (0.27)	−0.22 (0.19)

Table 3. *Cont.*

	Trait	10	11	12	13	14	15	16	17	18
1.	Milk yield	−0.42 (0.02)	0.42 (0.02)	−0.05 (0.03)	−0.02 (0.03)	0.05 (0.03)	0.02 (0.03)	0.00 (0.03)	−0.01 (0.02)	−0.05 (0.03)
2.	Fat yield	−0.21 (0.03)	−0.08 (0.03)	−0.45 (0.03)	−0.03 (0.03)	0.07 (0.03)	0.02 (0.03)	0.01 (0.03)	0.00 (0.02)	−0.01 (0.03)
3.	Protein yield	−0.10 (0.02)	0.12 (0.02)	0.07 (0.03)	−0.03 (0.03)	0.07 (0.03)	0.02 (0.03)	0.02 (0.03)	−0.01 (0.02)	−0.03 (0.03)
4.	Lactose yield	−0.14 (0.08)	0.44 (0.02)	−0.05 (0.03)	−0.02 (0.03)	0.06 (0.03)	0.03 (0.03)	0.01 (0.03)	−0.01 (0.02)	−0.04 (0.03)
5.	Somatic cell score [2]	0.28 (0.02)	−0.14 (0.03)	0.05 (0.03)	0.05 (0.03)	0.03 (0.03)	0.01 (0.03)	−0.02 (0.03)	0.01 (0.03)	−0.03 (0.03)
6.	Fat percentage	0.08 (0.04)	−0.50 (0.23)	−0.19 (0.06)	0.37 (0.08)	0.03 (0.03)	0.10 (0.04)	0.16 (0.04)	−0.03 (0.04)	0.16 (0.05)
7.	Protein percentage	0.85 (0.01)	−0.72 (0.01)	0.02 (0.03)	0.55 (0.21)	0.39 (0.03)	0.20 (0.04)	0.04 (0.03)	0.03 (0.03)	0.06 (0.03)
8.	Lactose percentage	−0.32 (0.03)	−0.18 (0.04)	−0.19 (0.07)	−0.34 (0.04)	0.07 (0.03)	0.39 (0.03)	0.05 (0.26)	0.03 (0.01)	0.29 (0.11)
9.	Protein:fat ratio [3]	0.25 (0.03)	0.38 (0.02)	−0.01 (0.03)	0.01 (0.03)	−0.02 (0.03)	0.00 (0.03)	0.01 (0.03)	−0.02 (0.03)	−0.06 (0.03)
10.	Protein:lactose ratio [3]		−0.80 (0.01)	0.01 (0.03)	0.00 (0.03)	−0.01 (0.03)	−0.04 (0.03)	0.01 (0.03)	0.00 (0.03)	0.01 (0.03)
11.	Lactose:fat + protein ratio [3]	−0.86 (0.02)		−0.02 (0.03)	0.01 (0.03)	0.00 (0.03)	0.03 (0.03)	−0.01 (0.03)	−0.01 (0.03)	−0.05 (0.03)
12.	Lameness susceptibility	0.10 (0.15)	0.09 (0.15)		0.84 (0.07)	0.01 (0.03)	0.06 (0.03)	0.26 (0.02)	0.29 (0.02)	0.03 (0.03)
13.	Lameness occurrence	0.11 (0.18)	0.02 (0.19)	-		0.00 (0.03)	0.06 (0.13)	0.22 (0.02)	0.23 (0.02)	0.05 (0.02)
14.	Horn sep. sus. [4]	−0.05 (0.13)	0.06 (0.13)	0.18 (0.22)	0.22 (0.28)		0.49 (0.02)	0.21 (0.02)	0.03 (0.02)	0.57 (0.02)
15.	Severe horn sep. sus. [4]	−0.22 (0.13)	0.14 (0.13)	0.36 (0.21)	0.23 (1.45)	0.63 (0.12)		0.24 (0.02)	0.03 (0.02)	0.25 (0.02)
16.	Rot susceptibility	0.04 (0.15)	0.01 (0.15)	0.70 (0.22)	0.97 (0.25)	0.11 (0.21)	0.02 (0.22)		0.26 (0.02)	0.20 (0.02)
17.	Granuloma susceptibility	0.25 (0.27)	−0.19 (0.28)	0.68 (0.39)	0.95 (0.25)	0.40 (0.40)	0.49 (0.39)	0.92 (0.48)		0.09 (0.02)
18.	Claw disorder occurrence	−0.19 (0.17)	0.09 (0.17)	−0.32 (0.27)	−0.28 (0.31)	0.92 (0.13)	0.45 (0.21)	−0.02 (0.29)	0.02 (0.49)	

Where '-' represents any non-estimable relationship. [1] None, weak, moderate, and strong correlations were <|0.3| (green), |0.30| to <|0.50| (blue), |0.50| to <|0.70| (yellow), and |0.70|–|1.00| (red), respectively. [2] Somatic cell score = average $\log_2$(somatic cell count/1000). [3] Ratios were calculated using percentages. [4] Horn sep. sus. = horn separation susceptibility; Severe horn sep. sus. = Severe horn separation susceptibility.

Generally, the phenotypic correlations between milk production traits were positive and strong or very strong. The two exceptions were for L:(F + P) with protein percentage and P:L, where the strong correlations were negative.

There were only three correlations between the lameness and claw disorder traits. There was one very strong correlation between lameness susceptibility and lameness occurrence, one moderate correlation between horn separation susceptibility and claw disorder occurrence, and one weak correlation between the two horn separation traits. Overall, there were no strong phenotypic correlations between lameness and claw disorder traits with milk traits. There was only a moderate correlation between milk protein percentage and lameness occurrence. There were also weak correlations for lameness occurrence with fat percentage and lactose percentage, between lameness susceptibility and fat yield, between horn separation susceptibility and protein percentage, and between severe horn separation and lactose percentage.

4. Discussion

4.1. Heritability Estimates

Results from our study indicate that lameness susceptibility, horn separation susceptibility, and rot susceptibility traits are heritable and could be selected against dairy goat breeding programs to improve animal welfare on farms. Our study reported the heritability of lameness susceptibility as a continuous trait rather than a binomial or ordinal trait. Previous studies have reported lameness traits as a binary trait (similar to the occurrence trait in this study) or as an ordinal trait. In dairy cows, the heritability of lameness as a binary trait has ranged from 0.01 to 0.04 using either a linear or threshold model. In contrast, the heritability of lameness as an ordinal trait has ranged from 0.03 to 0.15 [8]. The results in our study were consistent with the study of Weber et al. [21], which also used a 5-point locomotion scale and classified clinical lameness as a locomotion score ≥ 3. Heritability of lameness prevalence was 0.08 and 0.15 when using a linear and threshold model, respectively, similar to our study's 0.07 and 0.13 for lameness occurrence as a binary trait and lameness susceptibility as a continuous trait, respectively. When modeling binary or ordinal traits, threshold models can account for multiple cases over time, while linear models cannot [8]. Clinical lameness susceptibility had a higher heritability than clinical lameness occurrence. As susceptibility within our study considers multiple cases over a specific time period, it could explain why these heritabilities were similar to the results reported when using the threshold model.

Another factor that may explain the higher heritability estimate for clinical lameness susceptibility than for clinical lameness occurrence is that the calculation of lameness susceptibility includes some degree of the permanent environment effect. Vouraki et al. [14] indicated moderate to high (0.51) between-animal variation of lameness in Damascus dairy goats; however, they were unable to separate the genetic variation from the permanent environmental effects. A permanent environmental effect influences an individual's performance for a repeated trait. For example, if a doe is recorded as clinically lame in one visit, it could also be identified with clinical lameness in the subsequent visit because she was permanently damaged in the first event. Lameness susceptibility can capture this variation over time, whereas lameness occurrence cannot account for this effect.

Claw disorder susceptibility does not appear to have been studied previously in any species. In our study, horn separation susceptibility had the highest heritability estimate, while the estimate for granuloma susceptibility was the lowest. In the current study, horn separation could be comparable to severe white line disease in dairy cattle, while rot includes digital dermatitis, which has been extensively studied in dairy cattle. The heritability estimates for white line disease and digital dermatitis were higher in this study than those published for dairy cattle. In an extensive review of published heritabilities for claw disorders in dairy cows, traits were classified as binary variables, either presence or absence [8]. White line disease had estimates between 0.01 and 0.10, while digital dermatitis had estimates between 0.01 and 0.09 [8]. These differences in heritability estimates of the two claw disorders could be due to the different definitions of claw disorders. Horn separation is a more severe form of white line disease; therefore, the phenotype would be better defined, easier to identify, and more accurately recorded than white line disease. The more accurately recorded the phenotype is, the higher the heritability [22]. Rot in this study was broadly defined. It encompassed digital dermatitis and footrot because they are similar in their appearance in goats [23]. Due to digital dermatitis appearing differently in cows than goats, the goat's genetic response may differ from cows. More in line with our study's estimates were the heritabilities for horn separation and footrot in sheep. Heritability estimates for horn separation have been reported to be between 0.03 and 0.61 [6]. For footrot, estimates were between 0.09 and 0.33 [9–11]. These estimates varied depending on the animal's age (lamb or ewes), breed (Mule or Blackface ewes), number of feet scored, and the type of model used (threshold or linear). One of the traits defined by Nieuwhof et al. [9] had a similar definition to rot susceptibility in this study. Nieuwhof et al. [9] combined and then averaged the footrot scores for Mule ewes across four events. For this trait and using a linear model, they reported a heritability between 0.13 and 0.19 with standard errors between 0.09 and 0.10, which is consistent with the results for rot susceptibility in our study (0.13). Nieuwhof et al. [9] reported that the heritability estimate increased to between 0.21 and 0.23 when the number of affected feet was considered. They also reported that when multiple scores were accounted for over several events, the heritabilities were higher than if only one score was considered [9].

Heritability estimates reported in our study indicate underlying genetic variations for phenotypic traits within dairy goats. Consistent with previous studies, milk production traits in our study showed genetic variation [24]. Our study is the first to report heritability estimates for lameness and claw disorder traits in dairy goats and depicted genetic variation, a novel insight into goat health and breeding. This is in line with a recently published paper that indicated that limb problems and lameness had a moderate to high repeatability in dairy goats; therefore, the potential role of genetics on the occurrence of lameness is indicated [14]. Our study also suggests that lameness and claw disorders should be included in dairy goat breeding programs on commercial dairy goat farms.

4.2. Phenotypic Correlations

Currently, clinical lameness and claw disorder traits are not considered in breeding programs for the genetic improvement of dairy goats. One important aspect of introducing them into the breeding program is how the trait is phenotypically and genetically related to other measured traits. The phenotypic correlation between lameness susceptibility and clinical lameness occurrence was strongly positive (0.84), indicating that these two traits are related. However, the genetic correlation was not estimable, possibly due to a low number of observations. The genetic correlations between clinical lameness occurrence and susceptibility and between rot and granuloma susceptibility were positively strong (though relatively high standard errors were also noted to be present), while the phenotypic correlations were weak. Strong genetic correlations indicate that if there is genetic selection against lameness, there will also be genetic selection against rot and granulomas. Weber et al. [21] reported relatively high genetic correlations of 0.60 to 0.95 between lameness and claw disorders (including and excluding digital dermatitis), which was in line with the results of our study. When Weber et al. [21] included digital dermatitis within the claw disorder trait, the genetic correlation with lameness decreased, which is the opposite in our study, where digital dermatitis (included in 'rot') had a very strong relationship with clinical lameness. Our study suggests that digital dermatitis in cows may differ from digital dermatitis in dairy goats, which is supported by Groenevelt et al. [25]. The strong genetic correlations estimated in this study between lameness and claw disorders indicate that lameness can be used as an indicator trait for rot and granulomas or vice versa.

Our study determined that there were weak phenotypic correlations between claw disorders. The exception was between all horn separation and the most severe horn separation cases. The weak genetic correlation and strong phenotypic correlation between these two traits indicate that the environment is an important factor in determining the presence of horn separation.

4.3. Genotypic Correlations

There were underlying genetic correlations, varying in strength, between the claw disorders. Despite a high standard error, the strongest genetic relationship was between rot and granuloma. The mechanisms behind granuloma development are currently unknown. The granulomas' low heritability suggests that granulomas occur due to environmental factors. Winter [26] suggested granulomas were related to trimming injuries, although results from our study indicate that this may not be entirely correct. Our results suggest that the relationship between granulomas and hoof-trimming injuries remains unclear. Alternatively, there may be an indirect genetic relationship between rot healing and granulomas developing due to hoof trauma. In addition, horn separation (a form of white line disease) and rot (encompassing digital dermatitis and footrot) susceptibility were very weakly correlated. These were in line with previous research completed in dairy cows on digital dermatitis and white line disease, where genetic correlations ranged from −0.33 to 0.08 [8].

One important aspect of a breeding program for the genetic improvement of dairy goats is how health traits, such as lameness and claw disorders, are phenotypically and genetically related to milk production traits. In our study, clinical lameness susceptibility had a weak negative phenotypic correlation with fat percentage and a weak positive genotypic correlation with P:L ratio. Unlike previous dairy cattle studies, there was little correlation between lameness and milk yield. In our study, the correlation was less than 0.10, while in dairy cattle, it ranged from 0.17 to 0.44 [27,28].

The genetic correlation between the occurrence of clinical lameness and fat percentage was moderately positive (0.75). A reason for this relationship could be the goat's change in fat percentage in response to metabolic problems, as described for acidosis in dairy cattle. The fat percentage has been linked to negative energy balance and ruminal acidosis in dairy cattle [29]. Negative energy balance has indirectly been associated with the increased risk of lameness through loss of body condition [30,31]. Milk fat depression has been linked with subacute ruminal acidosis in cows [32]. The production of endotoxins resulting from acidosis, which are released into the bloodstream, has been associated with lameness-causing-laminitis [32]. Therefore, in dairy goats, events that may cause a goat to be in a negative energy balance, such as parturition, could increase fat percentage or P:F and indicate a higher risk of lameness in the herd. The results from this study also indicate that there may be some underlying genetic material that could be inherited. Within the breeding goal, selecting goats with higher estimated breeding values for fat percentage could indirectly select goats more susceptible to clinical lameness.

The opposite could be true for lactose concentrations. There was a strong negative genetic correlation between the occurrence of clinical lameness and lactose percentage (−0.94) and subsequently with milk production. It is possible that the control of lactose percentage shares some underlying biological mechanisms relating to the health status and potentially the immune system within dairy goats, as suggested in dairy cows [33,34]. In dairy cows, some studies have reported a negative association between lactose percentage and mastitis [35,36]. While there was not a strong genetic relationship between lameness occurrence and somatic cell score in our study, in dairy cows, one study observed a moderate and weak relationship between lactose yield, mastitis (0.52), and ketosis (0.42), respectively [36]. As Ingvartsen et al. [33] suggested, the effects of genotype, nutrition, time, and management could affect the immune system via the metabolic pathway, which increases the animal's susceptibility to disease, in this case, lameness. Our study results suggest that lactose percentage may be an important indicator of lameness occurrence within a lactation. This relationship should be further investigated with a larger dataset to validate these results. Other health traits and the relationship with lactose percentage should also be explored in the future.

Claw disorders have weak to moderate genetic correlations with milk production traits. Horn separation traits were negatively correlated with fat percentage and positively correlated with lactose percentage, while rot and granuloma susceptibility were positively correlated with milk yield. This suggests that genetic selection for higher yields of production traits may increase the goat's genetic susceptibility to rot and granulomas. In contrast, genetic selection for higher concentrations of fat could increase the goat's genetic susceptibility to horn separation. However, caution should be exercised for this interpretation due to the strength of the correlations. Estimates of genetic correlation between claw disorders and production traits are few in dairy cattle. Depending on the model, Koenig et al. [37] and König et al. [28] reported genetic correlations between digital dermatitis and milk yield of 0.07 to 0.24, which was similar to the values reported in this study (0.31). They also reported correlations of 0.17 to 0.28 between hoof wall disorders and milk yield, which were higher estimates than this study (0.05 to 0.07). Gernand et al. [38] reported weak genetic correlations between digital dermatitis and protein yield, fat and protein percentages (−0.11 to 0.05) which, considering the higher standard error, was still lower than the values calculated in this study (0.08 to 0.41). From the limited number of studies published, the relationship between milk production traits and claw disorder traits in dairy goats is still unclear. Based on dairy cow studies, a negative relationship exists between production traits and claw health. Our study tends to agree with this, although further investigation, with a larger cohort of animals, is needed in goats before conclusions can be made to determine the underlying genetic relationship between claw health and production traits.

4.4. Limitations

One of this study's limitations is the limited number of observations available. Many observations need to be collected on phenotypes of interest to precisely estimate genetic parameters (reducing the standard errors). The current methodology for measuring lameness and claw disorders in dairy goats is time consuming. Locomotion scoring would ideally need to be completed objectively and routinely. Introducing cameras that can record this information automatically would help the farmers immensely. An incentive must also be included for farmers and hoof trimmers to record information on lameness and claw disorders. The farmers would invest in equipment and additional services from the hoof trimmer. Dairy goat hoof trimmers and farmers can only be expected to record this information with reliable technology—preferably designed for dairy goats.

It would also be important to run a multivariate analysis rather than a bivariate analysis to investigate various trait combinations in one model so that covariances can be evaluated simultaneously. This requires more herds with genetic links among the sires. This would only be possible once artificial insemination is widely used within the dairy goat population of New Zealand.

Another approach to estimating heritabilities and variances is using threshold models rather than linear ones. Heritabilities of lameness and claw disorders traits in sheep and dairy cattle have previously been estimated using either threshold or linear models [6,8,9], while genetic and environmental variances can be obtained using linear or liability-threshold models. Threshold models can be better used when the trait of interest is binomial or ordinal and has an underlying continuous phenotype with a low heritability [39]. It is based on the probability of a phenotype being expressed depending on environmental or genetic conditions being met. However, depending on the information available and the trait being used, there may be little difference between the results of the two models [40]. As linear models (using either restricted maximum likelihood or least-squares) are simple and more widely used to estimate genetic parameters, therefore, linear instead of threshold models were chosen to model the estimates in our study [11,41]. However, Heringstad et al. [8] explained that, unlike linear models, threshold models are already adjusted for the time when there are repeated measures of ordinal traits for individuals. Therefore, instead of using the linear model in our study, where we modeled one trait over one lactation at the test-day level (to account for repeated measures over time), the threshold model may have been better for the two binomial traits in our study.

There are two alternatives to the current model that could be used to analyze lameness and claw disorder traits in the future. Firstly, including breed and the heterosis effect within our models. There are significant differences in milk production traits in different dairy goat breeds [24,41], and heterosis has been known to affect milk production traits significantly [15]. In dairy cows and sheep, the breed was a risk factor for developing lameness and claw disorders [6,42]. Therefore, heterosis should be included in the model to capture variation associated with the breed. To do this reliability, the data input of the breeds by the farmers needs to be accurate, which is challenging to achieve when only a pedigree is available.

Secondly, investigate the heritabilities of claw disorder and lameness traits at the test-day level rather than over one lactation. The genetic parameters for milk production traits have been reported to change the lactation [43], and this could also be the case for the genetic parameters of claw disorders and lameness. In dairy cattle, test-day milk yield has been associated with claw disorders [28]. Therefore, it should be established whether claw disorders and/or lameness traits are more heritable at the test-day level rather than at the lactation level.

This study is the first of its kind in dairy goats. It creates the foundation for future research on the genetic parameters of health traits besides the somatic cell score (related to mastitis). The objective would be to eventually include lameness and claw disorder traits in a breeding index to reduce lameness and claw disorder prevalence in herds and improve animal welfare in dairy goats. The economic values and weights need to be estimated before these traits are included in the breeding goal or selection index [44]. The model to estimate these economic values needs to include direct and indirect costs. Indirect costs, such as additional farm labor, should be included, as these costs are realized even when lameness scores are low. In contrast, the direct costs, such as those associated with loss of milk production, are not significant until lameness is severe [45]. Indirect costs may vary depending on the prevalence of lameness and the causes of lameness (which affects the duration) on the farm. The economic weights derived from the influence of other traits, such as milk production traits, claw disorders, and lameness, also need to be included. Finally, using the economic weight and including the desired gains value would establish a selection index for the breeding goal. The desired gains approach would capture costs that may not be monetary, such as animal welfare, and add value to the trait.

Improving claw health through genetics creates permanent gains [8]. Due to the low heritability of health traits, many observations are needed to produce reliable genetic evaluations. Agreeing on the definition of traits is important to be universally understood and have phenotypes accurately recorded.

5. Conclusions

This study reported heritability and phenotypic and genetic correlation estimates for claw disorder and lameness traits in dairy goats. The moderate heritability estimates suggest that selecting against clinical lameness, horn separation, and rot susceptibility is possible. There appear to be weak to strong genetic and phenotypic correlations between lameness and claw disorders traits with fat and lactose concentrations in milk. Further investigation into the relationship between these traits is required to determine the underlying biological relationships. If claw disorder and lameness traits are to be included in the breeding goal and selection index for dairy goats, research on the economic impact of these traits is needed in order to assign appropriate economic weights. Before this can be conducted, investing in better technology for dairy goat farmers and hoof trimmers to collect information on lameness and claw disorders is the next step forward.

Supplementary Materials: The following supporting information can be downloaded at: https://www.mdpi.com/article/10.3390/ani13081374/s1, Table S1: Locomotion scoring strategy used to measure lameness in dairy goats (adapted from Deeming et al. [18]).

Author Contributions: Conceptualization, N.J., N.L.-V., S.-A.T., E.V. and C.H.; methodology, N.J. and N.L.-V.; formal analysis, N.J. and N.L.-V.; investigation, N.J.; resources, N.J. and N.L.-V.; data curation, N.J. and N.L.-V.; writing—original draft preparation, N.J. and N.L.-V.; writing—review and editing, N.J., N.L.-V., S.-A.T., E.V. and C.H.; supervision, N.L.-V., S.-A.T., E.V. and C.H.; project administration, N.J.; funding acquisition, N.J. and N.L.-V. All authors have read and agreed to the published version of the manuscript.

Funding: This research was part of the project "Investigation of lameness and claw disorders in New Zealand dairy goats: a multidisciplinary approach". This research was funded by the Dairy Goat Co-Operative (NZ) Ltd. Natasha Jaques was funded and supported by Dairy Goat Co-Operative (NZ) Ltd. and the C. Alma Baker Trust (NZ) Ltd.

Institutional Review Board Statement: The animal study protocol was approved by the Ethics Committee of Massey University, New Zealand (MUAEC Protocol 19/51 and 29/05/2019).

Informed Consent Statement: Informed consent was obtained from all farmers involved in the study.

Data Availability Statement: Restrictions apply to the availability of these data. Data were obtained from New Zealand Dairy Goat Co-Operative Ltd. and are available from the authors with the permission of the New Zealand Dairy Goat Co-Operative Ltd.

Acknowledgments: We thank the farmers, farm workers, and hoof trimmers that were involved in this study for their time and respective expertise. We also acknowledge Vicki McLean, from the Dairy Goat Co-Operative, who gave us her expertise on dairy goat health in New Zealand and helped us with the design of the project.

Conflicts of Interest: The authors declare no conflict of interest. However, one of the authors, Sally-Anne Turner, from the Dairy Goat Co-Operative (NZ) Ltd., was involved in the design and collection of data of this study, as well as supervision of the PhD student, however, was not involved in the analyses or interpretation of data and in the original writing of the manuscript.

References

1. Prosser, C.; Stafford, K. Goat Production. In *Livestock Production in New Zealand*; Stafford, K., Ed.; Massey University Press: Palmerston North, New Zealand, 2017; pp. 123–126.
2. Deeming, L.; Beausoleil, N.; Stafford, K.; Webster, J.; Zobel, G. Brief Communication: Investigating the Pattern of Lameness in Dairy Goats across the First Two Years of Life. *N. Z. J. Anim. Sci. Prod.* **2021**, *81*, 35–38.
3. Chesterton, R.N.; Chesterton, S.J.; Laven, R.A. Lesions Found at Foot Trimming of Dairy Goats: Baseline Data for Comparing Lesions and Locomotion Scoring. *Vet. J.* **2022**, *290*, 105927. [CrossRef]
4. Vieira, A.; Oliveira, M.D.; Nunes, T.; Stilwell, G. Making the Case for Developing Alternative Lameness Scoring Systems for Dairy Goats. *Appl. Anim. Behav. Sci.* **2015**, *171*, 94–100. [CrossRef]
5. Matthews, J. Lameness in Adult Goats. In *Diseases of the Goat*, 4th ed.; Matthews, J., Ed.; John Wiley & Sons Limited: Chelmsford, Essex, UK, 2016; pp. 81–104.
6. Conington, J.; Nicoll, L.; Mitchell, S.; Bunger, L. Characterisation of White Line Degeneration in Sheep and Evidence for Genetic Influences on Its Occurrence. *Vet. Res. Commun.* **2010**, *34*, 481–489. [CrossRef]
7. Chesnais, J.P.; Cooper, T.A.; Wiggans, G.R.; Sargolzaei, M.; Pryce, J.E.; Miglior, F. Using Genomics to Enhance Selection of Novel Traits in North American Dairy Cattle1,2. *J. Dairy Sci.* **2016**, *99*, 2413–2427. [CrossRef]
8. Heringstad, B.; Egger-Danner, C.; Charfeddine, N.; Pryce, J.E.; Stock, K.F.; Kofler, J.; Sogstad, A.M.; Holzhauer, M.; Fiedler, A.; Muller, K.; et al. Invited Review: Genetics and Claw Health: Opportunities to Enhance Claw Health by Genetic Selection. *J. Dairy Sci.* **2018**, *101*, 4801–4821. [CrossRef]
9. Nieuwhof, G.J.; Conington, J.; Bunger, L.; Haresign, W.; Bishop, S.C. Genetic and Phenotypic Aspects of Foot Lesion Scores in Sheep of Different Breeds and Ages. *Animal* **2008**, *2*, 1289–1296. [CrossRef]
10. Skerman, T.M.; Johnson, D.L.; Kane, D.W.; Clarke, J.N. Clinical Footscald and Footrot in a New Zealand Romney Flock—Phenotypic and Genetic-Parameters. *Aust. J. Agric. Res.* **1988**, *39*, 907–916. [CrossRef]
11. Raadsma, H.W.; Egerton, J.R.; Wood, D.; Kristo, C.; Nicholas, F.W. Disease Resistance in Merino Sheep. 3. Genetic-Variation in Resistance to Footrot Following Challenge and Subsequent Vaccination with an Homologous RDNA Pilus Vaccine under Both Induced and Natural Conditions. *J. Anim. Breed. Genet.* **1994**, *111*, 367–390. [CrossRef]
12. O'Brien, A.C.; McHugh, N.; Wall, E.; Pabiou, T.; McDermott, K.; Randles, S.; Fair, S.; Berry, D.P. Genetic Parameters for Lameness, Mastitis and Dagginess in a Multi-Breed Sheep Population. *Animal* **2017**, *11*, 911–919. [CrossRef]

13. Hill, N.P.; Murphy, P.E.; Nelson, A.J.; Mouttotou, N.; Green, L.E.; Morgan, K.L. Lameness and Foot Lesions in Adult British Dairy Goats. *Vet. Rec.* **1997**, *141*, 412–416. [CrossRef]
14. Vouraki, S.; Gelasakis, A.I.; Fotiadou, V.; Banos, G.; Arsenos, G. Repeatability of Health and Welfare Traits and Correlation with Performance Traits in Dairy Goats Reared under Low-Input Farming Systems. *Vet. Sci.* **2022**, *9*, 289. [CrossRef]
15. Scholtens, M.R.; Lopez-Villalobos, N.; Garrick, D.; Blair, H.; Lehnert, K.; Snell, R. Genetic Parameters for Total Lactation Yields of Milk, Fat, Protein, and Somatic Cell Score in New Zealand Dairy Goats. *Anim. Sci. J.* **2020**, *91*, e13310. [CrossRef]
16. Scholtens, M.; Lopez-Lozano, R.; Rebecca, S. New Zealand Goat Industry. 2017. Available online: https://landusenz.org.nz/wp-content/uploads/2017/08/NZ_Goat_Industry_Report_to_Federated_Farmers_14_Mar_2017.pdf (accessed on 6 April 2023).
17. Scholtens, M.; Lopez-villalobos, N.; Garrick, D.; Blair, H.; Lehnert, K.; Snell, R. Estimates of Genetic Parameters for Lactation Curves for Milk, Fat, Protein and Somatic Cell Score in New Zealand Dairy Goats. *N. Z. J. Anim. Sci. Prod.* **2019**, *79*, 1–7.
18. Deeming, L.E.; Beausoleil, N.J.; Stafford, K.J.; Webster, J.R.; Zobel, G. Technical Note: The Development of a Reliable 5-Point Gait Scoring System for Use in Dairy Goats. *J. Dairy Sci.* **2018**, *101*, 4491–4497. [CrossRef]
19. Gilmour, A.; Gogel, B.; Cullis, B.; Welham, S.; Thompson, R. Asreml User Guide Release 4.1 Structural Specification. 2015. Available online: www.vsni.co.uk (accessed on 6 April 2023).
20. Weber, A.; Stamer, E.; Junge, W.; Thaller, G. Genetic Parameters for Lameness and Claw and Leg Diseases in Dairy Cows. *J. Dairy Sci.* **2013**, *96*, 3310–3318. [CrossRef]
21. Robinson, H.F.; Comstock, R.E.; Harvey, P.H. Estimates of Heritability and the Degree of Dominance in Corn. *Agron. J.* **1949**, *41*, 353–359. [CrossRef]
22. Schöpke, K.; Gomez, A.; Dunbar, K.A.; Swalve, H.H.; Döpfer, D. Investigating the Genetic Background of Bovine Digital Dermatitis Using Improved Definitions of Clinical Status. *J. Dairy Sci.* **2015**, *98*, 8164–8174. [CrossRef]
23. Groenevelt, M.; Anzuino, K.; Langton, D.A.; Grogono-Thomas, R. Association of Treponeme Species with Atypical Foot Lesions in Goats. *Vet. Rec.* **2015**, *176*, 626. [CrossRef]
24. Mucha, S.; Tortereau, F.; Doeschl-Wilson, A.; Rupp, R.; Conington, J. Animal Board Invited Review: Meta-Analysis of Genetic Parameters for Resilience and Efficiency Traits in Goats and Sheep. *Animal* **2022**, *16*, 100456. [CrossRef]
25. Groenevelt, M. Foot Lameness in Goats: Like Sheep, Like Cattle or Completely Different? *Livestock* **2017**, *22*, 98–101. [CrossRef]
26. Winter, A.C. Treatment and Control of Hoof Disorders in Sheep and Goats. *Vet. Clin. N. Am.-Food Anim. Pract.* **2011**, *27*, 187–192. [CrossRef]
27. Pryce, J.E.; Veerkamp, R.F.; Thompson, R.; Hill, W.G.; Simm, G. Genetic Aspects of Common Health Disorders and Measures of Fertility in Holstein Friesian Dairy Cattle. *Anim. Sci.* **1997**, *65*, 353–360. [CrossRef]
28. König, S.; Wu, X.L.; Gianola, D.; Heringstad, B.; Simianer, H. Exploration of Relationships between Claw Disorders and Milk Yield in Holstein Cows Via Recursive Linear and Threshold Models. *J. Dairy Sci.* **2008**, *91*, 395–406. [CrossRef]
29. de Vries, M.J.; Veerkamp, R.F. Energy Balance of Dairy Cattle in Relation to Milk Production Variables and Fertility. *J. Dairy Sci.* **2000**, *83*, 62–69. [CrossRef]
30. Collard, B.L.; Boettcher, P.J.; Dekkers, J.C.; Petitclerc, D.; Schaeffer, L.R. Relationships between Energy Balance and Health Traits of Dairy Cattle in Early Lactation. *J. Dairy Sci.* **2000**, *83*, 2683–2690. [CrossRef]
31. Huxley, J.N. Impact of Lameness and Claw Lesions in Cows on Health and Production. *Livest. Sci.* **2013**, *156*, 64–70. [CrossRef]
32. Plaizier, J.C.; Krause, D.O.; Gozho, G.N.; McBride, B.W. Subacute Ruminal Acidosis in Dairy Cows: The Physiological Causes, Incidence and Consequences. *Vet. J.* **2008**, *176*, 21–31. [CrossRef]
33. Ingvartsen, K.L.; Dewhurst, R.J.; Friggens, N.C. On the Relationship between Lactational Performance and Health: Is it Yield or Metabolic Imbalance That Cause Production Diseases in Dairy Cattle? A Position Paper. *Livest. Prod Sci.* **2003**, *83*, 277–308. [CrossRef]
34. Costa, A.; Egger-Danner, C.; Mészáros, G.; Fuerst, C.; Penasa, M.; Sölkner, J.; Fuerst-Waltl, B. Genetic Associations of Lactose and its Ratios to Other Milk Solids with Health Traits in Austrian Fleckvieh Cows. *J. Dairy Sci.* **2019**, *102*, 4238–4248. [CrossRef]
35. Ebrahimie, E.; Ebrahimi, F.; Ebrahimi, M.; Tomlinson, S.; Petrovski, K.R. A Large-Scale Study of Indicators of Sub-Clinical Mastitis in Dairy Cattle by Attribute Weighting Analysis of Milk Composition Features: Highlighting the Predictive Power of Lactose and Electrical Conductivity. *J. Dairy Res.* **2018**, *85*, 193–200. [CrossRef]
36. Costa, A.; Lopez-Villalobos, N.; Sneddon, N.W.; Shalloo, L.; Franzoi, M.; De Marchi, M.; Penasa, M. Invited Review: Milk Lactose—Current Status and Future Challenges in Dairy Cattle. *J. Dairy Sci.* **2019**, *102*, 5883–5898. [CrossRef]
37. Koenig, S.; Sharifi, A.R.; Wentrot, H.; Landmann, D.; Eise, M.; Simianer, H. Genetic Parameters of Claw and Foot Disorders Estimated with Logistic Models. *J. Dairy Sci.* **2005**, *88*, 3316–3325. [CrossRef]
38. Gernand, E.; Rehbein, P.; von Borstel, U.U.; König, S. Incidences of and Genetic Parameters for Mastitis, Claw Disorders, and Common Health Traits Recorded in Dairy Cattle Contract Herds. *J. Dairy Sci.* **2012**, *95*, 2144–2156. [CrossRef]
39. Dempster, E.R.; Lerner, M. Heritability of Threshold Characters. *Genetics* **1950**, *35*, 212–236. [CrossRef]
40. Varona, L.; Misztal, I.; Bertrand, J.K. Threshold-Linear Versus Linear-Linear Analysis of Birth Weight and Calving Ease Using an Animal Model: II. Comparison of Models. *J. Anim. Sci.* **1999**, *77*, 2003–2007. [CrossRef]
41. García-Peniche, T.B.; Montaldo, H.H.; Valencia-Posadas, M.; Wiggans, G.R.; Hubbard, S.M.; Torres-Vázquez, J.A.; Shepard, L. Breed Differences over Time and Heritability Estimates for Production and Reproduction Traits of Dairy Goats in the United States. *J. Dairy Sci.* **2012**, *95*, 2707–2717. [CrossRef]

42. Chawala, A.R.; Lopez-Villalobos, N.; Margerison, J.K.; Spelman, R.J. Genetic and Crossbreeding Parameters for Incidence of Recorded Clinical Lameness in New Zealand Dairy Cattle. *N. Z. Vet. J.* **2013**, *61*, 281–285. [CrossRef]
43. Swalve, H.H. The Effect of Test Day Models on the Estimation of Genetic Parameters and Breeding Values for Dairy Yield Traits. *J. Dairy Sci.* **1995**, *78*, 929–938. [CrossRef]
44. Byrne, T.J.; Santos, B.F.S.; Amer, P.R.; Martin-Collado, D.; Pryce, J.E.; Axford, M. New Breeding Objectives and Selection Indices for the Australian Dairy Industry. *J. Dairy Sci.* **2016**, *99*, 8146–8167. [CrossRef]
45. Jaques, N.; Turner, S.-A.; Vallée, E.; Heuer, C.; Lopez-Villalobos, N. *The Effect of Clinical Lameness on Milk Production of Dairy Goats;* Massey University: Palmerston North, New Zealand, 2023; *manuscript in preparation; to be submitted.*

animals

MDPI

Article

Orthopedic Diseases in the Pura Raza Española Horse: The Prevalence and Genetic Parameters of Angular Hoof Deviations

María Ripollés-Lobo [1], Davinia Isabel Perdomo-González [1], Pedro Javier Azor [2] and Mercedes Valera [1,*]

[1] Departamento de Agronomía, Escuela Técnica Superior de Ingeniería Agronómica, Universidad de Sevilla, 41013 Seville, Spain; marriplob@alum.us.es (M.R.-L.); dperdomo@us.es (D.I.P.-G.)
[2] Real Asociación Nacional de Criadores de Caballos de Pura Raza Española, 41014 Seville, Spain
* Correspondence: mvalera@us.es

Simple Summary: This study attempts to elucidate the prevalence and genetic factors associated with four angular hoof deviations in the Pura Raza Española horse: splay-footed forelimb (SFF), pigeon-toed forelimb (PTF), splay-footed rear limb (SFR), and pigeon-toed rear limb (PTR). From a total of 51,134 horses evaluated, only 15.75% did not have any of the four investigated angular hoof defects, while the rest presented varying degrees of defect. We identified the factors influencing these defects, including age, inbreeding coefficient, sex, and birth stud size, using a Bayesian multivariate animal model with the BLUPF90 software. The heritability estimates ranged from 0.11 (SFR) to 0.31 (PTR) for model A (three categories.), indicating the genetic influences of varying categories; and ranged from 0.11 (SFR) to 0.29 (PTR) for model B (two categories). Additionally, the study explored the genetic correlations between these defects and their relationship with the diameter of the superficial digital flexor tendon (SDFT) and the proportionality index (PI). The findings revealed that diameter of the SDFT is strongly correlated with inward-toe conditions (PTF and PTR), while PI is associated with outward defects (SFF and SFR). These results provide valuable insights into angular hoof deviations in PRE horses, shedding light on their prevalence, genetic factors, and associations with other anatomical features, which can have general implications for a horse's health and performance.

Abstract: Abnormalities in hoof shape are usually connected with limb conformation defects. The role of angular hoof deviations is important for longevity in sports competitions and is increasingly recognized as a factor associated with lameness in performance horses. In this paper, we measured the prevalence of four defects related to the angulation of the hoof in the Pura Raza Española horse (PRE): splay-footed forelimb (SFF), pigeon-toed forelimb (PTF), splay-footed rear limb (SFR), and pigeon-toed rear limb (PTR). A total of 51,134 animals were studied, of which only 15.75% did not have any of the four angular hoof defects investigated, while 26.61%, 23.76%, 79.53%, and 3.86% presented SFF, PTF, SFR, and PTR, respectively. Angular defects were evaluated using two different models; model A was a linear scale composed of three categories, where 0 corresponded to the absence of defects, 1 to a minor presence of the defect and 2 to the highest degree of the defect. Model B was composed of two categories, where 0 corresponded to the absence of defects and 1 to the presence of defects, joining classes 1 and 2. We measured the factors influencing the appearance of these defects: age, inbreeding coefficient, sex, and birth stud size. The heritability of each defect was also estimated using a multivariate animal model, using the Gibbsf90+ software from the BLUPF90 family, resulting in heritability estimates of 0.18 (s.d. = 0.009), 0.20 (s.d. = 0.010), 0.11 (s.d. = 0.009), and 0.31 (s.d. = 0.010) for SFF, PTF, SFR, and PTR defects, respectively, for model A, and 0.17 (s.d. = 0.008), 0.19 (s.d. = 0.009), 0.11 (s.d. = 0.009), and 0.29 (s.d. = 0.009) for SFF, PTF, SFR, and PTR defects, respectively, for model B. Finally, the genetic correlation between the diameter of the superficial digital flexor tendon (SDFT) and the proportionality index (PI) in relation to the higher or lower prevalence of the defects was analyzed. We concluded that diameter of SDFT development is strongly correlated with inward toe conditions (PTF, PTR; $P_{\neq 0} \geq 0.95$), while PI is associated with outward toe defects (SFF, SFR; $P_{\neq 0} \geq 0.95$).

Keywords: equine; forelimbs; genetic parameters; heritability; pigeon-toed forelimb; pigeon-toed rear limb; splay-footed forelimb; splay-footed rear limb

1. Introduction

The hoof structure plays a critical role in locomotion, weight distribution, and shock absorption during movement in equines [1], and disorders at the hoof level are considered to be a potential risk factor of lameness [2]. Defects in equine limbs usually consist of anomalies or irregularities in the structure or functionality of the joints in the limbs, including the hoof, the hock (tarsal joint), and the stifle [3]. Orthopedic diseases are of paramount importance in equine medicine, as they have a significant impact on a horse's health, performance, and longevity.

Angular hoof deviations encompass a spectrum of abnormalities that involve the alignment and orientation of a horse's hooves in relation to the limb axis. These deviations impact the biomechanics of the horse's limbs and, consequently, influence the horse's performance in various activities, including work and sports competitions, or simply its ability to move comfortably without pain. These areas' defects are associated with a series of factors, such as injuries, inflammation, bio-mechanical imbalances, abnormal development during the horse's growth, or even genetic factors. The main consequence is the effect on the horse's ability to move correctly [4,5]. Throughout history, horse breeders have developed routine trimming and shoeing techniques aimed at improving hoof balance by addressing factors such as hoof angle, length, mediolateral equilibrium, sole thickness, and the way the hoof interacts with the ground [6] to improve the horse movement. But also, a substantial body of research exists about the 'optimal' conformation of sport breed horses, such as Thoroughbreds [7–10], Warmbloods [11,12], Hanoverian warmblood horses [13], Trotting [14], Arabian [15], New Zealand Standardbred [16], Iceandic [17], and Pura Raza Española (PRE) [18]. Wilson et al. [6] identified a significant relationship between movement asymmetry and foot conformation in horses. In the same way, Ducro et al. [11] described that uneven feet have a moderate negative genetic relationship with foot conformation grade and this, in turn, has a genetic relationship with sporting performance in Warmblood riding horses. In addition, Ducro et al. [19] pointed out that uneven feet are known to negatively impact the competitive longevity of sport horses; given the observed prevalence of uneven feet in sports disciplines, it can be inferred that this is an undesirable characteristic, especially in elite jumping. Oosterlinck et al. [20] described that, while mild toed-in deviations have been found to have no significant impact on soundness or performance [21], there is empirical evidence suggesting that severe toed-in conformation is associated with an elevated risk of injury and a shortened lifespan for equine athletes. Nevertheless, more objective information regarding the biomechanical effects of toed-in and toed-out conformation during locomotion is necessary, even when, in horsemanship belief, the equine conformation is associated with performance and durability, as described in Thoroughbreds [9]. Thus, angular hoof deviations represent a specific area of interest, especially when considering the health, performance, and longevity on sport competition horses.

In this context, although all the above-mentioned risks produce themselves financial losses and a lower quality of life for the horse, from a population management point of view, knowledge about the genetic implications is of great interest. The identification and study of limb defects are essential for implementing the appropriate selection measures to reduce their incidence and improve the health and athletic performance of the breed, especially when considering sport competition horses such as the Pura Raza Española (PRE) horse. The PRE is a native Spanish breed, one of the oldest in Europe and one of the most important at the international level. This breed has gained international recognition in the field of Classical Dressage, taking part annually in a significant number of competitions, both in Spain and worldwide [22]. In general, PRE horses that excel in dressage performances have a higher economic value compared to other animals of the same breed, as this is one of the main objectives of their breeding program, which aims to enhance not only conformation and dressage functionality, but also, most importantly, gait quality, which is a key factor in optimizing dressage performance [23].

This study aims to provide a more comprehensive understanding of the factors affecting hoof health in PRE horses, with potential implications for the breeding program, management practices, and overall well-being of this equine breed. In this study, the incidence of hoof defects in the PRE horse population was estimated and the potential influencing risk factors in the presentation of angular hoof deviation were analyzed. Additionally, genetic parameters related to these defects were examined to determine if there is a genetic basis contributing to their occurrence. Finally, the relationship between the presence of hoof defects and morphological variables, specifically the diameter of the superficial digital flexor tendon (SDFT) and the proportionality index (PI), were explored.

2. Materials and Methods

2.1. Database and Description Traits

In our study, we analyzed hoof conformational data from 51,134 Pura Raza Española (PRE) horses (16,920 males and 34,214 females). The sample encompassed a variety of hoof conformational defects (Figure 1): hoof defects in the front legs were divided into 'splay-footed forelimb' (SFF) and 'pigeon-toed forelimb'(PTF), and those in the back legs into 'splay-footed rear limb' (SFR) and 'pigeon-toed rear limb' (PTR). Therefore, each angular hoof defect was analyzed independently. The average evaluation age of the horses was 4.86 (s.d. = 2.25) years. Data collection took place between 2012 and 2022, during the mandatory morphological assessment, which occurs once in a horse's lifetime before they are officially registered in the primary section of the PRE stud book. The evaluations were conducted by a total of 12 specially trained veterinarians responsible for performing standardized aptitude tests in this breed. Each horse was evaluated by a single evaluator. For the phenotypic evaluation of conformational defects, the horses were examined while standing on a firm, level surface with their limbs in a natural position. The fore and hind legs were placed in a parallel position as close to perpendicular as possible, with their hooves aligned. If the horse was still and there were doubts about the angular hoof defect, it was evaluated while the horse was in motion (walk or trot) to assign a defect category. For the evaluation of hoof deviation we considered, on the one hand, the forelegs, and, on the other hand, the hind legs for phenotyping angular defects in the limbs. This is because, in most cases, if one limb has an angular defect, it is common for the adjacent limb to also have one. Notably, no sedatives were administered during the evaluation process.

The four analyzed defects were as follows:

- Defects of the 'splay-footed forelimb' (SFF) and 'pigeon-toed forelimb' (PTF). These defects are related to the direction of the front hooves when seen from the front. An individual exhibits these defects when the hoof points either inwards or outwards, respectively, in relation to the vertical plumb line, as measured from the outer edge of the shoulder joint and the humero-cubital joint to the ground.
- Defects of the 'splay-footed rear' limb (SFR) and 'pigeon-toed rear limb' (PTR). These defects are related to the direction of the rear hooves when seen from the back. An individual has an SFR or PTR defect when the hoof tips either inwards or outwards, respectively, in relation to the vertical plumb line, as measured from the point of the hip and tibia-fibula joint to the ground, when seen from the rear.

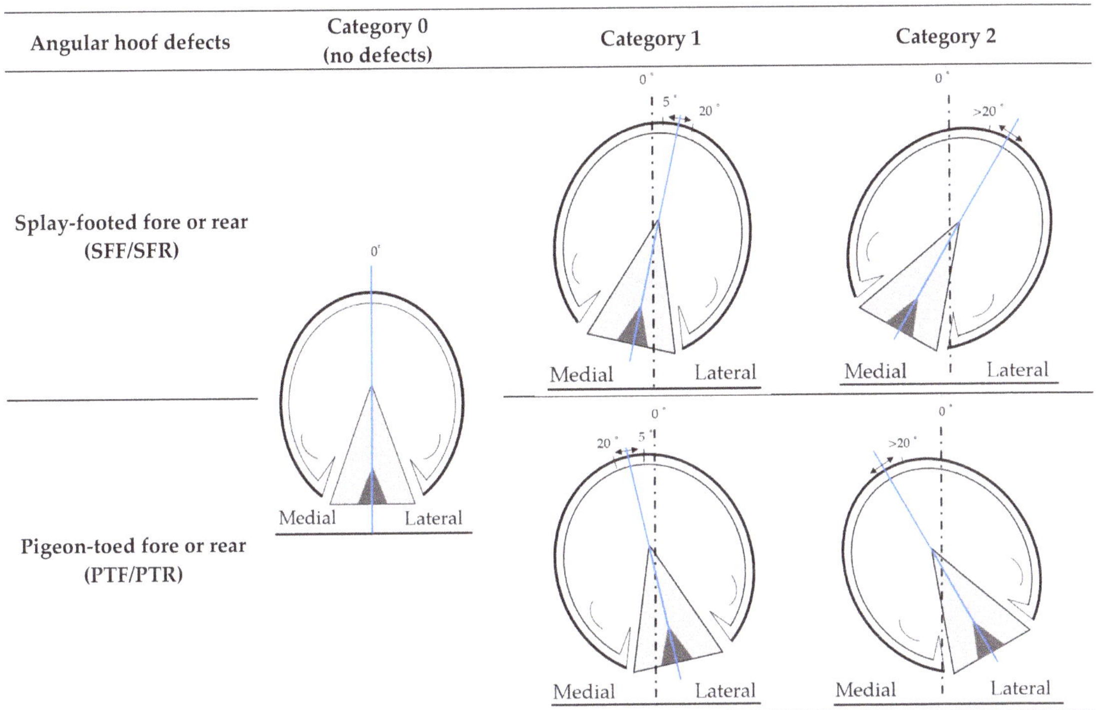

Figure 1. Conformational hoof defects (angular hoof deviations) in Pura Raza Española horses.

The estimation of genetic parameters was performed with two different models:

For model A, the four conformational hoof defects were recorded on a linear scale from 0 to 2. Category 0: represents a lack of the defect, the hooves are vertical, or slightly turned outwards (SFF or SFR) or inwards (PTF or PTR), with, at most, 5 degrees of angulation. Category 1: a minor presence of the defect, the hooves are turned outwards (SFF or SFR) or inwards (PTF or PTR) between 5 and 20 degrees of angulation. Category 2: the most significant level of the defect, the hooves are turned outwards (SFF or SFR) or inwards (PTF or PTR) with more than 20 degrees of angulation.

For model B, the four conformational hoof defects were recorded as a dichotomous model. Category 0: represents a lack of the defect, the hooves are vertical, or slightly turned outwards (SFF or SFR) or inwards (PTF or PTR), with, at most, 5 degrees of angulation. Category 1: presence of the defect, the hooves are turned outwards (SFF or SFR) or inwards (PTF or PTR), with more than 5 degrees of angulation.

In PRE horses, SFF and SFR defects are considered to be severe when they are in category 2, and PTR and PTF defects are considered to be very severe when they are in category 2. When a PRE horse, in any part of its body, has more than 2 very severe defects or more than 4 severe defects, the horse cannot be registered as a breeder of the PRE breed.

The covariates analyzed were the evaluation age (in years; from 2 to 23 years) and the classical inbreeding coefficient (F; from 0.00 to 0.47). The F value, following Wright [24], defined as the probability that the two alleles at any locus in an individual are identical by descent, was computed using the Endog software (v.3.0) [25]. The fixed effects analyzed were: sex (2 levels; male and female), birth stud size (3 levels; small: <3 foals born/year, medium: from 3 to 9 foals born/year, and large: >9 foals born/year); those effects are statistically significative for the four angular hoof defects analyzed. A comparison of the different methodological approaches was based on the deviance information criterion (DIC) parameters [26].

Additionally, heritabilities and genetic correlations between the hoof defects and two morphological variables, the diameter of the superficial digital flexor tendon (SDFT), and the proportionality index (PI) were analyzed. The SDFT extends from the elbow region to the distal interphalangeal joint in the forelimb or from the hock region in the hind limb. The diameter of the SDFT was measured on a linear scale from 1 to 9, depending on the thickness of the tendon of the four limbs as a whole, where 1 is very thin and 9 is highly developed (larger diameter); the evaluation is unique, grouping, in a single evaluation, both forelimb and rear limb diameter. This tendon is a fundamental part of the connective system that allows for the function of hoof flexion in the horse's limbs. The PI was defined as the height at the withers * 100 divided by the scapular-ischial length (distance of the straight segment from the union of the scapular-humerus joint to the point of the buttock). The PI is a variable widely used in equine breed characterization [27], and also because of its possible relationship with conformational defects [28].

2.2. Statistics and Genetic Analysis

A multivariate Generalized Non-linear Model (GLZ), with a multinomial logit distribution, was used to examine associations between the conformational hoof defects and their potential factors. All the statistical analyses were conducted using Statistica 11 for Windows software [29].

For the genetic evaluations, the genealogical information was obtained from the PRE StudBook managed by ANCCE [30], ensuring a minimum of 5 known generations for each animal in the performance control dataset. The pedigree file comprised total of 101,574 individuals. A multivariate model with the six variables (SFF, PTF, SFR, PTR, SDFT, and PI) was applied to study the genetic parameters of the conformational hoof defects. The equation in matrix notation used to solve the mixed model was:

$$\mathbf{y} = \mathbf{Xb} + \mathbf{Zu} + \mathbf{e}$$

$$\mathbf{u} \sim \mathrm{N}(0, \mathbf{A}\sigma_u^2), \mathbf{e} \sim \mathrm{N}\left(0,\ \mathbf{I}\sigma_e^2\right)$$

where **y** is the vector of observations, **X** is the incidence matrix of fixed effects, **Z** is the incidence matrix of the animal genetic effect, **b** is the vector of fixed effects, **u** is the vector of the direct animal genetic effect, **e** is the vector of residuals, σ_u^2 is the direct additive genetic variance, σ_e^2 is the residual variance, **I** is an identity matrix, and **A** is the numerator relationship matrix.

The genetic parameters were estimated using the Blupf90 software [31]. A Bayesian approach with threshold models was applied using GIBBSF90+. A total of 100,000 iterations were performed, with the initial 20,000 being considered as burn-in, after which, every 10th sample was saved. Therefore, 8000 chains were available for a post Gibbs analysis. Prior values were settled as similar to those reported for similar traits [3,32]. Finally, a post-Gibbs analysis was performed using POSTGIBBSf90 [33] to compute posterior means, additive and residual variances, and the standard deviations. Convergence was tested using the Z criterion of Geweke [34], and Monte Carlo sampling error was computed using time series procedures, as described in [35]. The Monte Carlo standard errors were small, ranging between 0.0004 (PTR) and 0.0010 (SFR) for model A, and ranging between 0.0004 (PTR) and 0.0020 (SFR) for model B. A lack of convergence was not detected by the Geweke test, ranging between −1.5130 (PTF) and 0.1670 (PTR) for model A, and ranging between −1.3929 (SFR) and 0.6989 (SFF) for model B. In order to identify the precision of the parameters, the 95% highest posterior density (HPD) intervals were determined from their marginal posterior distributions.

3. Results

The basic statistical results for angular hoof deviations in the PRE horses are shown in Table 1. The largest number of observations was for the SFR defect (50,717 records), with a mean value of 1.05, and the smallest number of observations was for the PTR defect (10,797 records), with a mean value of 0.05. Regarding the mode, all the defects had a mode of 1, except the SFR defect, with a mode of 0. Additionally, the highest 95% confidence interval was observed for PTF, and the smallest for PTR.

Table 1. Basic statistics of angular hoof defects in the Pura Raza Española horse.

Hoof Angular Deviations	N	Mean [1] (s.e.)	s.d.	Mode	Confidence [2] −95%	Confidence [2] +95%
SFF	41,614	0.326 (0.003)	0.583	0	0.320	0.332
PTF	40,061	0.467 (0.004)	0.842	0	0.460	0.476
SFR	50,717	1.051 (0.003)	0.677	1	1.046	1.058
PTR	10,797	0.045 (0.002)	0.240	0	0.041	0.050

[1] Mean: mean value of angular deviation score based on 0–2 ratings. [2] Confidence intervals of the mean. N: number of records, s.e.: standard error, s.d.: standard deviation, SFF: splay-footed forelimb, PTF: pigeon-toed forelimb, SFR: splay-footed rear limb, and PTR: pigeon-toed rear limb.

The percentage of affected animals (category 1 or 2) for each hoof defect can be observed in Figure 2. For the SFF defect, category 0 had a prevalence of 73.39%, while categories 1 and 2 had a prevalence of 26.61% (20.61% and 5.99%, respectively, for categories 1 and 2). In the case of the PTF defect, category 0 had a prevalence of 76.24%, while categories 1 and 2 had a prevalence of 23.76% (0.73% and 23.03%, respectively, for categories 1 and 2). On the other hand, for the SFR defect, category 0 had a prevalence of 20.47% compared to a total of 79.53% of animals presenting this defect (53.89% and 25.64%, respectively, for categories 1 and 2). Category 0 for the PTR defect showed a prevalence of 96.14%, with only 3.86% of the animals presenting this defect (3.16% and 0.70%, respectively, for category 1 and 2). Finally, the percentages of PRE horses simultaneously exhibiting category 2 for the SFF-SFR, SFF-PTR, PTF-SFR, and PTF-PTR defects were 2.92%, 0.02%, 5.49%, and 0.05%, respectively.

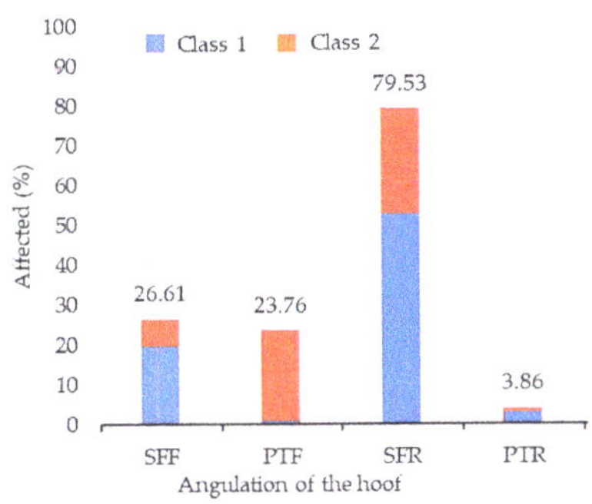

Hoof angular deviations	Prevalence C0	C1	C2
SFF	73.39	20.61	5.99
PTF	76.24	0.73	23.03
SFR	20.47	53.89	25.64
PTR	96.14	3.16	0.70

Figure 2. Percentage of affected animals for each of the angular hoof deviations in Pura Raza Española horse. SFF: splay-footed forelimb, PTF: pigeon-toed forelimb, SFR: splay-footed rear limb, and PTR: pigeon-toed rear limb.

In the analysis of angular hoof deviations in PRE horses, multiple effects were evaluated, including age, inbreeding, sex, and birth stud size (Table 2). A Generalized Non-Linear Model (GLZ) test revealed that, in the case of the PTF defect, all the effects studied were statistically significant. Regarding sex, significant differences were observed, with 20.34% of males and 25.30% of females being affected. Furthermore, birth stud size displayed statistically significant differences, with values of 23.90% (small), 23.46% (medium), and

23.99% (large) of animals being affected. In the case of the SFF and SFR defects, statistically significance differences were noted for age, but not for inbreeding. For SFF, statistically significant differences were found for sex, with 30.77% of males and 24.4% of females being affected. For SFR, significant differences were also found for birth stud size, with 78.44% of horses being affected in small studs, 79.92% in medium-sized studs, and 80.40% in large studs. Furthermore, for the PTR defect, no statistically significant differences were found for age or inbreeding, nor in sex. However, significant differences were observed for birth stud size, with 3.82% of animals being affected in small studs, 4.55% in medium studs, and 3.01% in large studs. These findings underscore the importance of considering age, inbreeding, sex, and birth stud size in genetic models when addressing angular hoof deviations and associated injuries in PRE horses.

Table 2. Generalized Non-linear Model (GLZ) between the angular hoof deviations with the risk factors for all PRE horses analyzed.

Hoof Angular Deviations	Age (*p*-Value)	Inbreeding Coefficient (*p*-Value)	Sex			Birth Stud Size			
			Male (%)	Female (%)	*p*-Value	Small (%)	Medium (%)	Large (%)	*p*-Value
SFF	0.050	0.477	30.77	24.40	<0.001	26.90	26.01	27.02	0.259
PTF	<0.001	0.019	20.34	25.30	<0.001	23.90	23.46	23.99	0.006
SFR	<0.001	0.568	79.64	79.40	<0.001	78.44	79.92	80.40	<0.001
PTR	0.829	0.652	4.21	3.69	0.301	3.82	4.55	3.01	0.007

SFF: splay-footed forelimb; PTF: pigeon-toed forelimb; SFR: splay-footed rear limb; PTR: pigeon-toed rear limb; small: <3 foals born/year; medium: 3 to 9 foals born/year; large: >9 foals born/year, and (%): percentage of animals affected.

The evolution of the four defects studied in the PRE horses over the last 20 years based on the horses' birth year is shown in Figure 3a. In the first 5 years (from 2000 to 2005), irregularities were observed in the percentages affected for all four defects, with fluctuations during that period. From 2006, the percentages of animals affected stabilized. Notably, the SFR defect appeared to be the most prevalent, with an incidence ranging from 83% to 77% over time. Conversely, the PTR defect appeared to be the least prevalent, with an incidence fluctuating between 3% and 6% across the time series.

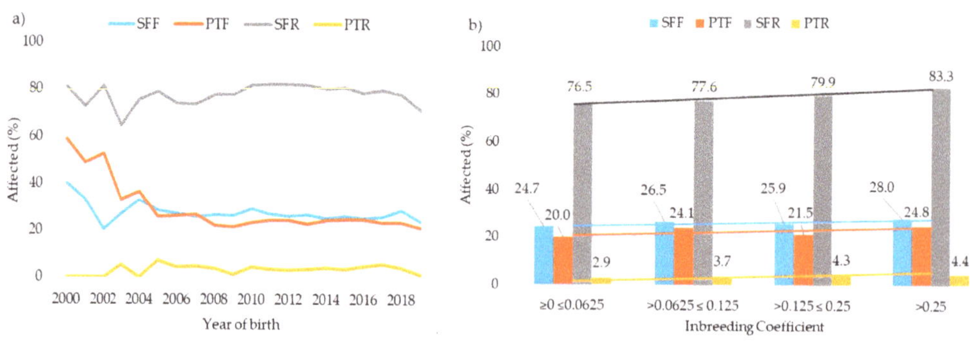

Figure 3. Evolution of the last 20 years of hoof defects in Pura Raza Española horses based on the year of birth (**a**) and incidence based on the level of consanguinity (**b**). SFF: splay-footed forelimb, PTF: pigeon-toed forelimb, SFR: splay-footed rear limb, and PTR: pigeon-toed rear limb.

Figure 3b illustrates the incidence of the four hoof defects based on the inbreeding coefficient of the PRE horses. A clear upward trend in the number of individuals affected was observed as consanguinity rose. The SFR defect saw the most significant increase, going from 76.5% (in the range from $\geq$0 to $\leq$0.0625) to 83.3% (in the range of >0.25). On the other hand, the PTR defect exhibited a greater stability, ranging from 2.9% (in the range from ≥0 to ≤0.0625) to 4.4% (in the range of >0.25).

The variance components and heritability values for the four hoof defects studied in PRE horses are shown in Table 3. In model A, for the defects in the front hoof, heritability values of 0.18 (s.d. = 0.009) (SFF) and 0.20 (s.d. = 0.010) (PTF) were obtained, compared to values of 0.11 (s.d. = 0.009) (SFR) and 0.31 (s.d. = 0.010) (PTR) for back hoof defects. In model B, for the defects in the front hoof, heritability values of 0.17 (s.d. = 0.008) (SFF) and 0.19 (s.d. = 0.009) (PTF) were obtained, compared to values of 0.11 (s.d. = 0.009) (SFR) and 0.29 (s.d. = 0.009) (PTR) for back hoof defects. According to the DIC, the best fitting model was B. Although the genetic parameters obtained with model A and B were very similar, note that the percentage of animals that coincided between both models, according to the genetic value, in the 80th percentile (animals with higher genetic values for each of the defects), coincided at 96.67% for the PTR defect, at 86.65% for the PTF defect, at 84.98% for the SFF defect, and at only 20.50% for SFR defect.

The genetic correlations among the four hoof defects are presented in Table 4. In model A, the genetic correlation between PTF-SFR was negative, with a value of −0.31 (s.d. = 0.046), indicating an inverse association between these two defects. The other genetic correlations were positive, with a low to moderate magnitude, ranging between 0.09 (s.d. = 0.033) for SFF-PTR and 0.27 (s.d. = 0.031) for PTF-PTR. This genetic correlations were different from zero, as indicated by the symmetric HPD at 95% that did not include zero and the high probability to be greater or lesser than zero ($P_{\neq 0} \geq 0.95$). On the other hand, for model B, the genetic correlations between SFF-PTR, SFF-SFR, and PTF-PTR were positive, with values of 0.14 (s.d. = 0.031), 0.35 (s.d. = 0.054) and 0.30 (s.d. = 0.030) for the three genetics correlations with a $P_{\neq 0} = 1$. The genetic correlation between PTF-SFR was small, with a value of −0.05 (s.d. = 0.055), and not different from zero, as indicated by the symmetric HPD at 95% that included zero (ranged from −0.053 to 0.159) and the low probability of being different from zero ($P_{\neq 0} = 0.17$).

Table 5 represents the means and heritability values for the diameter of the superficial digital flexor tendon (SDFT) and proportionality index (PI) in the dataset analyzed, as well as the genetic correlations between them and the four hoof defects, for both models, in the PRE horses. The means were 4.87 (s.d. = 1.083) for the SDFT and 99.80 (s.d. = 2.379) for the PI. For model A, the heritability of SDFT was 0.076 (s.d. = 0.014), and for PI, it was 0.348 (s.d.: 0.011), while for the model B, the heritability of SDFT was 0.077 (s.d. = 0.014), and for PI, it was 0.336 (s.d. = 0.010). Regarding the diameter of the superficial digital flexor tendon (SDFT), for model A, the PTF and PTR defect showed moderate, positive genetic correlations of 0.283 (s.d. = 0.060) and 0.276 (s.d. = 0.046), respectively. For model B, the PTF and PTR showed moderate, positive genetic correlations of 0.272 (s.d. = 0.051) and 0.266 (s.d. = 0.049), respectively. These genetic correlations were different from zero, as indicated by the symmetric HPD at 95% that did not include zero and the high probability to be greater than zero ($P_{\neq 0} = 1.00$) In model B, for the SFF and SFR defects, there was no evidence of a relationship between these factors ($P_{\neq 0} = 0.77$ and 0.03, respectively).

Regarding the proportionality index (PI), opposite correlations were observed compared to the SDFT. For model A, the defects SFF and SFR displayed low to moderate positive correlations of 0.212 (s.d. = 0.030) and 0.124 (s.d. = 0.041), respectively. For model B, the defect SFF displayed low to moderate positive correlation of 0.222 (s.d. = 0.031). In both models, for the PTF defect, there was no evidence of a relationship between these factors, not different from zero, as indicated by the symmetric HPD at 95% that included zero and the low probability to be lesser than zero ($P_{\neq 0} = 0.87$ in model A and $P_{\neq 0} = 0.54$ in model B). In model B, for the SFR defects, there was no evidence of a relationship between these factors, not different from zero, as indicated by its symmetric HPD at 95% that included zero and the probability to be lesser than zero ($P_{\neq 0} = 0.94$).

Table 3. Genetic parameters of the angular hoof deviations studied in Pura Raza Española horses.

Model	Hoof Angular Deviations	σ_u			σ_e			$b_{inbreeding}$ (s.d.)	b_{age} (s.d.)	h^2 (s.d.)	DIC
		Mean	Median	HPD 95%	Mean	Median	HPD 95%				
A	SFF	246.57	246.80	218.40–269.90	1140.70	1141.00	1118.00–1166.00	9.90 (4.418)	−0.57 (0.079)	0.18 (0.009)	−360,278,759.93
	PTF	325.95	325.40	293.60–359.20	1274.40	1274.00	1246.00–1304.00	−3.72 (4.680)	0.45 (0.083)	0.20 (0.010)	
	SFR	12.02	11.98	10.22–13.98	94.72	94.72	92.88–96.67	−1.36 (1.213)	−0.07 (0.021)	0.11 (0.009)	
	PTR	444.55	444.70	410.90–475.00	1000.20	619.10	975.10–1026.00	7.75 (4.647)	0.40 (0.076)	0.31 (0.010)	
B	SFF	6.93	6.92	6.27–7.73	34.81	34.80	34.03–35.62	1.80 (0.754)	−0.09 (0.014)	0.17 (0.008)	−781,593,187.00
	PTF	8.63	8.64	7.82–9.52	36.32	36.22	35.64–37.20	−0.28 (0.838)	0.06 (0.014)	0.19 (0.009)	
	SFR	1.05	1.05	0.80–1.29	8.49	8.48	8.26–8.70	−0.23 (0.392)	0.01 (0.006)	0.11 (0.009)	
	PTR	14.22	14.21	13.19–15.34	33.89	33.89	33.23–34.62	1.45 (0.850)	0.07 (0.018)	0.29 (0.009)	

SFF: splay-footed forelimb; PTF: pigeon-toed forelimb; SFR: splay-footed rear limb; PTR: pigeon-toed rear limb; σ_u: additive genetic variances; σ_e: residual variances; HPD 95%: 95% higher posterior density; h^2: heritabilities; s.d.: standard deviation; b: slope, and s.d.: standard deviation.

Table 4. Genetic correlations (r_g) among the angular hoof deviations studied in Pura Raza Española horses.

Model		r_g (s.d.) PTR	HPD 95%	$P_{\neq 0}$	r_g (s.d.) SFR	HPD 95%	$P_{\neq 0}$
A	SFF	0.09 (0.033)	0.025–0.154	1.00	0.09 (0.045)	0.003–0.180	0.97
	PTF	0.27 (0.031)	0.207–0.331	1.00	−0.31 (0.046)	(−0.389)–(−0.208)	1.00
B	SFF	0.14 (0.031)	0.086–0.200	1.00	0.35 (0.054)	0.228–0.444	1.00
	PTF	0.30 (0.030)	0.251–0.364	1.00	−0.05 (0.055)	−0.053–0.159	0.17

SFF: splay-footed forelimb, PTF: pigeon-toed forelimb, SFR: splay-footed rear limb, PTR: pigeon-toed rear limb, r_g: genetic correlations, s.d.: standard deviation, and HPD 95%: 95% higher posterior density. $P_{\neq 0}$ ($r_g > 0$): probability of genetic correlation being higher than zero. $P_{\neq 0}$ ($r_g < 0$): probability of genetic correlation being lower than zero.

Table 5. Mean and heritability (h^2) of the superficial digital flexor tendon development (SDFT) and proportionality index (IP) variables and their genetic correlations (r_g) with angular hoof deviations in Pura Raza Española horses.

		SDFT			PI		
Mean (s.d.)		4.87 (1.083)			99.80 (2.379)		
h^2 (s.d.)	Model A	0.076 (0.014)			0.348 (0.011)		
	Model B	0.077 (0.014)			0.336 (0.010)		
		r_g (s.d.)	HPD 95%	$P_{\neq 0}$	r_g (s.d.)	HPD 95%	$P_{\neq 0}$
Model A	SFF	−0.081 (0.054)	(−0.188)–0.028	0.93	0.212 (0.030)	0.151–0.270	1.00
	PTF	0.283 (0.060)	0.171–0.400	1.00	−0.034 (0.030)	(−0.093)–0.024	0.87
	SFR	−0.113 (0.064)	(−0.234)–0.003	0.96	0.124 (0.041)	0.039–0.201	1.00
	PTR	0.276 (0.046)	0.188–0.367	1.00	−0.088 (0.026)	(−0.142)–(−0.039)	1.00
Model B	SFF	−0.040 (0.052)	(−0.145)–0.057	0.77	0.222 (0.031)	0.165–0.293	1.00
	PTF	0.272 (0.051)	0.167–0.364	1.00	0.037 (0.031)	(−0.046)–0.069	0.54
	SFR	−0.118 (0.068)	(−0.014)–0.243	0.03	0.071 (0.045)	(−0.012)–0.160	0.94
	PTR	0.266 (0.049)	0.171–0.357	1.00	−0.088 (0.025)	(−0.138)–(−0.035)	1.00

SDFT: superficial digital flexor tendon; PI: proportionality index; SFF: splay-footed forelimb; PTF: pigeon-toed forelimb; SFR: splay-footed rear limb; PTR: pigeon-toed rear limb; s.d.: standard deviation; and HPD 95%: 95% higher posterior density. Highest posterior density interval at 95% in brackets. $P_{\neq 0}$ ($r_g > 0$): probability of genetic correlation being higher than zero. $P_{\neq 0}$ ($r_g < 0$): probability of genetic correlation being lower than zero.

4. Discussion

Addressing horse limb defects, especially hoof angular deviations, is essential to ensuring a horse's health, genetic quality, and performance, particularly in the context of sports competitions [36] and breeding programs. The uneven conformation of horses' hooves is frequently studied and is of significant clinical importance for most equine breeds [37]. Angular deviation in a horse's hoof leads to asymmetric limb loading, with subsequent implications for performance, injuries, and lameness [38]. Furthermore, Ducro et al. [19] demonstrated a link between uneven hoof conformation and a shorter competitive lifespan.

In the context of the basic statistical results, we observed significant variability in angular hoof deviations among the different defects analyzed (SFF, PTF, SFR, and PTR) in the PRE horses. The SFR defect exhibited the highest mean value and a wider confidence interval, indicating greater data dispersion compared to the PTR defect, which had the lowest mean value and the smallest confidence interval (Table 1). These intervals provide a measure of confidence in the variability of the mean estimates. Furthermore, this finding suggests the SFR defect was more prevalent compared to the PTR defect, as corroborated in Figure 2. The standard deviations turned out high, which is an indicator that further studies are needed to understand the mechanisms affecting the occurrence of these defects.

In the PRE horse, the outward deviation hoof defect can vary in severity and may appear in the forelimbs and/or rear limbs, although it is more common in the rear limbs (SFR) [39]. Horses, including racehorses, often exhibit some degree of inward-turning hocks [7], and as confirmed in the article by Ripolles et al. [3], in the case of PRE horses. This tendency can lead to the hoof turning slightly outward, contributing to a higher prevalence of SFR defect. The prevalence of this anomaly has been documented in various studies [40–42], underscoring this defect as one of the primary cases of angular deviation in the hindquarters of the equine breed. At times, the forelimbs may have good alignment, but are spoiled by a defect known as knock-knee. It can also occur that, while the metacarpal joint, carpal joint, and metatarsal bone are correctly aligned, the horse has SFF due to the outward twist of the fetlock, which usually causes the hoof to point in the same direction [43]. On the other hand, the inward deviation hoof defect can vary in severity, and can also appear in both pairs of limbs, although it is more common in the forelimbs (PTF). A very wide chest, conversely, results in widely separated front knee joints, leading to the hoof pointing inward [39]. In other cases, the defect exists when the horse has dropped heels, very long fetlocks, is bench-kneed, or has divergent hocks [43].

In each of the four hoof deviation defects, categories 1 and 2 represented the individuals affected, with significantly different percentages being observed among the defects (Figure 2). This indicates variations in the prevalence of the defects within the studied population. The prevalence of having any hoof deviation defect varies among breeds. In the PRE horse, it was 84.24%, which is similar to the prevalence in the Dutch Warmblood horse (85%) [44]. Moreover, this prevalence is lower in mules and Dutch Warmblood horses, with incidences of 37.1% [45] and 8% [11], respectively. It is important to focus on reducing the incidence because, as seen previously, horses that simultaneously exhibit PTR-PTF defects in category 2 (considering both defects are very severe for the PRE breed) are the animals most likely to be unsuitable as breeders for the breed, as they only require one more very severe defect, either in the limbs or elsewhere in the body, to be disqualified.

To achieve an accurate, comprehensive understanding of the patterns in hoof deviation defects, we need to consider that both environmental and genetic factors can influence the occurrence and severity of conformational defects. Although most widely published works have addressed the clinical perspective, no detailed genetic analyses for hoof deviation defects in PRE horses have been extensively conducted until now. In our study, we examined not only the influence of sex and age, but also the individual coefficient of inbreeding and the breeding stud size (Table 2). Overall, all these effects were statistically significant for most of the hoof deviation defects analyzed. The effects mentioned in this study have previously been implemented in estimations of the genetic parameters for morphological abnormalities in the PRE breed, such as cresty neck [46], ewe neck [28], or limb deformities [3], as well as those associated with medical conditions, like obesity [47] or the presence of cutaneous melanomas [48]. Additionally, they have also been included in the assessment of various PRE variables, including morphological [49] variables.

Hoof variations depend not only on the constant movements made during work [50] and the type of terrain where the horse treads, but also on the horse's age. The age of evaluation plays a crucial role in the development and conformation of a horse's limbs. Skeletal growth, hoof development, and the proper ossification of limb bones are critical factors to consider when evaluating hoof conformation. Newborn foals typically have a weaker conformation, narrow chests, and long legs relative to their size. This can lead to an outward hoof posture as they need to provide support to their forelimbs [51]. Over time, this may result in a correction of the inward hoof posture [52]. Events that affect a foal's intrauterine environment, such as placentitis during pregnancy or severe metabolic diseases in the mare, can compromise the proper ossification of these bones, but this only becomes a problem when hoof conformation defects persist with age [51].

Regarding sex, males exhibited a higher incidence of hoof deviation defects compared to females, which aligns with the findings in the lameness study conducted by Ross [53]. This sex disparity may be linked to factors such as differences in the intensity of the use of the horse. In sports competitions, there is often a higher participation of males, and pregnant mares generally undergo less intense exercise.

An association between the size of the breeding stud and the percentage of affected horses was also observed. Large-scale birth studs primarily focus on breeding, while smaller and medium-sized ones are mainly dedicated to competition. This distinction is crucial, because the type of livestock management significantly influences aspects such as the type of exercise horses undergo, the management regimen, the intensity of training, and other related factors. In this context, it has been documented that foals raised in large herds with extensive management may develop limb conformation irregularities due to uneven load patterns during grazing [54], as horses tend to systematically extend the same limb while grazing [55]. While exercise is obviously crucial for development, excessive exercise can lead to the development of angular deviations [56]. Another relevant consi-deration is nutritional imbalance, which can influence disproportionate growth in foals.

Regarding the evolution of defects related to consanguinity levels (Figure 3b), it is evident that there is a trend toward an increase in the percentage of animals affected as inbreeding values rise. These results support the theory that inbreeding plays a significant role in the occurrence and transmission of these genetic defects in the horse population. Increases in consanguinity are typically associated with a decrease in animal fitness and, consequently, a reduction in their performance in terms of capacity or physiological efficiency [57]. This finding underscores the importance of carefully considering genetic diversity and relatedness in horse breeding programs.

Conformation defects have been evaluated in numerous equine breeds using a linear scale that usually ranges from 5 classes [32,58] or 9–10 classes [12,59–61], where opposing defects are placed in the classes located at the extremes of the linear scale (for example, in a 5-category scale, where category 1 would be the SFF defect at its maximum grade, category 2 the SFF defect at its low grade, category 3 the animal does not present any defect, category 4 would be the PTF defect at its low grade, and category 5 would be the PTF defect at its maximum grade). However, recently, Ripollés et al. [3] showed that the best methodology for the genetic evaluation of conformation defects is to use an independent scale for each of the defects, where all the scale categories refer to assessing the grades of the defect presentation considered. Probably some of the genes involved in the presentation of a particular defect, even if they are related to the same anatomical area, may be different. Another issue when using a very large scale to evaluate a defect, especially when it is complex to assess through visual inspection, such as angular hoof defects, is that evaluators may not be able to utilize the entire scale and find it challenging to determine the differences between two consecutive categories. It is advisable to reduce the number of categories for phenotypically evaluating this type of defect. The heritabilities found in other studies have been higher than those obtained in the four defects studied in PRE horses (Table 3), with values of 0.45 in Icelandic horses [58], 0.40 in Brazilian sport horses [60], 0.12–0.27 found in KWPN breeds [11], and 0.26 (0.109) in the Pura Raza Menorquina [32]. The heritability of other hoof deformities was estimated to be 0.13 (0.022) for front hoof and 0.007 (0.018) for rear hoof in Warmblood foals [12] and 0.25 in Belgian horses [62]. In this study, despite the high residual variances, the heritabilities obtained suggest that selective breeding can be used to reduce the prevalence in the breed, even though the genetic component is smaller than the environmental one, with a moderate magnitude. It is also worth noting that the heritabilities from the two models (A and B) were very similar, so model B may also be effective when morphologically assessing angular hoof defects in horses. Although model B is the best fit, according to the DIC, it is important to work with three categories (model A), since, as has been seen, when we have a significant number of animals in category 1 and category 2 for a defect (example of the SFR defect; Figure 2), the percentage of animals that match according to the genetic value, at the 80th percentile, is very low (only 20.50% for the SFR defect). Therefore, model A allows us to diffe-rentiate the animals that transmit any of the defects studied to a greater degree. In this sense, it is very interesting for the breeder to be able to genetically identify the animals that are going to transmit some of the defects to a greater degree.

Regarding the genetic correlations between different hoof deviation defects, our results support that different sets of genes are involved in coordinating hoof deviation defects, depending on whether they occur in an outward or inward position (Table 4). In a study conducted in the Dutch Warmblood horse population, moderate correlations were observed between hoof deviation defects and traits of the distal limb [11]. Specifically, the heel height showed negative correlations with the cannon angle (−0.42) and hoof shape (−0.41), indicating that high heels are associated with narrow hooves and more vertical cannons.

The equine limb behaves like a large system of springs due to the extension of the metacarpophalangeal joint and the elastic recoil in tendons [63,64]; specifically, the SDFT is specialized in storing elastic energy [65]. It has been observed that the conformation of a horse's hoof may be related to predisposition for certain tendon defects [66,67]. Angular hoof defects can influence how a horse distributes its weight when bearing weight, affec-ting the load on associated tendons and ligaments, thus increasing its susceptibility to injuries [36,68,69]. Injuries to the SDFT are common in racehorses, especially in the forelimbs, with a prevalence ranging from 10% to 30% [66,70]. Additionally, it has been observed that the prevalence of these injuries tends to increase with a horse's age [71–73]. In our study, positive correlations were observed between the SDFT and angular hoof deviations (PTF and PTR; Table 5). These correlations may be related to the biomechanics of horse limbs. Excessive development of the SDFT could affect how a horse places its limbs during movements, which could influence their weight bearing and, in turn, contribute to certain inward hoof deviation defects. Additionally, Shade et al. [67] determined a correlation of −0.26 between static and dynamic angles of the metacarpophalangeal joint and the SDFT in Campeiro and Mangalarga Marchador horses.

On the other hand, the PI in horses seems to be a determining factor that influences movement and, consequently, plays an essential role in maintaining proper balance and hoof conformation [51]. This perspective underscores the importance of carefully considering the relationship between PI and hoof angular deviations. Furthermore, the need for hoof volume to be correctly adjusted to the horse's size and robustness, as suggested by Giles [74], fully justifies the inclusion of PI as a relevant study variable in this research. In contrast to SDFT, positive genetic correlations were observed between PI and outward angular deviations (SFF and SFR; Table 5), which suggests that horses with outward hoof defects tend to have a higher PI. Robust, broad-chested horses typically used for heavy pulling tasks have a lower withers height and often exhibit an inward hoof tread pattern, which is related to the occurrence of defects like PTF and PTR [1]. These types of horses tend to have larger hooves compared to saddle horses [51]. In contrast, horses destined for sports activities, such as dressage, as is the case with PREs, often exhibit a greater withers height and narrower chest conformation, resulting in a higher PI. These horses tend to bear weight in a way that directs the hooves outwards, which is associated with a more frequent occurrence of defects like SFF and SFR in breeds of this type. Furthermore, according to [75], there is a genetic correlation between height at withers and a hoof conformation of 0.15, which is directly related to PI.

It is possible that there may be limitations in this work, as it was not possible to stan-dardize the age of all the animals since the animals were not evaluated at the same age. Additionally, the presence of animals assessed both with and without horseshoes, as well as their participation in competitions, could have introduced biases in the assessment of angular defects, as footwear and exercise can impact the correction of these issues. Another limitation of phenotyping in the study is the assessment by a single evaluator, rather than using consensus from multiple evaluators for each horse, and variability between horses since there were different single evaluators across the large dataset. Despite this, this work had a significant representation of active PRE horses and it demonstrated that the angular hoof deviations affect limb biomechanics, influencing the horse's ability to perform, which may also have genetic implications, justifying the implementation of suitable selection measures to reduce their incidence and improve both the breed and equine sports performance. All these findings emphasize the importance of breeding programs that not only focus on the quality of movements and functionality in dressage, but also place a fundamental emphasis on conformational defects of the horses' limbs.

5. Conclusions

A high prevalence of angular hoof defects was found in the PRE population, with a higher occurrence in males and significant variations with an increasing inbreeding coe-fficient and breeder herd size. It was demonstrated that genetic factors play a role in the occurrence of these defects, with moderate heritability values. Furthermore, we observed genetic correlations between the defects analyzed, pointing to the existence of specific genes related to the presence of particular defects. Our study also underscored the signi-ficance of taking into account factors such as the proportionality index (PI) and development of the superficial digital flexor tendon (SDFT). It revealed correlations between these factors and the direction of angular hoof defects, with the SDFT being associated with inward defects and the PI with outward defects. Breeders and veterinarians should take into account, when organizing the mating of the stud, the genetic values of the breeding animals for each of the angular hoof deviation defects (SFF, PTF, SFR, and PTR), along with the genetic values of the rest of the morphological defects of the breed, both in the limbs and other areas of the body. From a reproductive perspective, breeders should prioritize horses whose genetic values, for each of the conformation defects analyzed in the breed (regardless of the anatomical region), have a high probability of not being transmitted to the offspring. Well-organized horse pairings based on the genetic value of the offspring will reduce defects in the population.

Author Contributions: Conceptualization: M.R.-L. and M.V.; methodology: M.V. and D.I.P.-G.; validation, M.R.-L. and D.I.P.-G.; formal analysis: M.R.-L. and D.I.P.-G.; research: all the authors; resources: P.J.A. and M.V.; data curation: M.R.-L.; writing—original draft preparation: M.R.-L. and D.I.P.-G.; writing—review and editing: D.I.P.-G. and M.V.; visualization: M.V.; supervision: M.V.; project administration: M.V. and P.J.A.; funding acquisition: M.V. and P.J.A. All authors have read and agreed to the published version of the manuscript.

Funding: This research received no external funding.

Institutional Review Board Statement: Not applicable.

Informed Consent Statement: Not applicable.

Data Availability Statement: The data used in this article is part of the performance control data carried out as part of the Pura Raza Español Breeding Program (BOE nº 175 of June 24, 2020. Sec. III Pág. 44191, https://www.mapa.gob.es/es/ganaderia/temas/zootecnia/ancceen_tcm30-540900.pdf, accessed on 1 June 2023), whose technical director of the Breeding Program is Dr. M. Valera (author of this work and director of Research Group PAIDI-AGR-273 the University of Seville, a qualified center for genetics https://servicio.mapa.gob.es/arca/flujos.html?_flowExecutionKey=c21e6192-8dd3-470b-8216-e0cc38ed413f_MTAzQjRDMDYxODMwNjgwQzIzNEI0RDc5NDBBMjAzODIuZTA0MTAwNjA=&_eventId=desplogarResponsa-ble&indiceResponsable=0, accessed on 1 June 2023).

Conflicts of Interest: The authors declare no conflict of interest.

References

1. Leśniak, K.; Whittington, L.; Mapletoft, S.; Mitchell, J.; Hancox, K.; Draper, S.; Williams, J. The Influence of Body Mass and Height on Equine Hoof Conformation and Symmetry. *J. Equine Vet. Sci.* **2019**, *77*, 43–49. [CrossRef] [PubMed]
2. Floyd, A.; Mansmann, R. Anatomy and Physiology of the Equine Foot. In *Equine Podiatry*; Elsevier: Maryland Heights, MO, USA, 2007; pp. 1–102.
3. Ripollés-Lobo, M.; Perdomo-González, D.I.; Azor, P.J.; Valera, M. Evaluation of Potential Effects and Genetic Parameters in Conformational Limb Defects in Pura Raza Española Horses. *Ital. J. Anim. Sci.* **2023**, *22*, 407–417. [CrossRef]
4. Collins, S.N.; Pollitt, C.; Wylie, C.E.; Matiasek, K. Laminitic Pain: Parallels with Pain States in Humans and Other Species. *Vet. Clin. N. Am.-Equine Pract.* **2010**, *26*, 643–671. [CrossRef]
5. Ireland, J.L.; Wylie, C.E.; Collins, S.N.; Verheyen, K.L.P.; Newton, J.R. Preventive Health Care and Owner-Reported Disease Prevalence of Horses and Ponies in Great Britain. *Res. Vet. Sci.* **2013**, *95*, 418–424. [CrossRef] [PubMed]

6. Wilson, A.; Agass, R.; Vaux, S.; Sherlock, E.; Day, P.; Pfau, T.; Weller, R. Foot Placement of the Equine Forelimb: Relationship between Foot Conformation, Foot Placement and Movement Asymmetry. *Equine Vet. J.* **2016**, *48*, 90–96. [CrossRef] [PubMed]
7. O'Donohue, D.D.; Smith, F.H.; Strickland, K.L. The Incidence of Abnormal Limb Development in the Irish Thoroughbred from Birth to 18 Months. *Equine Vet. J.* **1992**, *24*, 305–309. [CrossRef] [PubMed]
8. Kane, A.J.; Stover, S.M.; Gardner, I.A.; Bock, K.; Case, J.; Johnson, B.; Anderson, M.; Barr, B.; Daft, B.; Kinde, H.; et al. Hoof Size, Shape, and Balance as Possible Risk Factors for Catastrophic Musculoskeletal Injury of Thoroughbred Racehorses. *Am. J. Vet. Res.* **1998**, *59*, 1545–1552.
9. Love, S.; Wyse, C.A.; Stirk, A.J.; Stear, M.J.; Calver, P.; Voute, L.C.; Mellor, D.J. Prevalence, Heritability and Significance of Musculoskeletal Conformational Traits in Thoroughbred Yearlings. *Equine Vet. J.* **2006**, *38*, 597–603. [CrossRef]
10. Oki, H.; Miyake, T.; Kasashima, Y.; Sasaki, Y. Estimation of Heritability for Superficial Digital Flexor Tendon Injury by Gibbs Sampling in the Thoroughbred Racehorse. *J. Anim. Breed. Genet.* **2008**, *125*, 413–416. [CrossRef]
11. Ducro, B.J.; Bovenhuis, H.; Back, W. Heritability of Foot Conformation and Its Relationship to Sports Performance in a Dutch Warmblood Horse Population. *Equine Vet. J.* **2009**, *41*, 139–143. [CrossRef]
12. Novotna, A.; Birovas, A.; Vostra-Vydrova, H.; Vesela, Z.; Vostry, L. Genetic Parameters of Performance and Conformation Traits of 3-Year-Old Warmblood Sport Horses in the Czech Republic. *Animals* **2022**, *12*, 2957. [CrossRef] [PubMed]
13. Schröder, W.; Stock, K.F.; Distl, O. Genetic Evaluation of Hanoverian Warmblood Horses for Conformation Traits Considering the Proportion of Genes of Foreign Breeds. *Arch. Anim. Breed.* **2010**, *53*, 377–387. [CrossRef]
14. Clayton, H.M. The Effect of an Acute Hoof Wall Angulation on the Stride Kinematics of Trotting Horses. *Equine Vet. J.* **1990**, *22*, 86–90. [CrossRef]
15. Kolstrung, R.; Stachurska, A.; Miroslaw, P.; Silmanowicz, P.; Ussing, A.P. Hoof Wall Angulation in the Horse (*Equus Caballus*). *Vet. Med.* **2013**, *69*, 181–186.
16. Stowers, N.L.; Erdtsieck, B.; Rogers, C.W.; Taylor, T.B.; Firth, E.C. The Prevalance of Limb Deformities in New Zealand Standardbred Foals and Their Influence on Racing Success: A Preliminary Investigation. *Proc. N. Z. Soc. Anim. Prod.* **2010**, *70*, 140–142.
17. Herbrecht, V.; Waldern, N.M.; Mikkelsen, S.E.; Kjaer, M.; Dittmann, M.T.; Wiestner, T.; Weishaupt, M.A. Hoof Conformation in Icelandic Competition Horses and Its Interrelationship with Hoof Pathologies and Tölt Performance. *Vet. J.* **2020**, *259–260*, 105462. [CrossRef]
18. Sánchez Guerrero, M.J.; Cervantes, I.; Valera, M.; Gutiérrez, J.P. Modelling Genetic Evaluation for Dressage in Pura Raza Español Horses with Focus on the Rider Effect. *J. Anim. Breed. Genet.* **2014**, *131*, 395–402. [CrossRef]
19. Ducro, B.J.; Gorissen, B.; van Eldik, P.; Back, W. Influence of Foot Conformation on Duration of Competitive Life in a Dutch Warmblood Horse Population. *Equine Vet. J.* **2009**, *41*, 144–148. [CrossRef]
20. Oosterlinck, M.; Van der Aa, R.; Van de Water, E.; Pille, F. Preliminary Evaluation of Toe-Heel and Mediolateral Hoof Balance at the Walk in Sound Horses with Toed-In Hoof Conformation. *J. Equine Vet. Sci.* **2015**, *35*, 606–610. [CrossRef]
21. Holmstrom, M.; Magnusson, L.E.; Philipsson, J. Variation in Conformation of Swedish Warmblood Horses and Conformational Characteristics of Élite Sport Horses. *Equine Vet. J.* **1990**, *22*, 186–193. [CrossRef]
22. RFHE (Real Federación Hipica Española). Doma Clásica. Available online: https://rfhe.com/doma-clasica/ (accessed on 1 July 2023).
23. MAPA. Ministry of Agriculture. Fisheries and Food. Purebred Spanish Horse Breeding Program. 2023. Available online: https://www.mapa.gob.es/es/ganaderia/temas/zootecnia/ancceen_tcm30-540900.pdf (accessed on 17 October 2023).
24. Wright, S. Coefficients of Inbreeding and Relationship. *Am. Nat.* **1922**, *56*, 330–338. [CrossRef]
25. Gutiérrez, J.P.; Goyache, F. A Note on ENDOG: A Computer Program for Analysing Pedigree Information. *J. Anim. Breed. Genet.* **2005**, *122*, 172–176. [CrossRef] [PubMed]
26. Spiegelhalter, D.J.; Best, N.G.; Carlin, B.P.; Van Der Linde, A. Bayesian Measures of Model Complexity and Fit. *J. R. Stat. Soc. Ser. B Stat. Methodol.* **2002**, *64*, 583–616. [CrossRef]
27. Gómez, M.D.; Azor, P.J.; Alonso, M.E.; Jordana, J.; Valera, M. Morphological and Genetic Characterization of Spanish Heavy Horse Breeds: Implications for Their Conservation. *Livest. Sci.* **2012**, *144*, 57–66. [CrossRef]
28. Ripolles, M.; Sánchez-Guerrero, M.J.; Perdomo-González, D.I.; Azor, P.; Valera, M. Survey of Risk Factors and Genetic Characterization of Ewe Neck in a World Population of Pura Raza Español Horses. *Animals* **2020**, *10*, 1789. [CrossRef]
29. StatSoft Inc. STATISTICA (Data Analysis Software System). 2007. Available online: www.statsoft.com (accessed on 15 July 2023).
30. LG_ANCCE StudBook Pura Raza Español (PRE) Horse Managed by ANCCE (Real National Purebred Spanish Horse Breeders' Association). Available online: https://www.lgancce.com/lgpreancce/asp-publico/arbolGenealogicoPRE/BuscarArbolPRE.aspx?lang=ES (accessed on 9 January 2023).
31. Misztal, I.; Tsuruta, S.; Lourenco, D.; Masuda, Y.; Aguilar, I.; Legarra, A.; Vitezica, Z. Manual for BLUPF90 Family of Programs. Available online: http://nce.ads.uga.edu/wiki/lib/exe/fetch.php?media=blupf90.pdf (accessed on 1 July 2023).
32. Perdomo-González, D.I.; García de Paredes, R.d.l.A.; Valera, M.; Bartolomé, E.; Gómez, M.D. Morpho-Functional Traits in Pura Raza Menorquina Horses: Genetic Parameters and Relationship with Coat Color Variables. *Animals* **2022**, *12*, 2319. [CrossRef]
33. Tsuruta, S.; Misztal, I. THRGIBBS1F90 for Estimation of Variance Components with Threshold-Linear Models. In Proceedings of the 8th World Congress on Genetics Applied to Livestock Production, Belo Horizonte, Brazil, 13–18 August 2006.

34. Sorensen, D.; Gianola, D. *Likelihood, Bayesian, and MCMC Methods in Quantitative Genetics*; Springer Science and Business Media: New York, NY, USA, 2002.
35. Geyer, C.M. Practical Markov Chain Monte Carlo. *Stat. Sci.* **1992**, *7*, 467–511. [CrossRef]
36. Murray, R.C.; Walters, J.M.; Snart, H.; Dyson, S.J.; Parkin, T.D.H. Identification of Risk Factors for Lameness in Dressage Horses. *Vet. J.* **2010**, *184*, 27–36. [CrossRef]
37. Hagen, J.; Kojah, K.; Geiger, M. Correlations between the Equine Metacarpophalangeal Joint Angulation and Toe Conformation in Statics. *Open Vet. J.* **2018**, *8*, 96–103. [CrossRef]
38. Van Weeren, P.R.; Crevier-Denoix, N. Equine Conformation: Clues to Performance and Soundness? *Equine Vet. J.* **2006**, *38*, 591–596. [CrossRef]
39. Conde, J.; Peña, F.; Fernandez de la Vega, V.; Bravo, I. Morfología. In *Manual de Juzgamiento, Concurso Morfológico del Pura Raza Española*; Asociación Nacional de Criadores de Caballos de Pura Raza Española: Sevilla, Spain, 2016; pp. 65–88.
40. Greet, T.R.C. Managing Flexural and Angular Limb Deformities: The Newmarket Perspective. *AAEP Proc.* **2000**, *46*, 130–136.
41. Santschi, E.M.; Leibsle, S.R.; Morehead, J.P.; Prichard, M.A.; Clayton, M.K.; Keuler, N.S. Carpal and Fetlock Conformation of the Juvenile Thoroughbred from Birth to Yearling Auction Age. *Equine Vet. J.* **2006**, *38*, 604–609. [CrossRef] [PubMed]
42. Bramlage, L.R.; Auer, J.A. Diagnosis, Assessment, and Treatment Strategies for Angular Limb Deformities in the Foal. *Clin. Tech. Equine Pract.* **2006**, *5*, 259–269. [CrossRef]
43. Llamas, J. El Casco. In *El Caballo Español. Recopilación de Artículos Publicados en Extremadura PRE*; Asociación Extremeña de Criadores de Caballos de Pura Raza Española, Mercado Regional de Ganados: Cáceres, Spain, 2016; pp. 48–53.
44. Holzhauer, M.; Bremer, R.; Santman-Berends, I.; Smink, O.; Janssens, I.; Back, W. Cross-Sectional Study of the Prevalence of and Risk Factors for Hoof Disorders in Horses in The Netherlands. *Prev. Vet. Med.* **2017**, *140*, 53–59. [CrossRef]
45. Loiacono, B.Z.; Aranzales, J.R.M.; de Resende Faleiros, R.; Alves, G.E. Acquired Carpal Angular Limb Deformities in Mules: Diagnosis, Incidence and Treatment. *Ciênc. Rural.* **2012**, *42*, 1855–1861. [CrossRef]
46. Sánchez, M.J.; Azor, P.J.; Molina, A.; Parkin, T.; Rivero, J.L.L.; Valera, M. Prevalence, Risk Factors and Genetic Parameters of Cresty Neck in Pura Raza Español Horses. *Equine Vet. J.* **2017**, *49*, 196–200. [CrossRef] [PubMed]
47. Sánchez-Guerrero, M.J.; Ramos, J.; Valdés, M.; Rivero, J.L.L.; Valera, M. Prevalence, Environmental Risk Factors and Heritability of Body Condition in Pura Raza Español Horses. *Livest. Sci.* **2019**, *230*, 103851. [CrossRef]
48. Sánchez-Guerrero, M.J.; Solé, M.; Azor, P.J.; Sölkner, J.; Valera, M. Genetic and Environmental Risk Factors for Vitiligo and Melanoma in Pura Raza Español Horses. *Equine Vet. J.* **2019**, *51*, 606–611. [CrossRef]
49. Poyato-Bonilla, J.; Perdomo-González, D.I.; Sánchez-Guerrero, M.J.; Varona, L.; Molina, A.; Casellas, J.; Valera, M. Genetic Inbreeding Depression Load for Morphological Traits and Defects in the Pura Raza Española Horse. *Genet. Sel. Evol.* **2020**, *52*, 62. [CrossRef]
50. Butcher, M.T.; Ashley-Ross, M.A. Fetlock Joint Kinematics Differ with Age in Thoroughbred Racehorses. *J. Biomech.* **2002**, *35*, 563–571. [CrossRef]
51. Auer, J.A. Angular Limb Deformities. In *Equine Surgery*, 4th ed.; Elsevier: Amsterdam, The Netherlands, 2012; pp. 1201–1221, ISBN 9781437708677.
52. Finno, C.J.; Spier, S.J.; Valberg, S.J. Equine Diseases Caused by Known Genetic Mutations. *Vet. J.* **2009**, *179*, 336–347. [CrossRef]
53. Ross, M.W.; Acvs, D. Observations in Horses with Lameness Abolished by Palmar Digital Analgesia. *Methods* **1998**, *44*, 230–232.
54. Deuel, N.R.; Lawrence, L.M. Laterality in the Gallop Gait of Horses. *J. Biomech.* **1987**, *20*, 645–649. [CrossRef] [PubMed]
55. Van Heel, M.C.V.; Kroekenstoel, A.M.; Van Dierendonck, M.C.; Van Weeren, P.R.; Back, W. Uneven Feet in a Foal May Develop as a Consequence of Lateral Grazing Behaviour Induced by Conformational Traits. *Equine Vet. J.* **2006**, *38*, 646–651. [CrossRef] [PubMed]
56. García-López, J.M. Angular Limb Deformities: Growth Augmentation. *Vet. Clin. N. Am.-Equine Pract.* **2017**, *33*, 343–351. [CrossRef]
57. Leroy, G. Inbreeding Depression in Livestock Species: Review and Meta-Analysis. *Anim. Genet.* **2014**, *45*, 618–628. [CrossRef] [PubMed]
58. Albertsdóttir, E.; Eriksson, S.; Sigurdsson, Á.; Árnason, T. Genetic Analysis of "Breeding Field Test Status" in Icelandic Horses. *J. Anim. Breed. Genet.* **2011**, *128*, 124–132. [CrossRef]
59. Koenen, E.P.; Aldridge, L.; Philipsson, J. An Overview of Breeding Objectives for Warmblood Sport Horses. *Livest. Prod. Sci.* **2004**, *88*, 77–84. [CrossRef]
60. Medeiros, B.R.; Garbade, P.; Seixas, L.; Peripolli, V.; Mcmanus, C. Brazilian Sport Horse: Genetic Parameters for Approval of Brasileiro de Hipismo Stallions. *Trop. Anim. Health Prod.* **2020**, *52*, 1669–1680. [CrossRef]
61. Sánchez-Guerrero, M.J.; Cervantes, I.; Molina, A.; Gutiérrez, J.P.; Valera, M. Designing an Early Selection Morphological Linear Traits Index for Dressage in the Pura Raza Español Horse. *Animal* **2017**, *11*, 948–957. [CrossRef]
62. Rustin, M.; Janssens, S.; Buys, N.; Gengler, N. Multi-Trait Animal Model Estimation of Genetic Parameters for Linear Type and Gait Traits in the Belgian Warmblood Horse. *J. Anim. Breed. Genet.* **2009**, *126*, 378–386. [CrossRef] [PubMed]
63. McGuigan, M.P.; Wilson, A.M. The Effect of Gait and Digital Flexor Muscle Activation on Limb Compliance in the Forelimb of the Horse Equus Caballus. *J. Exp. Biol.* **2003**, *206*, 1325–1336. [CrossRef] [PubMed]
64. Patel, B.A. The Interplay between Speed, Kinetics, and Hand Postures during Primate Terrestrial Locomotion. *Am. J. Phys. Anthropol.* **2009**, *141*, 222–234. [CrossRef] [PubMed]

65. Dowling, B.A.; Dart, A.J. Mechanical and Functional Properties of the Equine Superficial Digital Flexor Tendon. *Vet. J.* **2005**, *170*, 184–192. [CrossRef]
66. Fugazzola, M.C.; Lancioni, I.; Duran, M.C.; Canonici, F.; Petrizzi, L. Correlation between the Conformation of the Distal Forelimb and Superficial Digital Flexor Tendon Lesions in Flat Racing Thoroughbreds. *J. Equine Vet. Sci.* **2015**, *35*, 264–270. [CrossRef]
67. Schade, J.; Fernando De Souza, A.; Vincensi, L.C.; Fonteque, J.H. The Influence of the Metacarpophalangeal Joint Angle on the Transversal Area and Mean Echogenicity of the Superficial Digital Flexor Tendon and Suspensory Ligament in Gaited Horses. *J. Equine Sci.* **2021**, 135–141. [CrossRef]
68. Weller, R.; Pfau, T.; Verheyen, K.; May, S.A.; Wilson, A.M. The Effect of Conformation on Orthopaedic Health and Performance in a Cohort of National Hunt Racehorses: Preliminary Results. *Equine Vet. J.* **2006**, *38*, 622–627. [CrossRef]
69. Lawson, S.E.M.; Chateau, H.; Pourcelot, P.; Denoix, J.M.; Crevier-Denoix, N. Effect of Toe and Heel Elevation on Calculated Tendon Strains in the Horse and the Influence of the Proximal Interphalangeal Joint. *J. Anat.* **2007**, *210*, 583–591. [CrossRef]
70. Genovese, R.L.; Simpson, B.S.; Simpson, D.M.; Rantanen, N.W. Clinical Experience with Quantitative Analysis of Superficial Digital Flexor Tendon Injuries in Thoroughbred and Standardbred Racehorses. *Vet. Clin. N. Am. Equine Pract.* **1990**, *6*, 129–145. [CrossRef]
71. Takahashi, T.; Kasashima, Y.; Ueno, Y. Association between Race History and Risk of Superficial Digital Flexor Tendon Injury in Thoroughbred Racehorses. *J. Am. Vet. Med. Assoc.* **2004**, *225*, 90–93. [CrossRef]
72. Kasashima, Y.; Takahashi, T.; Smith, R.K.W.; Goodship, A.E.; Kuwano, A.; Ueno, T.; Hirano, S. Prevalence of Superficial Digital Flexor Tendonitis and Suspensory Desmitis in Japanese Thoroughbred Flat Racehorses in 1999. *Equine Vet. J.* **2004**, *36*, 346–350. [CrossRef] [PubMed]
73. Williams, R.B.; Harkins, L.S.; Hammond, C.J.; Wood, J.L.N. Racehorse Injuries, Clinical Problems and Fatalities Recorded on British Racecourses from Flat Racing and National Hunt Racing during 1996, 1997 and 1998. *Equine Vet. J.* **2001**, *33*, 478–486. [CrossRef] [PubMed]
74. Giles, J.M. Partes Exteriores Del Caballo. In *Manual del Remontista o Sucinta Idea de los Conocimientos Necesarios para las Compras y Ventas de Caballos*; Imprenta Don Juan de la Vega: Madrid, Spain, 1842; pp. 1–36.
75. Ghavi Hossein-Zadeh, N. A Meta-Analysis of Genetic Parameter Estimates for Conformation Traits in Horses. *Livest. Sci.* **2021**, *250*, 104601. [CrossRef]

MDPI AG
Grosspeteranlage 5
4052 Basel
Switzerland
Tel.: +41 61 683 77 34

Animals Editorial Office
E-mail: animals@mdpi.com
www.mdpi.com/journal/animals

www.ingramcontent.com/pod-product-compliance
Lightning Source LLC
Chambersburg PA
CBHW061453210926
51850CB00008B/1707
* 9 7 8 3 7 2 5 8 4 4 4 5 6 *